Deepen Your Mind

序 Preface

在資訊大爆炸的當下，資訊超載已成為越來越多的人的負擔。

隨著 5G 時代的到來，物聯網和智慧城市將隨處可見，隨之而來的是資訊會更加複雜和龐大。如何掙脫資訊的束縛，高效率地找到自己需要的資訊呢？答案就是搜尋引擎，即借助搜尋引擎來尋找我們想要的資訊！

本書介紹的搜尋引擎是 Elasticsearch——一個開放原始碼的搜尋引擎。

目前，Elasticsearch 的功能已不侷限於搜索，它還在不斷地豐富和增強自己的生態。在 API 介面層面，除了基本的資料索引和資料搜索之外，Elasticsearch 還提供 Elasticsearch 服務監控介面、推薦相關介面，以及機器學習相關介面。

✤ 本書目的

與追求所有基礎知識都要論述但都泛泛而談的書不同，本書聚焦初學者的學習和實戰需要，將初學者接觸 Elasticsearch 從 0 到 1 過程中的必備基礎知識講透。只有學透基礎知識，才能學習更多有關 Elasticsearch 的進階知識。

這一點筆者在教育訓練 Elasticsearch 初學者時深有體會。因此，本書重點結合筆者在 Elasticsearch 上的沉澱、實戰、教育訓練和 Elasticsearch 最新版本內容，幫助 Elasticsearch 初學者突破這層限制！

正如王陽明在《傳習錄》中談為學之道時所言：「殊不知私欲日生，如地上塵，一日不掃便又有一層。著實用功，便見道無終窮，愈探愈深，必使精白無一毫不徹方可。」

對於知識與近代和現代高速發展的經濟之間的關係，管理學大師杜拉克有一段精闢論述。他認為二者的關係可以分為三個發展階段，即工業革命、生產力革命、管理革命。所謂工業革命，指的是知識應用於生產工具、生產流程和產品創新；所謂生產力革命，指的是知識以及被指定的含義開始被應用於工作中；所謂管理革命，指的是知識正被用於知識本身。而管理革命的核心在於連接。在知識領域，連接表示基礎知識連結。

很多人無法有效地將相似或連結的基礎知識進行連結，所以更談不上建置網狀知識系統。

因此，在本書行文過程中，筆者會以自己建置為基礎的知識系統向讀者進行必要的系統輸出，力求幫助讀者在快速上手的同時，建置搜尋引擎全景，洞悉 Elasticsearch 生態，建立連結知識網路。

本書基於 Elasticsearch 7.X 系列版本撰寫，內容由淺入深，先讓初學者會用、能用，再介紹背後的原理。這種方式在筆者主導過的 Elasticsearch 技術教育訓練中效果較好。

✤ 本書結構

本書分為三大部分，分別是 Elasticsearch 前傳、Elasticsearch 實戰和 Elasticsearch 生態。

Elasticsearch 前傳部分主要介紹搜索技術發展史和基礎，並介紹搜尋引擎技術原理，為讀者建置搜尋引擎全景。在技術發展史上，我們能看見多久的歷史，就能看見多遠的未來！

Elasticsearch 實戰部分主要介紹 Elasticsearch 的核心概念和架構設計，並重點介紹用戶端、文件、搜索、索引等實戰內容，待讀者能上手實戰後，再介紹這些內容的背後實現原理和連結知識，為讀者建置知識網路。

Elasticsearch 生態部分主要介紹外掛程式的使用和管理，以及 Elastic Stack 生態圈。

✤ 本書特色

特色 1：基於 Elasticsearch 7.X 系列版本撰寫。

特色 2：聚焦初學者學習和實戰需要，不求基礎知識都要論述，但求必備知識
　　　　透徹易懂。

特色 3：讓初學者快速上手的同時，幫助他們建置搜尋引擎全景、洞悉
　　　　Elasticsearch 生態、建立連結知識網路。

特色 4：由淺入深，先讓初學者會用，再介紹背後的原理。

在本書撰寫過程中，Elasticsearch 仍在升級版本，因此書中難免有理解和實作
不足之處。「卑辭俚語，不揣譾陋」，歡迎讀者和筆者交流學習，共同進步。

牛冬

Contents **目錄**

v

第二部分 Elasticsearch 實戰

03 初識 Elasticsearch

04 初級用戶端實戰

05 進階用戶端文件實戰一

06 進階用戶端文件實戰二

07 搜索實戰

08 索引實戰

第三部分　Elasticsearch 生態

09　Elasticsearch 外掛程式

10 Elasticsearch 生態圈

搜索技術發展史

人事有代謝
往來成古今

1.1 搜索技術發展史

「我們面前無所不有，我們面前一無所有。」

正如查理斯·狄更斯在《雙城記》中所述。在資訊大爆炸的當下，「我們面前無所不有」；而個人資訊超載已成為越來越多的人的負擔，「我們面前一無所有」。

如何掙脫超載的資訊的束縛，高效率地找到自己需要的資訊呢？——答案是搜尋引擎，借助搜尋引擎來實現！

本書介紹的搜尋引擎是 Elasticsearch——一個開放原始碼的搜尋引擎（簡稱 ES）。

我們每天都在某種場景下使用搜尋引擎，在電腦上、手機上，都可以找到自己慣用的搜尋引擎，例如百度搜索、搜狗搜索、神馬搜索、Google 搜索、360 搜索、頭條搜索，等等。

那麼，搜尋引擎是什麼呢，它是如何發展到今天的樣子呢？本章就介紹搜索技術發展，讓我們沿著技術發展的脈絡更深刻地認識搜索技術。

巨觀而言，搜尋引擎的發展經歷了五個階段和兩大分類。五個階段分別是 FTP 檔案檢索階段、分類目錄導覽階段、文字相關性檢索階段、網頁連結分析階段和使用者意圖識別階段。實際情況整理如下。

☑ FTP 檔案檢索階段

該階段的搜尋引擎只檢索多個 FTP 伺服器上儲存的檔案，代表作是 Archie。使用者搜索檔案時需輸入精確的檔案名稱來搜索尋找，搜尋引擎會告訴使用者從哪一個 FTP 位址可以下載被搜索的檔案。

☑ 分類目錄導覽階段

該階段的搜尋引擎就是一個導覽網站，網站中都是網址的分類陳列，使用者在網際網路上常用的網址在這裡一應俱全。

在使用該類別搜尋引擎時，使用者需要從各個分類目錄裡找到自己想要的網址，點擊其網站連結後進入對應的網站。

直到今天，這種搜尋引擎依然不過時，我們常用的網站如好 123、搜狗瀏覽器首頁、UC 導覽等均是這種導覽頁面。

☑ 文字相關性檢索階段

隨著網際網路內容不斷豐富，網頁的內容和形態也越來越多樣化，頁面中開始出現內容可能與網頁位址和網頁標題大相徑庭的情況。

為了解決這個問題，搜尋引擎引用全文檢索搜尋技術，來保障搜尋引擎檢索到的網頁標題與網頁全文內容強一致，摒棄了單純依靠網頁標題和網頁位址來判斷網頁內容的方法。

在使用這種搜尋引擎查詢資訊時，使用者將輸入的查詢資訊提交給搜尋引擎後台伺服器，搜尋引擎伺服器透過查閱已經索引好的網頁全文資訊，傳回一些相關程度高的頁面資訊。

計算輸入的查詢資訊與網頁內容相關性判斷的模型主要有布林模型、機率模型、向量空間模型等。

這個階段的搜尋引擎的主要代表作是 Alta Vista、Excite 等。

網頁連結分析階段

這個階段的搜尋引擎所使用的網站連結形式與目前大致相同。在該階段，外部連結表示推薦。

因此，透過計算每個網站的推薦連結的數量，就可以判斷一個網站的流行性和重要性。

於是，搜尋引擎透過結合網頁內容的重要性和相似程度來改善搜索的資訊品質。在這一階段，搜尋引擎的代表作是 Google 搜索。

這種模式是 Google 首創的，並且大獲成功，隨之引起了學術界和其他商業搜尋引擎的極度關注和效仿。目前，網頁連結分析演算法及其改進最佳化的版本在主流搜尋引擎中大行其道。

使用者意圖識別階段

這個階段的搜尋引擎以使用者為中心作為設計的初心，搜尋引擎力求了解每一位使用者的真正搜索訴求，力求做到千人千面，追求個性化識別和回饋。

在使用這種搜尋引擎時，即使是同一個查詢的請求關鍵字，不同的使用者可能也會獲得不同的查詢結果。例如輸入的是「小米」，那麼一個想要購買小米電子裝置的使用者和一個想要購買小米食用的使用者，他們的搜索意圖顯然天壤之別，因而獲得不同的搜索結果是順理成章的事情。不光是不同使用者之間，同一個使用者搜索同樣的關鍵字也會因時因地的不同而有所差異。例如當使用者在搜尋引擎上第一次輸入 "TAL" 時，可能是想尋找 TAL 股票程式對應的好未來公司的網站；當使用者在好未來的辦公區內搜索 "TAL" 時，有可能是想檢視 TAL 股票程式的即時股價。

其實在這兩個案例背後，搜尋引擎都在致力於解決同一個問題，即怎樣才能透過輸入的簡短的關鍵字來判斷使用者的真正查詢訴求。這也是我們將其歸類為使用者意圖識別的原因。這一階段的搜尋引擎典型代表就是百度。

在搜尋引擎技術不斷演進的過程中，為了更進一步地識別及滿足使用者的搜索需求，更多的新技術也在不斷引入，如 AI 技術、地理位置資訊、人物誌等。

兩大分類是指站內搜索和站外搜索。

站外搜索就是全網搜索，現在主流的搜尋引擎基本都是全網搜索，如 Google、百度。隨著技術的發展，搜索領域的生態圈搜索形態不斷擴大。以 Google 為代表的搜尋引擎推出了整合搜索、個人化搜索、即時搜索、地圖服務、線上檔案編輯、網站統計、瀏覽器、網管工具、超大容量電子郵件、即時通訊等。百度上線了百度百科、百度知道、百度貼吧等服務，這些服務中嵌入了文字搜索、語音搜索、影像搜索、地圖搜索等搜索形態。

站內搜索近幾年發展比較快速，各大網站平台紛紛上線了站內搜索，如 SNS 平台中的微博、人人網等，如電子商務平台中的京東、餓了麼、淘寶、美團等。

另外，區塊鏈內容搜索是近兩年新的站內搜索形式，如比特幣區塊鏈的搜索內容在比特幣公鏈上，但比特幣公鏈的節點所在地域卻是分散式的，和常見的站內搜索大相徑庭，如圖 1-1 所示。

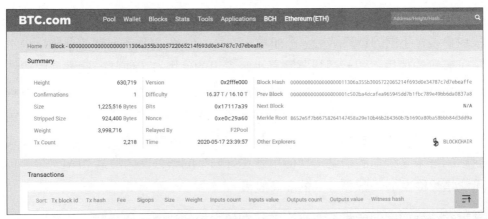

圖 1-1

在未來，搜尋引擎的發展會是什麼樣的呢？我們不妨天馬行空一下。隨著 5G 時代的到來，物聯網和智慧城市將隨處可見；AR/VR 技術會更加成熟，裝置更加普及和便宜。與之對應的,除現在的文字搜索、語音搜索、影像搜索外,還會出現 AR/VR 搜索等搜索形態。

在 5G 的加持下,搜尋引擎的搜索效率會更高;物聯網和區塊鏈中裝置和資訊搜索也會更加普遍,而搜尋引擎的商業模式也可能隨之升級,廣告的效果可能會更好。

1.2 Elasticsearch 簡介

Elasticsearch 是一個分散式、可擴充、近即時的高性能搜索與資料分析引擎。

Elasticsearch 提供了搜集、分析、儲存資料三大功能,其主要特點有:分散式、零設定、易裝好用、自動發現、索引自動分片、索引備份機制、RESTful 風格介面、多資料來源和自動搜索負載等。

Elasticsearch 並非從零起步,而是站在巨人的肩膀上。Elasticsearch 基於 Java 撰寫,其內部使用 Lucene 做索引與搜索。透過進一步封裝 Lucene,向開發人員隱藏了 Lucene 的複雜性。開發人員無須深入了解檢索的相關知識來了解它的運行原理,只需使用一套簡單一致的 RESTful API 即可,從此全文檢索搜尋變得非常簡單。

除此之外,Elasticsearch 還解決了檢索相關資料、傳回統計結果、回應速度等相關的問題。因此,Elasticsearch 能做到分散式環境下的即時文件儲存和即時分析搜索。即時儲存的文件,每個欄位都可以被索引與搜索。

最令人驚喜的是,Elasticsearch 能勝任成千上百個服務節點的分散式擴充,支援 PB 等級的結構化或非結構化巨量資料的處理。

2019 年 4 月 10 日,Elasticsearch 發佈了 7.0 版本。該版本的重要特性包含引用記憶體斷路器、引用 Elasticsearch 的全新叢集協調層──Zen2、支援更快的前 k 個查詢、引用 Function score 2.0 等。

其中記憶體斷路器可以更精準地檢測出無法處理的請求,並防止它們使單一節點不穩定；Zen2 是 Elasticsearch 的全新叢集協調層，加強了可用性、效能和使用者體驗，使 Elasticsearch 變得更快、更安全，且更易於使用。

1.3 Lucene 簡介

Lucene 是一個免費、開放原始碼、高性能、純 Java 撰寫的全文檢索引擎。

在業務開發場景中，Lucene 幾乎適用任何需要全文檢索的場景，各種程式語言的 Lucene 版本不斷湧現。目前，Lucene 先後發展出了 C++、C#、Perl 和 Python 等語言的版本，Lucene 逐漸成為開放原始程式碼中最好的全文檢索引擎工具套件。

2005 年，Lucene 升級成為 Apache 頂級專案。

Lucene 包含大量相關專案，核心專案有 Lucene Core、Solr 和 PyLucene。

需要指出的是，Lucene 僅是一個工具套件，它並非一個完整的全文檢索引擎，這和 Lucene 的初衷有關。Lucene 是軟體開發人員的一個簡單好用的工具套件，主要提供倒排索引的查詢結構，以方便軟體開發人員在其業務系統中實現全文檢索的功能。這也是我們常説全文檢索引擎主要是 Solr 和 Elasticsearch 的原因，雖然二者均是以 Lucene 為基礎所建立。

Lucene 作為一個全文檢索引擎工具套件，具有以下優點。

☑ 索引檔案格式獨立於應用平台
Lucene 定義了一套以 8 位位元組為基礎的索引檔案格式，使得相容系統或不同平台的應用能夠共用建立的索引檔案。

☑ 索引速度快
在傳統全文檢索引擎的倒排索引的基礎上，實現了分段索引，能夠針對新的檔案建立小檔案索引，提升索引速度。然後透過與原有索引的合併，達到最佳化的目的。

☑ 簡單易學

優秀的物件導向的系統架構，降低了 Lucene 擴充的學習難度，方便擴充新功能。

☑ 跨語言

設計了獨立於語言和檔案格式的文字分析介面，索引子透過接收 Token 流完成索引檔案的創立，使用者擴充新的語言和檔案格式，只需實現文字分析的介面即可。

☑ 強大的查詢引擎

Lucene 預設實現了一套強大的查詢引擎，使用者無須自己撰寫程式即可透過系統獲得強大的查詢能力。Lucene 預設實現了布林操作、模糊查詢、分組查詢等。

Lucene 的主要模組有 Analysis 模組、Index 模組、Store 模組、QueryParser 模組、Search 模組和 Similarity 模組，各模組的功能分別整理如下。

（1）Analysis 模組：主要負責詞法分析及語言處理，也就是我們常說的斷詞，透過該模組可最後形成儲存或搜索的最小單元 Term。

（2）Index 模組：主要負責索引的建立工作。

（3）Store 模組：主要負責索引的讀和寫，主要是對檔案的一些操作，其主要目的是抽象出和平台檔案系統無關的儲存。

（4）QueryParser 模組：主要負責語法分析，把查詢敘述產生 Lucene 底層可以識別的條件。

（5）Search 模組：主要負責對索引的搜索工作。

（6）Similarity 模組：主要負責相關性評分和排序的實現。

在 Lucene 中，還有一些核心術語，主要有關 Term、詞典（Term Dictionary，也叫作字典）、倒排表（Posting List）、正向資訊和段（Segment），這些術語的含義整理如下。

（1）Term：索引中最小的儲存和查詢單元。對於英文語境而言，一般是指一個單字；對於中文語境而言，一般是指一個斷詞後的詞。

（2）詞典：是 Term 的集合。詞典的資料結構有很多種，各有優缺點。例如可以透過排序陣列（透過二分尋找來檢索資料）、HashMap（雜湊表，檢索速度更快，屬於空間換時間的模式）、FST（Finite-State Transducer，有很好的壓縮率）等來實現。

（3）倒排表：一篇文章通常由多個詞組成，倒排表記錄的是某個詞在哪些文章中出現過。

（4）正向資訊：原始的文件資訊。可以用來做排序、聚合、展示等。

（5）段：索引中最小的獨立儲存單元。一個索引檔案由一個或多個段組成。在 Lucene 中，段有不變性，段一旦產生，在段上只能讀取、不能寫入。

1.4 基礎知識連結

站在巨人的肩膀上

從上文中我們可以看到，Elasticsearch 並非從 0 起步，重複製造輪子，而是站在巨人 Lucene 的肩膀上建置。其實很多新技術都是以已有的技術為基礎發展演進而來，例如最近幾年火熱的區塊鏈技術。

■ REST 協定

RESTful 程式設計風格是近年來前後端流行的互動模式，在工作和面試中，有人會將 REST 協定與 HTTP 混為一談，因此這裡介紹 REST 協定的由來。

下面從幾個維度來展開敘述：

（1）Web 技術發展與 REST 的由來——講歷史。

（2）REST 架構風格的推導過程——講過程。

（3）REST 定義——講定義。

（4）REST 關鍵原則——講原則。

（5）歸納 REST 風格的架構特點——講特點。

（6）REST 架構風格的優點和缺點——講辯證。

2 Web 技術發展與 REST 的由來

Web 技術基礎有哪些？從技術架構層面來看，Web 技術架構包含四個基礎：URI、HTTP、HyperText、MIME（Multipurpose Internet Mail Extensions，多用途網際網路郵件擴充類型）。這四個基礎相互支撐，相互幫助，共同推動著 Web 這座宏偉的大廈以幾何級數的速度快速發展起來。

在 Web 開發技術歷程中，Web 開發技術經歷了 5 個階段，分別是靜態內容階段、CGI（Common Gateway Interface）程式階段、指令碼語言階段、精簡型用戶端應用階段、RIA（Rich Internet Application）應用階段和行動 Web 應用階段。其中：

在靜態內容階段，Web 由靜態 HTML 檔案組成，Web 伺服器有點像支援超文字的共用檔案伺服器。

在 CGI 程式階段，Web 伺服器增加了程式設計 API，研發人員可以透過 CGI 協定完成程式開發，這種應用程式被稱作 CGI 程式。

在指令碼語言階段，伺服器端先後出現了 ASP、PHP、JSP、ColdFusion 等支援 session 的指令碼語言技術，瀏覽器端則出現了 Java Applet、JavaScript 等技術。以這些技術為基礎，網站可以提供更加豐富的動態內容。

在精簡型用戶端應用階段，伺服器端出現了獨立於 Web 伺服器的應用伺服器，以及 Web MVC 開發模式。以這些架構開發為基礎的 Web 應用，通常都是精簡型用戶端應用。

在 RIA 應用階段，出現了多種 RIA 技術，大幅改善了 Web 應用的使用者體驗，例如最為廣泛的 RIA 技術 DHTML+Ajax，Ajax 技術支援前端在不更新頁面的情況下動態更新頁面中的局部內容，顯著提升了使用者體驗。這一時期也誕生了大量的 Web 前端 DHTML 開發函數庫，如 Prototype、Dojo、ExtJS、 jQuery/jQuery UI。其他的 RIA 技術還有 Adobe 公司的 Flex、微軟公司的 Silverlight、Sun 公司的 JavaFX（現在為 Oracle 公司所有），等等。

在行動 Web 應用階段，出現了大量針對行動裝置的 Web 應用程式開發技術。除 Android、iOS、Windows Phone 等作業系統平台原生的開發技術外，以 HTML5 為基礎的開發技術也變得非常流行。

從上述 Web 開發技術的發展過程可以得出結論，Web 從最初其設計者所構思的主要支援靜態文件的階段，逐漸變得越來越動態化；Web 應用的互動模式變得越來越複雜，從靜態文件發展到以內容為主的入口網站、電子商務網站、搜尋引擎、社交網站，再到以娛樂為主的大型多人線上遊戲、手機遊戲。

Web 開發技術和 REST 之間有什麼關係呢？上面提及 Web 技術架構包含四個基礎，其中之一就是 HTTP，下面簡介 HTTP 的發展歷程。

1995 年，CGI、ASP 等技術出現後，沿用多年、針對靜態文件的 HTTP/1.0 已無法滿足 Web 應用的開發需求，因此需要設計新版本的 HTTP。於是，HTTP/1.0 協定專案小組成立。在這些專家之中，有一位名叫 Roy Fielding 的年輕人脫穎而出，顯示出了不凡的洞察力，後來 Roy Fielding 成為 HTTP/1.1 協定專案小組的負責人。而 Roy Fielding 還是 Apache HTTP 伺服器的核心開發者，更是 Apache 軟體基金會的合作創始人。

HTTP/1.1 的第一個草稿於 1996 年 1 月發佈，後經三年多的反覆修訂和論證，於 1999 年 6 月成為 IETF 的正式標準。

2000 年，Roy Fielding 在其博士學位論文 *Architectural Styles and the Design of Network-based Software Architectures* 中，有系統、嚴謹地說明了這套理論架構，還使用這套理論架構推導出一種新的架構風格，並且為這種架構風格取了一個令人輕鬆愉快的名字 "REST"——Representational State Transfer（表述性狀態傳輸）。

2007 年 1 月，支援 REST 開發的 Ruby on Rails 1.2 版發佈。特別的是，Rails 將支援 REST 開發作為其未來發展的優先內容。Ruby on Rails 的創始人 DHH

做了一場名為 "World of Resources" 的精彩演講。DHH 在 Web 開發技術社區中的強大影響力，使得 REST 一下子處在 Web 開發技術舞台的聚光燈之下。隨著 Rails 等新興 Web 開發技術的推波助瀾，Web 開發技術社區掀起了一場重歸 Web 架構設計本源的運動。REST 架構風格獲得了越來越多的關注，現在各種流行的 Web 開發架構幾乎都支援 REST 開發。

而廣大的開發人員也紛紛學習 REST 架構風格。當然，大多數 Web 開發者都是透過閱讀 REST 開發架構的文件，以及透過一些實例程式來學習 REST 開發。不過，透過實例程式來學習 REST 有非常大的限制，因為 REST 並不是一種實際的技術，也不是一種實際的標準，而是一種內涵非常豐富的架構風格。

雖然透過實例程式來學習 REST，能「短平快」地學習到一種有趣的 Web 開發技術，但並不能全面深入地了解 REST 究竟是什麼，甚至還會誤以為這些簡單的實例程式就是 REST 本身，以為 REST 不過是一種簡單的 Web 開發技術而已。這有點類似瞎子摸象，有的人摸到了象鼻子、有的人摸到了象耳朵、有的人摸到了象腿、有的人摸到了象尾巴。摸象的眾人都堅信自己感覺到的大象才是最真實的大象，而其他人的感覺都是錯誤的，但現實常常恰巧相反。

3 REST 架構風格的推導過程——講過程

Roy Fielding 在批判性繼承前人研究成果的基礎上，建立起了一整套研究和評價軟體架構的方法論。這套方法論的核心是「架構風格」這個概念。

架構風格是一種研究和評價軟體架構設計的方法，它是比架構更加抽象的概念。一種架構風格是由一組相互協作的架構約束所定義。架構約束是指軟體的執行環境施加在架構設計之上的約束。

Roy Fielding 的 REST 架構風格的推導過程如圖 1-2 所示。

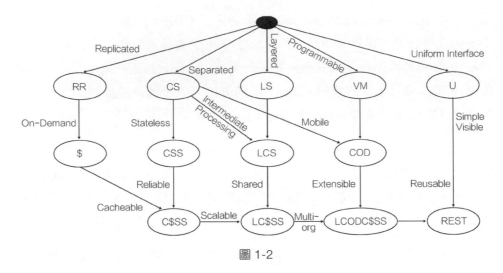

圖 1-2

每一個橢圓形裡面的縮寫詞代表了一種架構風格,而每一個箭頭旁邊的單字代表了一種架構約束。其中:

- RR(Replicated Repository)可譯為複製倉庫。RR 利用多個處理程序提供相同的服務,來改善資料的可存取和可伸縮。
- COD(Code-On-Demand),可譯為隨選程式。
- $ 表示 Cache,指的是複製個別請求的結果,以便可以被後面的請求重用。
- CSS(Client-Stateless-Server)可譯為客戶無狀態伺服器。
- LCS(Layered-Client-Server)可譯為分層客戶伺服器。
- VM(Virtual Machine)可譯為虛擬機器風格。

透過推導,Roy Fielding 得出了 REST 架構風格中最重要的 6 個架構約束,即客戶─伺服器(Client-Server)、無狀態(Stateless)、快取(Cache)、統一介面(Uniform Interface)、分層系統(Layered System)和隨選程式(Code-on-Demand)。其中:

- 客戶 - 伺服器是指通訊只能由用戶端單方面發起,表現為請求 - 回應的形式。
- 無狀態是指通訊的階段狀態(Session State)應該全部由用戶端負責維護。
- 快取是指回應內容可以在通訊鏈的某處被快取,以改善網路效率。

- 統一介面是指通訊鏈的元件之間透過統一的介面相互通訊，以加強互動的可見性。
- 分層系統是指透過限制元件的行為，將架構分解為許多等級的層。
- 隨選程式是指支援透過下載並執行一些程式，對用戶端的功能進行擴充。

HTTP/1.1 身為 REST 架構風格的架構實例，其架構演變經歷了圖 1-3 和圖 1-4 所示的過程。

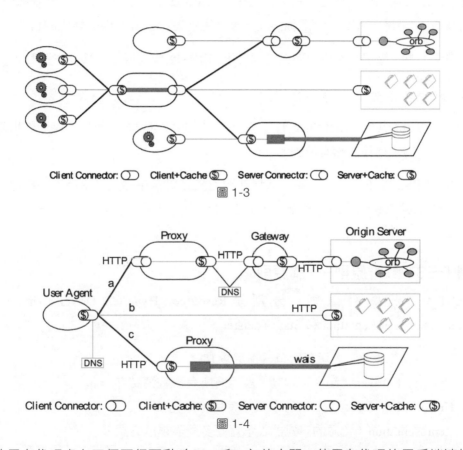

Client Connector: ◯ Client+Cache ⑤ Server Connector: ◯ Server+Cache: ⑤

圖 1-3

Client Connector: ◯ Client+Cache: ⑤ Server Connector: ◯ Server+Cache: ⑤

圖 1-4

使用者代理處在三個平行互動（a、b 和 c）的中間，使用者代理的用戶端連接器快取無法滿足請求，因此它根據每個資源識別符號的屬性和用戶端連接器的設定，將每個請求路由到資源的來源。

請求 a 被發送到一個本機代理，代理隨後存取一個透過 DNS 尋找發現的快取閘道，該閘道將這個請求轉發到一個能夠滿足該請求的原始伺服器，原始伺服器的內部資源由一個封裝過的物件請求代理（Object Request Broker）架構來定義。

請求 b 直接發送到一個原始伺服器，它能夠透過自己的快取滿足這個請求。

請求 c 被發送到一個代理，它能夠直接存取 WAIS（一種與 Web 架構分離的資訊服務），並將 WAIS 的回應翻譯為一種通用的連接器介面能夠識別的格式。

在圖 1-4 中，每個元件只知道與它們自己的用戶端或伺服器連接器的互動。

■4 REST 定義——講定義

REST（REpresentational State Transfer，表述性狀態傳輸）是所有 Web 應用都應該遵守的架構設計指導原則。

當然，REST 並不是法律法規。如果在軟體開發實作中違反了 REST 的指導原則，雖然能夠實現應用的功能，但是會付出很多代價，特別是對於高流量的網站。

■5 REST 關鍵原則——講原則

REST 有 5 個關鍵原則，分別是 Resource、Hypertext Driven、Uniform Interface、Representation 和 State Transfer。

- Resource（資源）表示為所有事物定義 ID。
- Hypertext Driven（超文字驅動）表示將所有事物連結在一起。
- Uniform Interface（統一介面）表示使用標準方法。
- Representation（資源的表述）表示資源多重表述的方式。
- State Transfer（狀態傳輸）表示無狀態的通訊。

註：在翻譯外文文獻時，可參考嚴復在《天演論》中的「譯例言」講到的「譯事三難：信、達、雅」。這三個原則的核心在於意譯。

☑ 資源（Resource）

資源其實可以看作一種看待伺服器的方式。

與物件導向設計類似，資源是以名詞為核心組織而成。

一個資源可以由一個或多個 URI 來標識。URI 既是資源的名稱，也是資源在 Web 上的地址。對某個資源有興趣的用戶端應用，可以透過資源的 URI 與其進行互動存取並取得資源。

將伺服器看作由很多離散的資源群組成，每個資源是伺服器上一個可命名的抽象概念。資源可以是伺服器檔案系統中的檔案，也可以是資料庫中的一張表。

上文提到資源表示為所有事物定義 ID，也就是說，每個事物都可以被標示，都會擁有一個 ID 識別符號。在 Web 開發中，代表 ID 的統一概念就是 URI。URI 組成了一個全域命名空間，使用 URI 標識你的關鍵資源表示它們獲得了一個唯一、全域的 ID。

這裡要注意區分 URI、URL 和 URN。

- URI，即 Uniform Resource Identifier：統一資源識別項，包含 URL 和 URN。
- URL，即 Uniform Resource Locator：統一資源定位器，常見的有 Web URL 和 FTP URL。
- URN，即 Uniform Resource Name：統一資源名稱，例如 tel:+1-888-888-8888。

設計 URI 的 4 個原則是：

- 它們是名詞。
- 區分單複數。
- URI 有長度限制，建議小於 1KB。
- 在 URI 中不要放未經加密的敏感資訊。

在 URI 中，有幾個符號需要特別注意：

（1）/ 展現層次關係，例如 https://www.×××/cn/products/elasticsearch。

（2）;, 表 示 並 列 關 係 ， 例 如 https://www.×××.com/axis;x=0,y=9 sip:user @domain.com; foo=bar;x=y。

（3）- 用來加強可讀性，最好全用小寫，例如 https://www.×××-fan.com/。

（4）用參數或 HTTP Range Header 來限定範圍，例如 https://www.×××.com/ sortbyAsc=name&fileds=email,title&limit=10&start=20。

超文字驅動

超文字驅動表示將所有事物連結在一起。超媒體被當作應用狀態引擎（Hypermedia As The Engine of Application State，HATEOAS），其核心是超媒體概念，換句話說是連結的思想。

引用連結後，我們可以將 Web 應用看作一個由很多狀態組成的有限狀態機。資源之間透過超連結相互連結，超連結既代表資源之間的關係，也代表可執行的狀態遷移。

使用 URI 表示連結時，連結可以指向由不同應用、不同伺服器甚至位於不同地域、不同洲際的不同公司提供的資源——因為 URI 命名標準是全球標準，所以組成 Web 的所有資源都可以互連互通。

超媒體原則還表示應用「狀態」，伺服器端為用戶端提供一組連結，使用戶端能夠透過連結將應用從一個狀態改變為另一個狀態。而連結剛好是組成動態應用的非常有效的方式。

因此連結成就了現在的 Web。

統一介面

統一介面表示使用標準方法。

按 REST 要求，必須透過統一介面對資源執行各種操作。對於每個資源只能執行一組有限的操作。

在 RESTful HTTP 方式中，一般開發人員透過組成 HTTP 應用協定的通用介面存取服務程式。在 HTTP/1.1 中，已經定義了一個操作資源的統一介面，主要包含以下內容。

- 7 個 HTTP 方 法：POST、GET、PUT、DELETE、PATCH、HEAD 和 OPTIONS。
- HTTP 標頭資訊（可自訂）。
- HTTP 回應狀態碼（可自訂）。
- 一套標準的內容協商機制。
- 一套標準的快取機制。
- 一套標準的用戶端身份認證機制。

對開發人員而言，程式碼將從圖 1-5 中轉變。

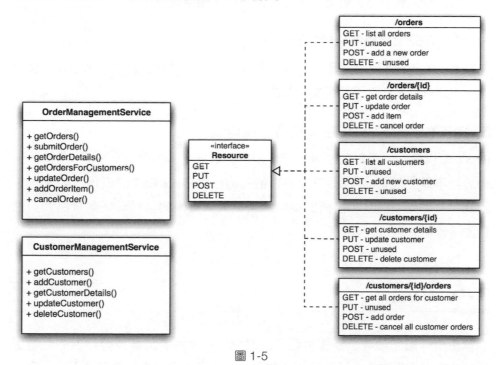

圖 1-5

▨ 資源的表述

資源的表述是一段對資源在某個特定時刻的狀態的描述。資源可以在用戶端一伺服器端之間傳輸。

資源的表述有多種格式，如 HTML、XML、JSON、文字、圖片、視訊、音訊等。資源的表述格式可以透過協商機制來確定。請求一回應的表述通常使用不同的格式。

▨ 狀態傳輸

狀態傳輸表示無狀態通訊。這裡的狀態傳輸是指在用戶端和伺服器端之間傳輸。透過傳輸和操作資源的表述，間接實現操作資源的目的。

無狀態表示用戶端看不見伺服器的變化。這主要是為了確保架構設計的可伸縮性和可擴充性。試想一下，如果伺服器需要儲存每個用戶端狀態，那麼大量的用戶端互動會嚴重影響伺服器的記憶體可用空間。

▨ 歸納 REST 風格的架構特點

判斷 REST 架構設計是否優秀的標準有 6 個：分別是針對資源、可定址、連通性、無狀態、統一介面和超文字驅動。

▨ REST 架構風格的優點 & 缺點——講辯證

REST 架構設計的優點是簡單、可伸縮、鬆散耦合。需要指出的是，REST 架構設計不是「銀彈」。在即時性要求很高的應用中，REST 的表現不如 RPC。因此，在做架構設計和技術選型時，需要根據實際的執行環境對實際問題進行實際分析。

1.5 小結

本章主要介紹了搜尋引擎技術發展的歷史，幫助讀者沿著技術發展的脈絡更深刻地認識搜索技術，並天馬行空想像未來搜尋引擎技術的發展情況。

在生活中，搜索無處不在，搜索隨時可在——在查字典時、在圖書館找書時。其實人類社會的發展史也是一部資訊高效整理、高效查閱的歷史，而資訊的高效搜索與高效查閱，剛好是搜尋引擎的本質。

搜索技術基礎

騏驥一躍，不能十步
駑馬十駕，功在不舍

在正式介紹 Elasticsearch 的相關內容之前，先介紹搜尋引擎的相關工作原理。了解通用搜尋引擎工作原理之後，再學習 Elasticsearch 時會輕鬆很多。

2.1 資料搜索方式

在第 1 章中，我們已經了解到：搜尋引擎主要是對資料進行檢索。而在研發過程中不難發現，資料有兩種類型，即結構化資料和非結構化資料。

對軟體研發人員來說，在做資料持久化時，對資料的結構化感知會特別強烈。如結構化資料一般我們會放入關聯式資料庫（如 MySQL、Oracle 等），這是因為結構化資料有固定的資料格式和有限個數的欄位，因此可以透過二維化的表結構來承載。

而非結構化資料一般會放入 MongoDB 中，這是因為非結構化的資料長度不定且無固定資料格式，顯然在關聯式資料庫中儲存這種資料較為困難。

與資料形態相對應，資料的搜索分為兩種，即結構化資料搜索和非結構化資料搜索。

因為結構化資料可以基於關聯式資料庫來儲存，而關聯式資料庫常常支援索引，因此結構化資料可以透過關聯式資料庫來完成搜索和尋找。常用的方式有順序掃描、關鍵字精確比對、關鍵字部分比對等，對於較為複雜的關鍵字部分比對，通常需要借助 like 關鍵字來實現，例如左比對關鍵字 "TAL" 時需要使用 like "TAL%"，右比對關鍵字 "TAL" 時需要使用 like "%TAL"，完全模糊比對關鍵字 "TAL" 時需要使用 like "%TAL%"。

對於非結構化資料，資料的搜索主要有順序掃描和全文檢索兩種方法。顯然，對於非結構化資料而言，順序掃描是效率很低的方法，因此全文檢索技術應運而生，而全文檢索搜尋就是本書所說的搜尋引擎要做的事情。

在實現全文檢索的過程中，一般都需要分析非結構化資料中的有效資訊，重新組織資料的承載結構形式。而搜索資料時，需要以新結構化的資料為基礎展開，進一步達到加強搜索速度的目的。顯而易見，全文檢索是一種空間換時間的做法──前期進行資料索引的建立，需要花費一定的時間和空間，但能顯著加強後期搜索的效率。

下面介紹搜尋引擎的工作原理。

2.2　搜尋引擎工作原理

通用搜尋引擎的工作原理如圖 2-1 所示。

搜尋引擎的工作原理分為兩個階段，即網頁資料爬取和索引階段、搜索階段。其中網頁資料爬取和索引階段包含網路爬蟲、資料前置處理、資料索引三個主要動作，搜索階段包含搜索關鍵字、輸入內容前置處理、搜索關鍵字查詢三個主要動作。

其中，網路爬蟲用於爬取網際網路上的網頁，爬取到一個新網頁後還要繼續透過該頁面中的連結來爬取其他網頁。因此網路爬蟲是一個不停歇的工作，一般需要自動化方法來實施。網路爬蟲的主要工作就是盡可能快、盡可能全面地發現和抓取網際網路上的各種網頁。

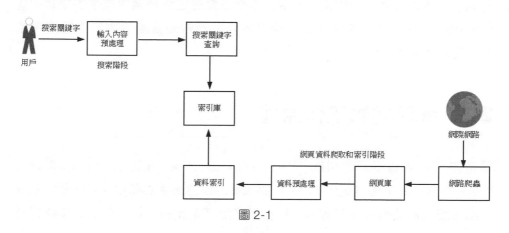

圖 2-1

網頁被網路爬蟲爬取後，會被存入網頁函數庫，以備下一階段進行資料的前置處理。需要指出的是，網頁函數庫裡儲存的網頁資訊與我們在瀏覽器看到的頁面內容相同。此外，由於網際網路上的網頁有一定的重複性，因此在把新網頁真正插入網頁函數庫之前，需要檢查是否重複。

網頁資料前置處理程式不斷地從網頁函數庫中取出網頁進行必要的前置處理。常見的前置處理動作有去除雜訊內容、關鍵字處理（如中文斷詞、去除停止詞）、網頁間連結關係計算等。其中，去除雜訊內容包含版權宣告文字、導覽列、廣告等。網頁經過前置處理後，會被濃縮成以關鍵字為核心的內容。

此外，網際網路上的內容除正常的頁面外，還有各種檔案（如 PDF、Word、WPS、XLS、PPT、TXT 等）、多媒體檔案（如圖片、視訊）等，這些內容均需進行對應的資料前置處理動作。

資料前置處理後，要進行資料索引過程。索引過程先後經歷正向索引和倒排索引階段，最後建立索引函數庫。隨著新的網頁等內容不斷地被加入網頁函數庫，索引函數庫的更新和維護常常也是增量進行的。

以上是網頁資料爬取和索引階段的核心工作，下面介紹搜索階段的核心工作。

使用者輸入的關鍵字同樣會經過前置處理，如刪除不必要的標點符號、停用詞、空格、字元拼字錯誤判等，隨後進行相關的斷詞，斷詞後搜尋引擎系統將向索引函數庫發出搜索請求。索引函數庫會將包含搜索關鍵字的相關網頁從索引函數庫中找出來，搜尋引擎根據索引函數庫傳回的內容進行排序處理，最後傳回給使用者。

2.3 網路爬蟲工作原理

網路爬蟲有多個不同的稱謂，如網路探測器、Crawler 爬蟲、Spider 蜘蛛、Robot 機器人等，其中網路爬蟲或網路蜘蛛的叫法會更具體更生動一些，取意為網頁爬取程式像蟲子和蜘蛛一樣在網路間爬來爬去，從一個網頁連結爬到另一個網頁連結。世界上第一個網路爬蟲是 MIT Matthew Gray 的 World Wide Web Wanderer，Wanderer 主要用於追蹤網際網路發展規模。

網路蜘蛛在工作時，透過種子爬取網頁的連結位址來尋找目標網頁。隨後從網站的 1 個頁面，如首頁，開始讀取網頁的內容和網頁中其他網頁的連結位址，然後透過這些連結位址繼續尋找下一個網頁。如此循環，直到所有內容都被抓取完成。

在網路爬蟲爬取過程中，為了加強爬取效率，一般採用平行爬取的方式。多個網路爬蟲在平行爬取過程中，不重複爬取同一個網頁尤為重要，這將會大幅加強爬取效率。

正常的做法如圖 2-2 所示。

網路爬蟲在爬取網頁時，搜尋引擎會建立兩張不同的表，一張表記錄已經造訪過的網址，另一張表記錄沒有存取過的網址。當網路爬蟲爬取某個外部連結頁面 URL 時，需把該網站的 URL 下載回來分析，當網路爬蟲處理好這個 URL 後，將該 URL 存入已經造訪過的表中。當另一個網路爬蟲從其他網站或頁面中又發現了這個 URL 時，它會在已存取列表中比較檢視有沒有該 URL 的存取記錄，如果有，則網路爬蟲會自動捨棄該 URL，不再存取。

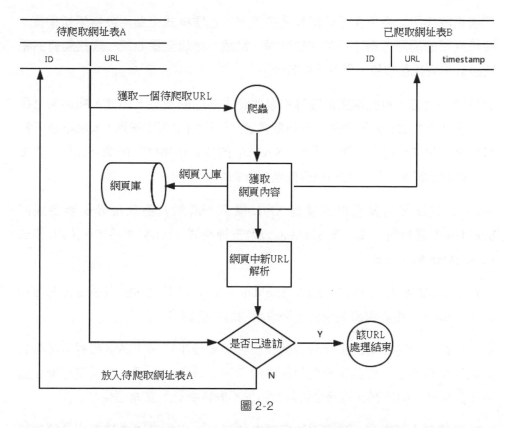

圖 2-2

網路爬蟲在按照連結爬取網頁的過程中，網頁之間的關係類似有向圖。在有向圖的節點檢查過程中，我們可以按照「先深度後廣度」的方式檢查，也可以按照「先廣度後深度」的方式檢查。同樣地，在網路爬蟲爬取網頁的過程中，網路爬蟲需要根據一定的策略來爬取網頁，一般採用「先深度後廣度」的方式。

此外，在網路爬蟲爬取網頁的過程中，還需要注意網頁的收錄模式，一般有兩種，即增量收集和全量搜集。全量搜集，顧名思義，每次爬取網頁都更新全部資料內容。這種模式一般定期展開，因為全量搜集模式的資源負擔大、付出成本高、內容更新的時效性低、網路寬頻消耗多，而且更新全量資料所需時間也比較長。

而增量收集可以避免全量收集模式的弊端，這種模式主要用於搜集新網頁、搜集更新的網頁，刪除不存在的頁面。當然，比起全量收集，網路爬蟲的系統設計也會比較複雜，但時效性較佳。

對於網站而言，被各家主流搜尋引擎收錄是共同的夙願。因此，網站常常採取一些技術方法告知搜尋引擎來抓取內容。一般網站可以透過 SiteMap 告知搜尋引擎網站中可提供抓取的網址，SiteMap 的核心作用就是向網路抓取工具提供一些提示訊息，以便它們更有效地抓取網站。

SiteMap 最簡單的實現形式就是 XML 檔案。當然，各家搜尋引擎定義的 SiteMap 不盡相同，如百度 SiteMap 分為三種格式：txt 文字格式、XML 格式和 SiteMap 索引格式。

對網站的維護人員而言，除了 SiteMap，還可以結合 SEO（Search Engine Optimization，搜尋引擎最佳化）來改善網站的被抓取效果。

綜上所述，網路爬蟲的工作核心就在於在網頁搜集效率、品質和對目標網站的人性化程度上。網路爬蟲要用最少的資源、最少的時間，搜集盡可能多的高品質網頁；同時對目標網站的內容抓取不影響網站的正常運轉和使用。

對軟體開發人員來說，我們可以用現有的爬蟲架構為基礎來實現對網路資料的爬取。Java 語言堆疊的讀者可以使用 WebMagic、Gecco；Python 語言堆疊的讀者可以使用 Scrapy；Go 語言堆疊的讀者可以使用 YiSpider。

2.4 網頁分析

網路爬蟲將網頁資料爬取後，會將網頁內容儲存到網頁函數庫。隨後，網頁分析程式將自動對網頁進行分析。主要的分析動作有網頁內容摘要、連結分析、網頁重要程度計算、關鍵字分析 / 斷詞、去除雜訊等。經過網頁分析後，網頁資料將變成網頁中關鍵字組、連結與關鍵字的相關度、網頁重要程度等資訊。

網頁內容中的去除雜訊主要是去除如廣告、無關的導覽列、版權資訊、調查問卷等和文章主體內容無關的內容。這些內容如果納入網頁分析中，常常會讓網頁的主題發生偏移；同時更多的無關內容的索引，還會使得索引結構規模變大，拖累搜索的準確性，降低搜索速度，而這一點非常重要，因為搜索的準確性和檢索速度是衡量搜尋引擎的主要標準之一。在實作中，去除雜訊可以基於 Doc View 模型實現。

網頁內容摘要一般由網頁正文產生，摘要一般會顯示在搜索結果的展示區，如圖 2-3 中框形區域所示。

圖 2-3

一般來説，摘要的產生方式有靜態產生方式和動態產生方式兩種。其中靜態產生方式比較簡單，在網頁分析階段即可從網頁內容中分析。雖然這種方式「短、平、快」，但缺點也很明顯，即當呈現搜索結果時，展示的摘要可能與搜索的關鍵字無關。

現在的搜尋引擎常常採用動態產生模的方式產生，即根據查詢關鍵字在文件中的位置，分析其週邊的文字，並反白顯示。

網頁重要程度計算用於衡量網站的權威性。一般越權威的網站，越容易被其他網站主動連結；換言之，網站被參考的次數越多，説明該網站越重要。對搜尋引擎而言，在傳回相關性強的內容時，應該儘量先傳回權威網站的內容；對搜尋引擎的使用者而言，這樣常常更能比對他們的需要。因此這也是評價搜尋引擎體驗好壞的核心指標之一。

網站之間連結的關係，其實可以從演算法的維度來解析。連結其實是一種投票、一種信任。網站被主動連結的次數越多，説明網際網路環境下其他網站對該網站的投票越多、信任越多，該網站在網際網路中也越流行。本質上這就是一種分散式系統下的共識投票。如圖 2-4 所示，網頁 A 在多個網頁間被主動連結、被主動投票的次數最多，因此我們認為網頁 A 更權威。

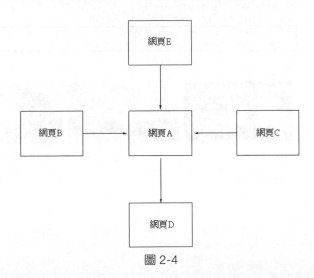

圖 2-4

如果將網頁間的連結關係視為有向圖，則網頁的連結關係就會變成內分支度和外分支度。這裡的內分支度指的是網頁能透過其他網頁的連結來存取；外分支度指的是網頁中連結了其他網頁。因此，內分支度大的網頁，説明其被多個網頁參考，這也表示該網頁比較權威、比較流行和熱門，如圖 2-4 中的網頁 A。

在各家搜尋引擎的實作中，Google 提出了 PageRank 演算法，該演算法也是 Google 搜尋引擎的重要法寶。

在關鍵字分析 / 斷詞環節，基礎技術是斷詞，因此斷詞是各家搜尋引擎中十分重要的技術。不論中英文網站，在建立索引之前都需要對內容進行斷詞。斷詞不僅是關鍵字分析的前提，也是後續文字採擷的基礎。

在斷詞方面，中英文斷詞有天然的差異性。相較而言，英文斷詞會簡單一些，因為英文有天然的空格作為單字的分隔符號。中文不僅沒有天然空格來分隔中文字，而且中文字的片語大部分由兩個及以上的中文字組成，中文敘述也習慣連續性撰寫，因此增加了中文斷詞的難度。

此外，在中英文中，都有一種詞稱之為 "Stop Word"，即「停用詞」。停用詞一般是指無內容指示意義的詞。

在中文斷詞時，常用的演算法可以分為兩大類，一種是以字典為基礎的機械式斷詞，另一種是以統計為基礎的斷詞。

以字典為基礎的斷詞方法，一般會按照一定的策略將待分析的中文字串與一個足夠大的詞典的詞條進行比對，若在詞典中找到某個字串詞條，則比對成功。因此以字典為基礎的斷詞方法的核心字串的比對。

在比對字串時分為正向比對和逆向比對兩種。正向比對指的是在比對字串時從左向右比對。如「中國」、「中國人」兩個詞條字串在比對時，從左向右可以依次比對成功「中」和「國」。而逆向比對與之相反，一般是從右向左比對。同樣，當「中國」、「中國人」兩個詞條字串從右向左比對時，「中國」的「國」會與「中國人」的「人」去比對，顯而易見，比對失敗；此時需要符合的字就是「中國人」中「人」字左側的「國」字，這時，兩個「國」字可以比對成功，依此類推，兩個字串詞條中的「中」字也能比對成功。

由於中文的特性，多個詞條常常有相關的字首或副檔名，如「中國」、「中國人」、「中國話」三個詞條都有「中國」這個公共字首。因此在某方向，如正向或逆向比對過程中，按比對長度的不同，還可以細分為最大 / 最長比對和最

小 / 最短比對。我們仍然以「中國」、「中國人」、「中國話」三個詞條舉例，正向最短符合的結果是「中」，而正向最長符合的結果是「中國」。

因此，以字典為基礎的斷詞演算法一般常用正向最大比對、逆向最大比對，或是組合模式。不過以字典為基礎的斷詞演算法對斷詞詞典的更新有較強依賴，特別是需要對新詞敏感的場景。

第二種斷詞演算法是以統計為基礎的斷詞演算法。該演算法無須詞典，一般會根據中文字與中文字相鄰出現的機率來進行斷詞。因此以統計為基礎的斷詞演算法常常需要建置一個語料庫，並且不斷地更新。在斷詞前，演算法需要進行前置處理，即對語料庫中相鄰出現的各個字的組合進行統計，計算兩個中文字間的組合機率。

其實不論是哪一種斷詞演算法，都會對新詞不敏感，都會分出一些經常出現但並非有效的片語，如「有點」、「有的」、「好像」、「非常」等。因此有必要對這種詞語進行過濾，畢竟斷詞演算法最主要的指標就是斷詞的準確率。

在開發實作中，各個語言堆疊常用的斷詞中介軟體整理如下。Python 語言中的中文片語組件有 jieba 中文斷詞，Java 語言中常用 Jcseg、Ansj 和庖丁斷詞，Go 語言中常用 sego。

2.5 倒排索引

在中文資訊檢索領域，索引的發展經歷了以字為基礎的索引和以詞為基礎的索引兩種。不論是基於字做索引，還是基於詞做索引，在建立索引過程中，均有關正排索引和倒排索引兩個資料結構。

正排索引的資料結構如圖 2-5 所示。

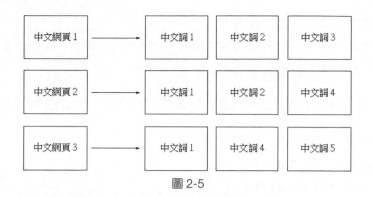

圖 2-5

如圖 2-5 所示，在正排索引中，以網頁或文章對映關係為 Key、以斷詞的列表為 Value。而在實際搜索網頁或文章時，剛好與此結構相反，即在搜索時是以查詢敘述的斷詞列表為 Key 來進行搜索的。因此，為了加強搜索效率，我們需要對正排索引進行轉化，將其轉化為以斷詞為 Key、以網頁或文章清單為 Value 的結構，而這個結構就是倒排索引，如圖 2-6 所示。

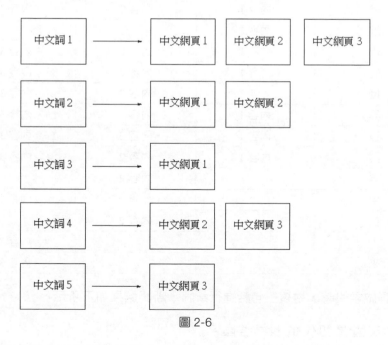

圖 2-6

在介紹完正排索引和倒排索引的結構後，下面我們以三句話為例，展示以字為基礎的索引和以詞為基礎的索引的區別。

內容 1：英國是近代的世界強國。

內容 2：美國是當代的世界強國。

內容 3：中國是未來的世界強國。

當以字做索引時，對上述三句話進行字為基礎的分解，分解結果如下所示。

內容 1：英 國 是 近 代 的 世 界 強 國。

內容 2：美 國 是 當 代 的 世 界 強 國。

內容 3：中 國 是 未 來 的 世 界 強 國。

倒排索引結構如表 2-1 所示。

表 2-1

字	內容清單		
英	內容 1		
國	內容 1	內容 2	內容 3
是	內容 1	內容 2	內容 3
近	內容 1		
代	內容 1	內容 2	內容 3
的	內容 1	內容 2	內容 3
世	內容 1	內容 2	內容 3
界	內容 1	內容 2	內容 3
強	內容 1	內容 2	內容 3
國	內容 1	內容 2	內容 3
美		內容 2	
中			內容 3
當		內容 2	
未			內容 3
來			內容 3

當基於詞做索引時，對這三句話進行斷詞，斷詞結果如下所示。

內容 1：英國 是 近代 的 世界 強國。

內容 2：美國 是 當代 的 世界 強國。

內容 3：中國 是 未來 的 世界 強國。

倒排索引結構如表 2-2 所示。

表 2-2

詞	內容清單		
英國	內容 1		
是	內容 1	內容 2	內容 3
近代	內容 1		
的	內容 1	內容 2	內容 3
世界	內容 1	內容 2	內容 3
強國	內容 1	內容 2	內容 3
美國		內容 2	
當代		內容 2	
中國			內容 3
未來			內容 3

透過比較不難發現，以詞做索引為基礎的索引內容明顯比以字做索引為基礎的索引內容要少，因而查詢時會更加高效。

此外，在倒排索引中一般還會記錄更多的資訊，如詞彙在網頁或文章中出現的位置、頻率和加權等，這些資訊在查詢階段會用到。

在倒排索引中，有詞條（Term）、詞典（Term Dictionary）、倒排表（Post List）三個名詞，下面我們一一介紹。

詞條是索引裡面最小的儲存和查詢單元。一般來說，在英文語境中詞條是一個單字，在中文語境中詞條指的是斷詞後的片語。

詞典又稱字典，是詞條的集合。單字詞典一般是由網頁或文章集合中出現過的所有詞組成的字串集合。

倒排表如圖 2-2 所示，倒排表記錄的是詞出現在哪些文件裡、出現的位置和頻率等。

在倒排表中，每筆記錄被稱為一個倒排項。

前文中提及，Elasticsearch 是以 Lucene 實現作基礎。在 Lucene 中，詞典和倒排表是實現快速檢索的重要基礎。另外，詞典和倒排表分兩部分儲存，詞典儲存在記憶體中，倒排表儲存在磁碟上。

2.6 結果排序

搜尋引擎除了對網路爬蟲爬取的網頁進行處理，將它們結構化成倒排索引之外，還有一項重要工作就是回應使用者的查詢需求。

搜尋引擎系統接收使用者提交的查詢字串後，對字串進行斷詞，去除不必要的停用詞等無意義詞彙後，進行倒排索引的查詢。多個關鍵字的倒排索引查詢結果的交集即為搜索的結果。

而搜索的結果常常需要進一步處理，例如一般都會進行排序。搜索結果排序是搜尋引擎查詢服務的核心所在。排序結果決定了搜尋引擎體驗的好與壞、使用者的使用滿意度和搜尋引擎的口碑。

搜索結果的排序演算法也在不斷反覆運算發展，早期主要以查詢詞出現的頻率來排序，隨後出現了 PageRank 和相關性等演算法排序。

一般而言，相關性演算法主要考慮的因素有關鍵字的使用頻率、關鍵字在網頁中的詞頻、關鍵字出現在所在網頁的位置、關鍵字間的距離、網頁連結及重要性。其中，關鍵字的使用頻率指的是日常生活使用的頻率，如「有的」、「有點」、「可能」、「非常」這些詞經常出現在日常生活會話中，但在搜尋引擎看來，這些詞彙的意義並不大。

關鍵字在網頁中的詞頻越高，表示出現次數越多，說明頁面與搜索詞的關係越密切。

關鍵字出現在所在網頁的位置是指關鍵字是否出現在比較重要的位置，例如標題。關鍵字出現在所在網頁的位置越重要，說明頁面與關鍵字越相關。一般而言，倒排索引函數庫在建立時，會將關鍵字出現在所在網頁的位置記錄起來。

關鍵字間的距離指的是多個關鍵字在頁面上出現的位置的接近程度，關鍵字間越接近，說明在該網頁與搜索詞字串的相關度越高。

網頁連結及重要性指的是頁面有越多以搜索詞為關鍵字的匯入連結，說明頁面的相關性越強；連結分析還包含了連結來源頁面本身的主題、目標文字周圍的文字等。

當然，在目前的搜尋引擎中，還都不約而同地引用了使用者行為分析、資料採擷等技術，來提升搜索結果的品質。

2.7 中文斷詞實戰

前文提及了 Ansj 和 Jcseg 兩種輕量級 Java 中文斷詞器，本節將介紹這兩種片語組件的使用方法。

2.7.1 Ansj 中文斷詞

Ansj 中文斷詞基於 n-Gram+CRF+HMM 演算法，用 Java 實現。Ansj 中文斷詞工具的斷詞速度可達到大約 200 萬字 /s，準確率達 96% 以上。

目前 Ansj 中文斷詞工具已經實現了中文斷詞、中文姓名識別、使用者自訂字典、關鍵字分析、自動摘要、關鍵字標記等功能，可以應用到自然語言處理等方面，適用於對斷詞效果要求較高的各種專案。

Ansj 中文斷詞工具支援的斷詞方式有 ToAnalysis（精準斷詞）、DicAnalysis（使用者自訂字典優先策略的斷詞）、NlpAnalysis（帶有新詞發現功能的斷詞）、IndexAnalysis（針對索引的斷詞）、BaseAnalysis（最小顆粒度的斷詞）。

其中，ToAnalysis 是 Ansj 斷詞方式的偏好。該斷詞方式在便利性、穩定性、準確性及斷詞效率上，都獲得了一個不錯的平衡。如果初次嘗試使用 Ansj，想「開箱即用」，那麼建議使用 ToAnalysis。

在 DicAnalysis 中，使用者可以自訂字典優先策略。如果自訂字典足夠好，或使用者對自訂字典的要求比較高，那麼強烈建議使用 DicAnalysis。在此種情景下，很多方面它會優於 ToAnalysis 的斷詞結果。

NlpAnalysis 可以識別出未登入詞。它的缺點是速度比較慢、穩定性比較差。如果需要進行未登入詞識別、對文字進行分析，則首推該斷詞方法。

IndexAnalysis 適合在 Lucene 等文字檢索中使用，其召回率較高，準確率也較高。

BaseAnalysis 則確保了最基本的斷詞，詞語顆粒度非常小，所涉及的詞大約是 10 萬（個）左右。這種斷詞方式的斷詞速度非常快，可達到 300 萬字 /s，同時準確率也很高。不過，對於新詞，這種斷詞方式的效果不太理想。

下面展示上述斷詞方式的斷詞 API 的使用和斷詞效果。斷詞 API 的使用如下所示：

```
package com.niudong.esdemo.util;
import org.ansj.splitWord.analysis.BaseAnalysis;
import org.ansj.splitWord.analysis.DicAnalysis;
import org.ansj.splitWord.analysis.IndexAnalysis;
import org.ansj.splitWord.analysis.NlpAnalysis;
import org.ansj.splitWord.analysis.ToAnalysis;
public class AnsjSegUtil {
  public static void processString(String content) {
    // 最小顆粒度的斷詞
    System.out.println(BaseAnalysis.parse(content));
    // 精準斷詞
    System.out.println(ToAnalysis.parse(content));
    // 使用者自訂字典優先策略的斷詞
    System.out.println(DicAnalysis.parse(content));
    // 針對索引的斷詞
    System.out.println(IndexAnalysis.parse(content));
    // 帶有新詞發現功能的斷詞
    System.out.println(NlpAnalysis.parse(content));
  }
```

```
public static void main(String[] args) {
    String content ="15 年來首批由深海探險家組成的國際團隊五次潛入大西洋海底 3800
公尺深處,對鐵達尼號沉船殘骸進行調查。探險隊發現,雖然沉船的部分殘骸狀況良好,也有一些
部分已消失在大海中。強大的洋流、鹽蝕和細菌正不斷侵蝕著這艘沉船。英媒曝光的高畫質圖片可
見,沉船的部分殘骸遭腐蝕情況嚴重。";
    processString(content);
}
}
```

在目前類別中執行 main 方法,各個斷詞方式的執行結果如下所示。

▨ 最小顆粒度的斷詞

15/m, 年來 /t, 首批 /n, 由 /p, 深海 /n, 探險家 /n, 組成 /v, 的 /u, 國際 /n, 團隊 /n, 五 /m, 次 /q, 潛入 /v, 大西洋 /ns, 海底 /n,3800/m, 米 /q, 深處 /s, , /w, 對 /p, 鐵達尼號 /nz, 沉船 /n, 殘骸 /n, 進行 /v, 調查 /vn。/w, 探險隊 /n, 發現 /v, , /w, 雖然 /c, 沉船 /n, 的 /u, 部分 /n, 殘骸 /n, 狀況 /n, 良好 /a, , /w, 也 /d, 有 /v, 一些 /m, 部分 /n, 已 /d, 消失 /v, 在 /p, 大海 /n, 中 /f。/w, 強大 /a, 的 /u, 洋流 /n、/w, 鹽 /n, 蝕 /vg, 和 /c, 細菌 /n, 正 /d, 不斷 /d, 侵蝕 /vn, 著 /u, 這 /r, 艘 /q, 沉船 /n。/w, 英 /j, 媒 /ng, 曝光 /v, 的 /u, 高畫質 /n, 圖片 /n, 可見 /v, , /w, 沉船 /n, 的 /u, 部分 /n, 殘骸 /n, 遭 /v, 腐蝕 /v, 情況 /n, 嚴重 /a。/w

該方式累計有效斷詞 71 個,其中單字斷詞結果 24 個。

▨ 精準斷詞

15/m, 年來 /t, 首批 /n, 由 /p, 深海 /n, 探險家 /n, 組成 /v, 的 /u, 國際 /n, 團隊 /n, 五次 /mq, 潛入 /v, 大西洋 /ns, 海底 /n,3800 公尺 /mq, 深處 /s, , /w, 對 /p, 鐵達尼號 /nz, 沉船 /n, 殘骸 /n, 進行 /v, 調查 /vn。/w, 探險隊 /n, 發現 /v, , /w, 雖然 /c, 沉船 /n, 的 /u, 部分 /n, 殘骸 /n, 狀況 /n, 良好 /a, , /w, 也 /d, 有 /v, 一些 /m, 部分 /n, 已 /d, 消失 /v, 在 /p, 大海 /n, 中 /f。/w, 強大 /a, 的 /u, 洋流 /n、/w, 鹽 /n, 蝕 /vg, 和 /c, 細菌 /n, 正 /d, 不斷 /d, 侵蝕 /vn, 著 /u, 這 /r, 艘 /q, 沉船 /n。/w, 英 /j, 媒 /ng, 曝光 /v, 的 /u, 高畫質 /n, 圖片 /n, 可見 /v, , /w, 沉船 /n, 的 /u, 部分 /n, 殘骸 /n, 遭 /v, 腐蝕 /v, 情況 /n, 嚴重 /a。/w

該方式累計有效斷詞 68 個，其中單字斷詞結果 21 個。

☑ 使用者自訂字典優先策略的斷詞

15/m, 年來 /t, 首批 /n, 由 /p, 深海 /n, 探險家 /n, 組成 /v, 的 /u, 國際 /n, 團隊 /n, 五次 /mq, 潛入 /v, 大西洋 /ns, 海底 /n,3800 公尺 /mq, 深處 /s,，/w, 對 /p, 鐵達尼號 /nz, 沉船 /n, 殘骸 /n, 進行 /v, 調查 /vn,。/w, 探險隊 /n, 發現 /v,，/w, 雖然 /c, 沉船 /n, 的 /u, 部分 /n, 殘骸 /n, 狀況 /n, 良好 /a,，/w, 也 /d, 有 /v, 一些 /m, 部分 /n, 已 /d, 消失 /v, 在 /p, 大 /a, 海中 /s,。/w, 強大 /a, 的 /u, 洋流 /n,、/w, 鹽 /n, 蝕 /vg, 和 /c, 細菌 /n, 正 /d, 不斷 /d, 侵蝕 /vn, 著 /u, 這 /r, 艘 /q, 沉船 /n,。/w, 英 /j, 媒 /ng, 曝光 /v, 的 /u, 高畫質 /n, 圖片 /n, 可見 /v,，/w, 沉船 /n, 的 /u, 部分 /n, 殘骸 /n, 遭 /v, 腐蝕 /v, 情況 /n, 嚴重 /a,。/w

該方式累計有效斷詞 68 個，其中單字斷詞結果 21 個。

☑ 針對索引的斷詞

15/m, 年來 /t, 首批 /n, 由 /p, 深海 /n, 探險家 /n, 組成 /v, 的 /u, 國際 /n, 團隊 /n, 五次 /mq, 潛入 /v, 大西洋 /ns, 海底 /n,3800 公尺 /mq, 深處 /s,，/w, 對 /p, 鐵達尼號 /nz, 沉船 /n, 殘骸 /n, 進行 /v, 調查 /vn,。/w, 探險隊 /n, 發現 /v,，/w, 雖然 /c, 沉船 /n, 的 /u, 部分 /n, 殘骸 /n, 狀況 /n, 良好 /a,，/w, 也 /d, 有 /v, 一些 /m, 部分 /n, 已 /d, 消失 /v, 在 /p, 大海 /n, 中 /f,。/w, 強大 /a, 的 /u, 洋流 /n,、/w, 鹽 /n, 蝕 /vg, 和 /c, 細菌 /n, 正 /d, 不斷 /d, 侵蝕 /vn, 著 /u, 這 /r, 艘 /q, 沉船 /n,。/w, 英 /j, 媒 /ng, 曝光 /v, 的 /u, 高畫質 /n, 圖片 /n, 可見 /v,，/w, 沉船 /n, 的 /u, 部分 /n, 殘骸 /n, 遭 /v, 腐蝕 /v, 情況 /n, 嚴重 /a,。/w

該方式累計有效斷詞 68 個，其中單字斷詞結果 21 個。

☑ 帶有新詞發現功能的斷詞

15 年 /t, 來 /v, 首 /m, 批 /q, 由 /p, 深海 /n, 探險家 /n, 組成 /v, 的 /u, 國際 /n, 團隊 /n, 五次 /mq, 潛入 /v, 大西洋 /ns, 海底 /n,3800 公尺 /mq, 深處 /s,，/w, 對 /p, 鐵達尼號 /nz, 沉船 /n, 殘骸 /n, 進行 /v, 調查 /vn,。/w, 探險隊 /n, 發現 /v,，/w, 雖然 /c, 沉船 /n, 的 /u, 部分 /n, 殘骸 /n, 狀況 /n, 良好 /a,，/w, 也 /d, 有 /v, 一些 /m,

部分 /n, 已 /d, 消失 /v, 在 /p, 大海 /n, 中 /f,。/w, 強大 /a, 的 /u, 洋流 /n,、/w, 鹽蝕 /nw, 和 /c, 細菌 /n, 正 /d, 不斷 /d, 侵蝕 /vn, 著 /u, 這 /r, 艘 /q, 沉船 /n,。/w, 英媒 /nw, 曝光 /v, 的 /u, 高畫質 /n, 圖片 /n, 可見 /v,,/w, 沉船 /n, 的 /u, 部分 /n, 殘骸 /n, 遭 /v, 腐蝕 /v, 情況 /n, 嚴重 /a,。/w

該方式累計有效斷詞 67 個，其中單字斷詞結果 20 個。

透過上述斷詞結果我們可以看到，精準斷詞在整體的斷詞效果上確實較好，在新詞識別上，如上面文字中的「鹽蝕」、「英媒」的識別，是帶有新詞發現功能的斷詞，即 NlpAnalysis 的效果更好。

2.7.2 Jcseg 輕量級 Java 中文斷詞器

Jcseg 是以 MMSEG 演算法為基礎的輕量級中文斷詞器，整合了關鍵字分析、關鍵子句分析、關鍵句子分析和文章自動摘要等功能，並且提供了一個以 Jetty 為基礎的 Web 伺服器，方便各大語言直接 HTTP 呼叫，同時提供了最新版本的 Lucene、Solr 和 Elasticsearch 的斷詞介面。

此外，Jcseg 還附帶了一個 jcseg.properties 檔案，用於快速設定，進一步獲得適合不同場合的斷詞應用，如最大比對詞長、是否開啟中文人名識別、是否追加拼音、是否追加同義字等。

Jcseg 支援的中文斷詞模式有六種，即簡易模式、複雜模式、檢測模式、檢索模式、分隔符號模式、NLP 模式，各自特點如下。

（1）簡易模式：以 FMM 演算法實現，適合斷詞速度要求較高的場合。

（2）複雜模式：基於 MMSEG 四種過濾演算法實現，可去除較高的問題，斷詞準確率達 98.41%。

（3）檢測模式：只傳回詞函數庫中已有的詞條，很適合某些應用場合。

（4）檢索模式：轉為細粒度切分，專為檢索而生。除中文處理外（不具備中文的人名、數字識別等智慧功能），其他與複雜模式一致（英文，組合詞等）。

（5）分隔符號模式：按照指定的字元切斷詞條，預設是空格，特定場合的應用。

（6）NLP 模式：繼承自複雜模式，更改了數字、單位等詞條的組合方式，增加了電子郵件、手機號碼、網址、人名、地名、貨幣等無限種自訂實體的識別與傳回。

Jcseg 的核心功能有中文斷詞、關鍵字分析、關鍵子句分析、關鍵句子分析、文章自動摘要、自動詞性標記和命名實體標記，並提供了 RESTful API。各個功能的介紹如下。

（1）中文斷詞：以 MMSEG 演算法和 Jcseg 獨創為基礎的最佳化演算法實現，有四種切分模式。

（2）關鍵字分析：基於 textRank 演算法實現。

（3）關鍵子句分析：基於 textRank 演算法實現。

（4）關鍵句子分析：基於 textRank 演算法實現。

（5）文章自動摘要：基於 BM25+textRank 演算法實現。

（6）自動詞性標記：基於詞函數庫實現。目前效果不是很理想，對詞性標記結果要求較高的應用不建議使用。

（7）命名實體標記：基於詞函數庫實現，可以識別電子郵件、網址、手機號碼、地名、人名、貨幣、datetime 時間、長度、面積、距離單位等。

（8）RESTful API：嵌入 Jetty 並提供了一個絕對高性能的 Server 模組，包含全部功能的 HTTP 介面，標準化 JSON 輸出格式，方便各種語言用戶端直接呼叫。

Jcseg 支援自訂詞函數庫。在 jcseg\vendors\lexicon 資料夾下，可以隨便增加、刪除、更改詞函數庫和詞函數庫內容，並且對詞函數庫進行了分類。此外，Jcseg 還支援詞函數庫多目錄載入，在設定 lexicon.path 時使用 ';' 隔開多個詞函數庫目錄即可。

詞函數庫中的中文字類型可以分為簡體、繁體、簡繁體混合類型，Jcseg 可以專門適用於簡體切分、繁體切分、簡繁體混合切分，並且可以利用同義字實現簡繁體的相互檢索。Jcseg 同時提供了兩個簡單的詞函數庫管理工具來進行簡繁體的轉換和詞函數庫的合併。

Jcseg 還支援中英文同義字追加、同義字比對、中文詞條拼音追加。Jcseg 的詞函數庫中整合了《現代中文詞典》和 cc-cedict 辭典中的詞條，並且依據 cc-cedict 詞典為詞條標上了拼音。使用時，可以更改 jcseg.properties 設定文件，在斷詞的時候，把拼音和同義字加入斷詞結果中。

Jcseg 還支援中文數字和中文分數識別，如「一百五十個人都來了，四十分之一的人。」中的「一百五十」和「四十分之一」均可有效識別。在輸出結果時，Jcseg 會自動將其轉為阿拉伯數字加入斷詞結果中，如 150、1/40。

Jcseg 支援中英混合詞和英中混合詞的識別，如 B 超、X 射線、卡拉 OK、奇都 KTV、哆啦 A 夢等。此外，Jcseg 還支援阿拉伯數字 / 小數 / 中文數字基本單字單位的識別，如 2012 年、1.75 公尺、38.6℃、五折（Jcseg 會將其轉為「5 折」加入斷詞結果中）。

Jcseg 對智慧圓角半形、英文大小寫轉換、特殊字母識別（如 I、II、①、⑩）等均能有效支援。

Jcseg 可以設定 Elasticsearch 使用，實際步驟如下。

（1）拉取 Jcseg 專案程式並進行編譯包裝。其中，拉取程式的指令為 git clone https://gitee.com/lionsoul/jcseg.git，如圖 2-7 所示。

```
牛冬@LAPTOP-1S8BALK3 MINGW64 ~/Desktop/ES/code
$ git clone https://gitee.com/lionsoul/jcseg.git
Cloning into 'jcseg'...
remote: Enumerating objects: 6378, done.
remote: Counting objects: 100% (6378/6378), done.
remote: Compressing objects: 100% (2317/2317), done.
remote: Total 6378 (delta 2868), reused 6056 (delta 2575)
Receiving objects: 100% (6378/6378), 8.91 MiB | 901.00 KiB/s, done.
Resolving deltas: 100% (2868/2868), done.
```

圖 2-7

將 Jcseg 專案程式拉取到本機後切換到 Jcseg 目錄，即在 DOS 視窗中執行指令 cd jcseg，隨後在該目錄下執行編譯包裝指令 mvn package，編譯執行的輸出結果如圖 2-8 所示。

```
牛冬@LAPTOP-1S8BALK3 MINGW64 ~/Desktop/ES/code/jcseg (master)
$ mvn package
[INFO] Scanning for projects...
[INFO] ------------------------------------------------------------------------
[INFO] Reactor Build Order:
[INFO]
[INFO] jcseg
[INFO] jcseg-core
[INFO] jcseg-analyzer
[INFO] jcseg-elasticsearch
[INFO] jcseg-server
[INFO]
[INFO] ------------------------------------------------------------------------
[INFO] Building jcseg 2.4.1
[INFO] ------------------------------------------------------------------------
[INFO]
[INFO] ------------------------------------------------------------------------
[INFO] Building jcseg-core 2.4.1
[INFO] ------------------------------------------------------------------------
```

圖 2-8

編譯包裝成功後，輸出的內容如圖 2-9 所示。

```
[INFO] ------------------------------------------------------------------------
[INFO] Reactor Summary:
[INFO]
[INFO] jcseg .............................................. SUCCESS [  0.279 s]
[INFO] jcseg-core ......................................... SUCCESS [ 26.282 s]
[INFO] jcseg-analyzer ..................................... SUCCESS [  4.896 s]
[INFO] jcseg-elasticsearch ................................ SUCCESS [  6.274 s]
[INFO] ------------------------------------------------------------------------
[INFO] BUILD SUCCESS
[INFO] ------------------------------------------------------------------------
[INFO] Total time: 37.968 s
[INFO] Finished at: 2019-08-23T22:49:04+08:00
[INFO] Final Memory: 44M/618M
[INFO] ------------------------------------------------------------------------
Picked up JAVA_TOOL_OPTIONS: -Dfile.encoding=UTF-8
```

圖 2-9

（2）在 Elasticsearch 專案中設定 Jcseg 斷詞，實際方法如下。

先在 \elasticsearch-7.2.0\plugins 目錄下新增資料夾 jcseg，如圖 2-10 所示。

| jcseg | 2019/8/24 15:49 |

圖 2-10

將（1）中包裝的 3 個 jar 檔案 jcseg-analyzer-2.4.1.jar、jcseg-core-2.4.1.jar 和 jcseg-elasticsearch-2.4.1.jar 複製到 {ES_HOME}/plugins/jcseg 目錄下，如圖 2-11 所示。

jcseg-elasticsearch-2.4.1.jar	2019/8/23 22:49	Executable Jar File	14 KB
jcseg-analyzer-2.4.1.jar	2019/8/23 22:48	Executable Jar File	7 KB
jcseg-core-2.4.1.jar	2019/8/23 22:48	Executable Jar File	1,847 KB

圖 2-11

將 Jcseg 專案中的設定檔 \jcseg-core\jcseg.properties 複製到 {ES_HOME}/plugins/jcseg 目錄下，如圖 2-12 所示。

jcseg.properties	2019/8/23 22:29	PROPERTIES 文件	3 KB
jcseg-analyzer-2.4.1.jar	2019/8/23 22:48	Executable Jar File	7 KB
jcseg-core-2.4.1.jar	2019/8/23 22:48	Executable Jar File	1,847 KB
jcseg-elasticsearch-2.4.1.jar	2019/8/23 22:49	Executable Jar File	14 KB

圖 2-12

將 Jcseg 專案中的設定檔 jcseg-elasticsearch/plugin/plugin-descriptor.properties 複製到 {ES_HOME}/plugins/jcseg 目錄下，如圖 2-13 所示。

jcseg.properties	2019/8/23 22:29	PROPERTIES 文件	3 KB
jcseg-analyzer-2.4.1.jar	2019/8/23 22:48	Executable Jar File	7 KB
jcseg-core-2.4.1.jar	2019/8/23 22:48	Executable Jar File	1,847 KB
jcseg-elasticsearch-2.4.1.jar	2019/8/23 22:49	Executable Jar File	14 KB
plugin-descriptor.properties	2019/8/23 22:29	PROPERTIES 文件	1 KB

圖 2-13

將 Jcseg 專案中的 lexicon 資料夾複製到 {ES_HOME}/plugins/jcseg 目錄下，如圖 2-14 所示。

lexicon	2019/8/24 15:55	文件夾	
jcseg.properties	2019/8/23 22:29	PROPERTIES 文件	3 KB
jcseg-analyzer-2.4.1.jar	2019/8/23 22:48	Executable Jar File	7 KB
jcseg-core-2.4.1.jar	2019/8/23 22:48	Executable Jar File	1,847 KB
jcseg-elasticsearch-2.4.1.jar	2019/8/23 22:49	Executable Jar File	14 KB
plugin-descriptor.properties	2019/8/23 22:29	PROPERTIES 文件	1 KB

圖 2-14

隨後設定 jcseg.properties，主要是設定 lexicon.path，以便指向正確的詞函數庫，筆者本機的設定如圖 2-15 所示。

```
####about the lexicon
#abusolte path of the lexicon file.
#Multiple path support from jcseg 1.9.2, use ';' to split different path.
#example: lexicon.path = /home/chenxin/lex1;/home/chenxin/lex2 (Linux)
#        : lexicon.path = D:/jcseg/lexicon/1;D:/jcseg/lexicon/2 (WinNT)
#lexicon.path=/Code/java/JavaSE/jcseg/lexicon
#lexicon.path = {jar.dir}/lexicon ({jar.dir} means the base directory of jcseg-core-{version}.jar)
#@since 1.9.9 Jcseg default to load the lexicons in the classpath
lexicon.path = C:/elasticsearch-7.2.0-windows-x86_642/elasticsearch-7.2.0/plugins/jcseg/lexicon →

#Wether to load the modified lexicon file auto.
lexicon.autoload = 0

#Poll time for auto load. (seconds)
lexicon.polltime = 300



####lexicon load
#Wether to load the part of speech of the entry.
jcseg.loadpos = 1

#Wether to load the pinyin of the entry.
jcseg.loadpinyin = 0

#Wether to load the synoyms words of the entry.
jcseg.loadsyn = 1

#wether to load the entity of the entry
jcseg.loadentity = 1
```

圖 2-15

啟動 Elasticsearch，在啟動過程中會載入 Jcseg，如圖 2-16 所示。

```
[2019-08-24T17:58:53,712][INFO ][o.e.p.PluginsService     ] [LAPTOP-1S8BALK3] loaded plugin [jcseg]
[2019-08-24T17:59:02,210][INFO ][o.e.x.s.a.s.FileRolesStore] [LAPTOP-1S8BALK3] parsed [0] roles from file [C:\elasticsea
rch-7.2.0-windows-x86_642\elasticsearch-7.2.0\config\roles.yml]
[2019-08-24T17:59:04,341][INFO ][o.e.x.m.p.l.CppLogMessageHandler] [LAPTOP-1S8BALK3] [controller/28728] [Main.cc@110] co
ntroller (64 bit): Version 7.2.0 (Build 65aefcbfce449b) Copyright (c) 2019 Elasticsearch BV
[2019-08-24T17:59:05,737][DEBUG][o.e.a.ActionModule       ] [LAPTOP-1S8BALK3] Using REST wrapper from plugin org.elastic
search.xpack.security.Security
[2019-08-24T17:59:06,471][INFO ][o.e.d.DiscoveryModule    ] [LAPTOP-1S8BALK3] using discovery type [zen] and seed hosts
providers [settings]
[2019-08-24T17:59:08,465][INFO ][o.e.n.Node               ] [LAPTOP-1S8BALK3] initialized
[2019-08-24T17:59:08,466][INFO ][o.e.n.Node               ] [LAPTOP-1S8BALK3] starting ...
[2019-08-24T17:59:10,563][INFO ][o.e.t.TransportService   ] [LAPTOP-1S8BALK3] publish_address {127.0.0.1:9300}, bound_ad
dresses {127.0.0.1:9300}, {[::1]:9300}
[2019-08-24T17:59:10,641][WARN ][o.e.b.BootstrapChecks    ] [LAPTOP-1S8BALK3] the default discovery settings are unsuita
ble for production use; at least one of [discovery.seed_hosts, discovery.seed_providers, cluster.initial_master_nodes] m
ust be configured
[2019-08-24T17:59:10,657][INFO ][o.e.c.c.Coordinator      ] [LAPTOP-1S8BALK3] cluster UUID [6UkGmpPrSwyX1zPHgSaivg]
[2019-08-24T17:59:10,693][INFO ][o.e.c.c.ClusterBootstrapService] [LAPTOP-1S8BALK3] no discovery configuration found, wi
ll perform best-effort cluster bootstrapping after [3s] unless existing master is discovered
[2019-08-24T17:59:11,135][INFO ][o.e.c.s.MasterService    ] [LAPTOP-1S8BALK3] elected-as-master ([1] nodes joined)[{LAPT
OP-1S8BALK3}{3gNRbEL_TJSocpgRnc2wjA}{6Pqmhu5_QbC-AFSkD6kc2w}{127.0.0.1}{127.0.0.1:9300}{ml.machine_memory=8467296256, xp
ack.installed=true, ml.max_open_jobs=20} elect leader, _BECOME_MASTER_TASK_, _FINISH_ELECTION_], term: 7, version: 63, r
eason: master node changed {previous [], current [{LAPTOP-1S8BALK3}{3gNRbEL_TJSocpgRnc2wjA}{6Pqmhu5_QbC-AFSkD6kc2w}{127.
0.0.1}{127.0.0.1:9300}{ml.machine_memory=8467296256, xpack.installed=true, ml.max_open_jobs=20}]}
[2019-08-24T17:59:11,295][INFO ][o.e.c.s.ClusterApplierService] [LAPTOP-1S8BALK3] master node changed {previous [], curr
ent [{LAPTOP-1S8BALK3}{3gNRbEL_TJSocpgRnc2wjA}{6Pqmhu5_QbC-AFSkD6kc2w}{127.0.0.1}{127.0.0.1:9300}{ml.machine_memory=8467
296256, xpack.installed=true, ml.max_open_jobs=20}]}, term: 7, version: 63, reason: Publication{term=7, version=63}
[2019-08-24T17:59:12,298][INFO ][o.e.l.LicenseService     ] [LAPTOP-1S8BALK3] license [bbcef3ff-378f-4c13-bb43-962a9dff2
841] mode [basic] - valid
[2019-08-24T17:59:12,345][INFO ][o.e.g.GatewayService     ] [LAPTOP-1S8BALK3] recovered [2] indices into cluster_state
[2019-08-24T17:59:15,005][INFO ][o.e.h.AbstractHttpServerTransport] [LAPTOP-1S8BALK3] publish_address {127.0.0.1:9200},
bound_addresses {127.0.0.1:9200}, {[::1]:9200}
[2019-08-24T17:59:15,006][INFO ][o.e.n.Node               ] [LAPTOP-1S8BALK3] started
[2019-08-24T17:59:15,790][INFO ][o.e.c.r.a.AllocationService] [LAPTOP-1S8BALK3] Cluster health status changed from [RED]
to [YELLOW] (reason: [shards started [[ultraman1][0], [ultraman][0]] ...]).
```

圖 2-16

在 DOS 視窗執行以下指令，測試斷詞效果。

```
curl 'http://localhost:9200/_analyze?pretty=true' -H 'Content-Type:
application/ json' -d'
{
    "analyzer": "jcseg_search",
    "text": " 一百美金等於多少人民幣 "
}'
```

斷詞結果列印如下所示。

```
{
  "tokens" : [
    {
      "token" : " 一 ",
      "start_offset" : 0,
      "end_offset" : 1,
      "type" : "word",
      "position" : 0
    },
    {
      "token" : " 一百 ",
      "start_offset" : 0,
      "end_offset" : 2,
      "type" : "word",
      "position" : 1
    },
    {
      "token" : " 百 ",
      "start_offset" : 1,
      "end_offset" : 2,
      "type" : "word",
      "position" : 2
    },
    {
      "token" : " 美 ",
      "start_offset" : 2,
      "end_offset" : 3,
      "type" : "word",
      "position" : 3
```

```
    },
    {
      "token" : "美金",
      "start_offset" : 2,
      "end_offset" : 4,
      "type" : "word",
      "position" : 4
    },
    {
      "token" : "元",
      "start_offset" : 3,
      "end_offset" : 4,
      "type" : "word",
      "position" : 5
    },
    {
      "token" : "等",
      "start_offset" : 4,
      "end_offset" : 5,
      "type" : "word",
      "position" : 6
    },
    {
      "token" : "等於",
      "start_offset" : 4,
      "end_offset" : 6,
      "type" : "word",
      "position" : 7
    },
    {
      "token" : "於",
      "start_offset" : 5,
      "end_offset" : 6,
      "type" : "word",
      "position" : 8
    },
    {
```

```
    "token" : " 多 ",
    "start_offset" : 6,
    "end_offset" : 7,
    "type" : "word",
    "position" : 9
},
{
    "token" : " 多少 ",
    "start_offset" : 6,
    "end_offset" : 8,
    "type" : "word",
    "position" : 10
},
{
    "token" : " 少 ",
    "start_offset" : 7,
    "end_offset" : 8,
    "type" : "word",
    "position" : 11
},
{
    "token" : " 人 ",
    "start_offset" : 8,
    "end_offset" : 9,
    "type" : "word",
    "position" : 12
},
{
    "token" : " 人民 ",
    "start_offset" : 8,
    "end_offset" : 10,
    "type" : "word",
    "position" : 13
},
{
    "token" : " 人民幣 ",
    "start_offset" : 8,
```

```
        "end_offset" : 11,
        "type" : "word",
        "position" : 14
      },
      {
        "token" : "民",
        "start_offset" : 9,
        "end_offset" : 10,
        "type" : "word",
        "position" : 15
      },
      {
        "token" : "幣",
        "start_offset" : 10,
        "end_offset" : 11,
        "type" : "word",
        "position" : 16
      }
    ]
}
```

2.8 基礎知識連結

倒排索引的思想在日常生活中十分常見。例如在某個公車站或捷運站，我們會以該站為起始點做路線查詢與規劃，而不是以公車路線或捷運路線──顯然該站就是倒排索引中的片語，而公車路線或捷運路線就是倒排索引中的文章。

同樣，我們在做讀書筆記的過程中，通常是將書中的內容拆解為基礎知識，以基礎知識複習和思考──顯然基礎知識就是倒排索引中的片語，而以書名為單位的書就是倒排索引中的文章。

在本章開篇，我們提及了結構化資料和非結構化資料。結構化不僅是一種資料表現形式，更是一種思維方式。

結構化思維的核心是邏輯性。一般常用的邏輯有因果關係、時空關係、優先順序關係等。而對這些邏輯性內容，我們其實並不陌生，並且從小就接觸。例如在小學的作文練習中，「總分總」結構是必練的思維之一；在解答應用題時，老師教的「策略」也是結構化思維的一種外化形式；而在工作中，常用的思維導圖工具更是結構化思維的產物。對程式設計師而言，架構設計一般採用從上往下的模式也是結構化思維的一種表現。

結構化思維不僅表現在日常的溝通表達上、在敲程式的過程中，更滲透在我們分析問題、解決問題的點滴過程裡。

2.9 小結

本章以資料的檢索為切入點，主要介紹了搜尋引擎的工作原理，對搜尋引擎的核心模組如網路爬蟲、網頁分析、倒排索引、結果排序和中文斷詞等進行了詳細說明，並介紹了近幾年開放原始碼片語組件中 Java 語言堆疊的優秀代表 Ansj 和 Jcseg。

初識 Elasticsearch

與君初相識
猶如故人歸

本章主要介紹 Elasticsearch 的基礎知識，如 Elasticsearch 的安裝、設定，另外，還會介紹 Elasticsearch 的相關術語及架構設計，以方便讀者學習後續章節。

3.1 Elasticsearch 簡介

Elasticsearch 是一個分散式、可擴充、近即時的高性能搜索與資料分析引擎。Elasticsearch 基於 Apache Lucene 建置，採用 Java 撰寫，並使用 Lucene 建置索引、提供搜索功能。Elasticsearch 的目標是讓全文檢索搜尋功能的落地變得更簡單。

Elasticsearch 的特點和優勢如下：

（1）分散式即時檔案儲存。Elasticsearch 可將被索引文件中的每一個欄位存入索引，以便欄位可以被檢索到。

（2）即時分析的分散式搜尋引擎。Elasticsearch 的索引分拆成多個分片，每個分片可以有零個或多個備份。叢集中的每個資料節點都可承載一個或多個分片，並且協調和處理各種操作；負載再平衡和路由會自動完成。

（3）高可擴充性。大規模應用方面，Elasticsearch 可以擴充到上百台伺服器，處理 PB 等級的結構化或非結構化資料。當然，Elasticsearch 也可以執行在單台 PC 上。

（4）可抽換外掛程式支援。Elasticsearch 支援多種外掛程式，如斷詞外掛程式、同步外掛程式、Hadoop 外掛程式、視覺化外掛程式等。

根據最新的資料庫引擎排名顯示，Elasticsearch、Splunk 和 Solr 分別佔據了資料庫搜尋引擎的前三位，如圖 3-1 所示。

	Rank					Score		
Aug 2019	Jul 2019	Aug 2018	DBMS		Database Model	Aug 2019	Jul 2019	Aug 2018
					351 systems in ranking, August 2019			
1.	1.	1.	Oracle ✚		Relational, Multi-model ℹ	1339.48	+18.22	+27.45
2.	2.	2.	MySQL ✚		Relational, Multi-model ℹ	1253.68	+24.16	+46.87
3.	3.	3.	Microsoft SQL Server ✚		Relational, Multi-model ℹ	1093.18	+2.35	+20.53
4.	4.	4.	PostgreSQL ✚		Relational, Multi-model ℹ	481.33	-1.94	+63.83
5.	5.	5.	MongoDB ✚		Document	404.57	-5.36	+53.59
6.	6.	6.	IBM Db2 ✚		Relational, Multi-model ℹ	172.95	-1.19	-8.89
7.	7.	↑8.	Elasticsearch ✚		Search engine, Multi-model ℹ	149.08	+0.27	+10.97
8.	8.	↓7.	Redis ✚		Key-value, Multi-model ℹ	144.08	-0.18	+5.51
9.	9.	9.	Microsoft Access		Relational	135.33	-1.98	+6.24
10.	10.	10.	Cassandra ✚		Wide column	125.21	-1.80	+5.63
11.	11.	11.	SQLite ✚		Relational	122.72	-1.91	+8.99
12.	12.	↑13.	Splunk		Search engine	85.88	+0.39	+15.39
13.	13.	↑14.	MariaDB ✚		Relational, Multi-model ℹ	84.95	+0.52	+16.66
14.	14.	↑18.	Hive ✚		Relational	81.80	+0.93	+23.86
15.	15.	↓12.	Teradata ✚		Relational, Multi-model ℹ	76.64	-1.18	-0.77
16.	16.	↓15.	Solr		Search engine	59.12	-0.52	-2.78
17.	17.	↑19.	FileMaker		Relational	58.02	+0.12	+1.96
18.	↑20.	↑21.	Amazon DynamoDB ✚		Multi-model ℹ	56.57	+0.15	+4.91
19.	↓18.	↓17.	HBase		Wide column	56.54	-1.00	-2.27
20.	↓19.	↓16.	SAP Adaptive Server		Relational	55.86	-0.79	-4.57
21.	21.	↓20.	SAP HANA ✚		Relational, Multi-model ℹ	55.43	-0.11	+3.50
22.	22.	22.	Neo4j ✚		Graph	48.39	-0.59	+7.47
23.	23.	23.	Couchbase ✚		Document, Multi-model ℹ	33.83	+0.12	+0.88
24.	24.	↑29.	Microsoft Azure Cosmos DB ✚		Multi-model ℹ	29.94	+0.85	+10.41

圖 3-1

3.2 Elasticsearch 的安裝與設定

常言道：工欲善其事，必先利其器。因此在使用 Elasticsearch 之前，我們需要安裝 Elasticsearch。下面介紹 Elasticsearch 在 Windows 環境下和在 Linux 環境下的安裝方法。由於 Elasticsearch 依賴 Java 環境，因此首先介紹 Java 環境的安裝方法。

3.2.1 安裝 Java 環境

首先下載並安裝 JDK（JAVA Development Kit）。JDK 是整個 Java 開發的核心，它包含了 Java 的執行環境、Java 工具和 Java 基礎類別庫。

本書使用版本是 JDK 12.0。

1 在 Windows 環境下安裝

下面以 jdk-12.0.2_windows-x64_bin.zip 為例，展示 JDK 的安裝過程。

首先，將 jdk-12.0.2_windows-x64_bin.zip 下載並解壓縮後放到某一目錄下，如在 C:\Program Files\ 下新增 Java 資料夾，並將壓縮檔放到該目錄下，即 C:\Program Files\Java 資料夾下。

然後，進入 C:\Program Files\Java\jdk-12.0.2_windows-x64_bin\jdk-12.0.2 ，即可看到解壓縮後的檔案目錄，如圖 3-2 所示。

bin	2019/7/16 9:01	文件夾
conf	2019/7/16 9:01	文件夾
include	2019/7/16 9:01	文件夾
jmods	2019/7/16 9:01	文件夾
legal	2019/7/16 9:01	文件夾
lib	2019/7/16 9:01	文件夾
release	2019/7/16 9:01	文件

圖 3-2

jdk-12.0.2 資料夾下包含 bin、conf、include、jmods、legal、lib 資料夾和 release 檔案。

- bin 資料夾下儲存的是可執行程式──Java 執行時期環境（JRE）的實現。JRE 包含 Java 虛擬機器（JVM）、類別庫和支援 Java 程式語言撰寫程式執行的其他檔案。該目錄還包含工具和應用程式，這些工具和應用程式將幫助軟體開發人員開發、執行、偵錯和記錄用 Java 撰寫的程式。
- conf 資料夾下儲存的是設定檔，包含可設定選項的檔案。讀者可以編輯此目錄中的檔案以更改 JDK 的存取權限，還可以設定安全演算法，設定可用於限制 JDK 密碼強度的 Java 加密擴充策略檔案。

- include 資料夾下儲存的是 C 語言撰寫的標頭檔——基於 Java 本機介面和 Java 虛擬機器（JVM）偵錯器介面，以支援本機程式設計的 C 語言標頭檔。
- jmods 資料夾下儲存的是編譯好的 Java 模組，讀者可以用 J-link 建立自訂執行時期的編譯模組。
- legal 資料夾下儲存的是版權和許可檔案，其中包含 JDK 每個模組的授權和版權檔案，以及作為 .md 檔案的協力廠商使用須知。
- lib 資料夾下儲存的是其他類別庫，主要是 JDK 所需的附加類別庫和支援檔案，這些檔案不供外部使用。
- release 檔案中包含的是 JDK 版本相關資訊。

隨後，開始設定 JDK 所需的環境變數。筆者所用的電腦的環境為 Windows 10，相關環境變數設定過程如下。

首先，開啟一個資料夾，在左側導覽視窗選取「這台電腦」，點擊滑鼠右鍵，在出現的快顯功能表中點擊「內容」選項，出現「控制台」視窗。

在「控制台」視窗中，點擊「進階系統設定」→「進階」→「環境變數 (N)…」，出現「環境變數設定」對話方塊，如圖 3-3 所示。

JAVA_HOME	C:\Program Files\Java\jdk-12.0.2
OneDrive	C:\Users\牛冬\OneDrive
Path	C:\Program Files\Java\jdk-12.0.2\bin;C:\Users\牛冬\AppData\Local...
TEMP	C:\Users\牛冬\AppData\Local\Temp
TMP	C:\Users\牛冬\AppData\Local\Temp

ES_HOME	C:\Program Files\Elastic\Elasticsearch\
JAVA_HOME	C:\Program Files\Java\jdk-12.0.2
JAVA_TOOL_OPTIONS	-Dfile.encoding=UTF-8
MAVEN_HOME	C:\apache-maven-3.5.0-bin\apache-maven-3.5.0
NUMBER_OF_PROCESSORS	4
OS	Windows_NT
Path	%JAVA_HOME%\bin;C:\Program Files (x86)\RBTools\bin;C:\Progra...
PATHEXT	.COM;.EXE;.BAT;.CMD;.VBS;.VBE;.JS;.JSE;.WSF;.WSH;.MSC

圖 3-3

點擊「新增 (W)…」按鈕，即可開啟「新增使用者變數」對話方塊，如圖 3-4 所示。

圖 3-4

在該對話方塊中，我們分別設定 JAVA_HOME 和 Path，實際參數詳見圖 3-3。設定後，開啟 Windows 10 附帶的 PowerShell 視窗，輸入指令 "java -version"，並按 "Enter" 鍵，即可驗證 JDK 是否安裝成功。如果安裝成功，則執行 "java -version" 指令，視窗中會顯示如圖 3-5 所示內容。

```
Microsoft Windows [版本 6.1.7601]
Copyright (c) 2009 Microsoft Corporation.  All rights reserved.

C:\Users\joshhu>java -version
java version "1.8.0_172"
Java(TM) SE Runtime Environment (build 1.8.0_172-b11)
Java HotSpot(TM) 64-Bit Server VM (build 25.172-b11, mixed mode)

C:\Users\joshhu>
```

圖 3-5

2 在 Linux 環境下安裝

首先將 jdk-12.0.2_linux-x64_bin.tar.gz 下載到本機。在本機開啟 SecureCRT 軟體，連接 Linux 伺服器。切換到 java 目錄下之後，用 rz 指令將 jdk-12.0.2_linux-x64_bin.tar.gz 上傳到該目錄。

隨後使用指令 tar -zxvf jdk-12.0.1_linux-x64_bin.tar.gz 將壓縮檔解壓縮到目前的目錄下。解壓縮成功之後，使用 ll 指令檢視檔案列表，此時壓縮檔解壓縮成了 jdk-12.0.1 資料夾。jdk-12.0.1 資料夾中的檔案和 Windows 環境下的相同。

隨後設定環境變數，實際方法為使用 vim 指令修改 /etc/profile 檔案。在 profile 檔案中增加以下環境變數：

```
JAVA_HOME=/usr/java/jdk-12.0.1
CLASSPATH=$JAVA_HOME/lib/
PATH=$PATH:$JAVA_HOME/bin
export PATH JAVA_HOME CLASSPATH
```

儲存修改後，需重新載入 /etc/profile 設定檔，實際指令為 source /etc/profile。
指令執行成功後，檢查 JDK 是否安裝成功。檢查指令的方法與在 Windows 環
境下的相同，不再贅述。

3.2.2 Elasticsearch 的安裝

Elasticsearch 支援多平台，我們可以在 Elasticsearch 官網找到 Elasticsearch 官
方支援的作業系統和 JVM 的矩陣，如圖 3-6 所示。

	CentOS/RHEL 6.x/7.x	Oracle Enterprise Linux 6/7 with RHEL Kernel only	Ubuntu 14.04	Ubuntu 16.04	Ubuntu 18.04	SLES 11 SP4**	SLES 12	openSUSE Leap 42	Windows Server 2012/R2	Windows Server 2016
Elasticsearch 5.0.x	✔	✔	✔	✔	✘	✔	✔	✔	✔	✘
Elasticsearch 5.1.x	✔	✔	✔	✔	✘	✔	✔	✔	✔	✘
Elasticsearch 5.2.x	✔	✔	✔	✔	✘	✔	✔	✔	✔	✘
Elasticsearch 5.3.x	✔	✔	✔	✔	✘	✔	✔	✔	✔	✘
Elasticsearch 5.4.x	✔	✔	✔	✔	✘	✔	✔	✔	✔	✘
Elasticsearch 5.5.x	✔	✔	✔	✔	✘	✔	✔	✔	✔	✔
Elasticsearch 5.6.x	✔	✔	✔	✔	✘	✔	✔	✔	✔	✔
Elasticsearch 6.0.x	✔	✔	✔	✔	✘	✘	✔	✔	✔	✔
Elasticsearch 6.1.x	✔	✔	✔	✔	✘	✔	✔	✔	✔	✔
Elasticsearch 6.2.x	✔	✔	✔	✔	✘	✘	✔	✔	✔	✔
Elasticsearch 6.3.x	✔	✔	✔	✔	✘	✔	✔	✔	✔	✔
Elasticsearch 6.4.x	✔	✔	✔	✔	✘	✔	✔	✔	✔	✔
Elasticsearch 6.5.x	✔	✔	✔	✔	✔	✔	✔	✔	✔	✔
Elasticsearch 6.6.x	✔	✔	✔	✔	✔	✔	✔	✔	✔	✔
Elasticsearch 6.7.x	✔	✔	✔	✔	✔	✘	✔	✔	✔	✔
Elasticsearch 6.8.x	✔	✔	✔	✔	✔	✘	✔	✔	✔	✔
Elasticsearch 7.0.x	✔	✔	✘	✔	✔	✔	✔	✔	✔	✔
Elasticsearch 7.1.x	✔	✔	✘	✔	✔	✔	✔	✔	✔	✔

圖 3-6

軟體開發人員既可以選擇自行安裝 Elasticsearch 來建置搜索服務，也可以選擇使用雲端託管的 Elasticsearch 服務。Elasticsearch 服務可以在 AWS 和 GCP 上使用。

下面介紹如何安裝 Elasticsearch，本書主要介紹在 Windows 系統和常用的 Linux 系統下的安裝方法。

1 在 Windows 系統下安裝 Elasticsearch

在 Windows 系 統 下， 我 們 可 以 用 Windows 中 的 zip 安 裝 套 件 來 建 置 Elasticsearch 服務。該 zip 安裝套件附帶了一個 elasticsearch-service.bat 指令檔案，執行該指令檔案，即可將 Elasticsearch 作為服務執行。

下載 Elasticsearch V7.2.0 的 .zip 安裝套件，解壓縮後，在同級目錄下將建立一個名為 elasticsearch-7.2.0 的資料夾，我們稱之為 %ES_HOME%，如圖 3-7 所示。

bin	2019/6/20 16:02
config	2019/6/20 16:02
jdk	2019/6/20 16:02
lib	2019/6/20 16:02
logs	2019/6/20 15:56
modules	2019/6/20 16:02
plugins	2019/6/20 15:56
LICENSE.txt	2019/6/20 15:50
NOTICE.txt	2019/6/20 15:56
README.textile	2019/6/20 15:50

圖 3-7

在 elasticsearch-7.2.0 資 料 夾 中 有 bin、config、jdk、lib、logs、modules、plugins 和資料夾。

- bin 資料夾下儲存的是二進位指令稿，包含啟動 Elasticsearch 節點和安裝的 Elasticsearch 外掛程式。
- config 資料夾下儲存的是包含 elasticsearch.yml 在內的設定檔。
- jdk 資料夾下儲存的是 Java 執行環境。
- lib 資料夾下儲存的是 Elasticsearch 本身所需的 jar 檔案。

- logs 資料夾下儲存的是記錄檔。
- modules 資料夾下儲存的是 Elasticsearch 的各個模組。
- plugins 資料夾下儲存的是設定外掛程式，每個外掛程式都包含在一個子目錄中。

啟動 Elasticsearch 服務。

首先切換到終端視窗，如 PowerShell 視窗，在命令列視窗下執行 cd 指令 cd c:\elasticsearch-7.2.0，以便切換到 %ES_HOME% 目錄。

然後從命令列啟動 Elasticsearch，啟動指令如下所示。

```
PS C:\elasticsearch-7.2.0-windows-x86_642\elasticsearch-7.2.0> .\bin\elasticsearch.bat
```

指令執行後，我們可以在視窗中看到 Elasticsearch 的啟動過程，如下所示。當看到節點 started 的輸出後，說明 Elasticsearch 服務已經啟動。

```
[2019-07-29T19:16:12,579][INFO ][o.e.c.m.MetaDataIndexTemplateService] [LAPTOP-1S8BALK3] adding template [.watch-history
-9] for index patterns [.watcher-history-9*]
[2019-07-29T19:16:12,646][INFO ][o.e.c.m.MetaDataIndexTemplateService] [LAPTOP-1S8BALK3] adding template [.triggered_wat
ches] for index patterns [.triggered_watches*]
[2019-07-29T19:16:12,720][INFO ][o.e.c.m.MetaDataIndexTemplateService] [LAPTOP-1S8BALK3] adding template [.monitoring-lo
gstash] for index patterns [.monitoring-logstash-7-*]
[2019-07-29T19:16:12,797][INFO ][o.e.c.m.MetaDataIndexTemplateService] [LAPTOP-1S8BALK3] adding template [.monitoring-es
] for index patterns [.monitoring-es-7-*]
[2019-07-29T19:16:12,875][INFO ][o.e.c.m.MetaDataIndexTemplateService] [LAPTOP-1S8BALK3] adding template [.monitoring-be
ats] for index patterns [.monitoring-beats-7-*]
[2019-07-29T19:16:12,934][INFO ][o.e.c.m.MetaDataIndexTemplateService] [LAPTOP-1S8BALK3] adding template [.monitoring-al
erts-7] for index patterns [.monitoring-alerts-7]
[2019-07-29T19:16:13,003][INFO ][o.e.c.m.MetaDataIndexTemplateService] [LAPTOP-1S8BALK3] adding template [.monitoring-ki
bana] for index patterns [.monitoring-kibana-7-*]
[2019-07-29T19:16:13,071][INFO ][o.e.x.i.a.TransportPutLifecycleAction] [LAPTOP-1S8BALK3] adding index lifecycle policy
[watch-history-ilm-policy]
[2019-07-29T19:16:13,286][INFO ][o.e.l.LicenseService     ] [LAPTOP-1S8BALK3] license [bbcef3ff-378f-4c13-bb43-962a9dff2
841] mode [basic] - valid
[2019-07-29T19:16:13,420][INFO ][o.e.h.AbstractHttpServerTransport] [LAPTOP-1S8BALK3] publish_address {127.0.0.1:9200},
bound_addresses {127.0.0.1:9200}, {[::1]:9200}
[2019-07-29T19:16:13,421][INFO ][o.e.n.Node               ] [LAPTOP-1S8BALK3] started
```

我們可以看到名為 "LAPTOP_1S8BALK3" 的節點（不同的電腦顯示不同）已經啟動，並且選舉它自己作為單一叢集中的 Master 主節點。

Elasticsearch 啟動後，在預設情況下，Elasticsearch 將在前台執行，並將其記錄檔列印到標準輸出（stdout）。可以按 Ctrl+C 組合鍵停止執行 Elasticsearch。

在 Elasticsearch 執行過程中，如果需要將 Elasticsearch 作為守護處理程序執行，則需要在命令列上指定指令參數 "- d"，並使用 "- p" 選項將 Elasticsearch

的處理程序 ID 記錄在檔案中。此時的啟動指令如下：

```
. / bin / elasticsearch - d - p pid
```

此時 Elasticsearch 的記錄檔訊息可以在 $ ES_HOME / logs / 目錄中找到。

在啟動 Elasticsearch 的過程中，我們可以透過命令列對 Elasticsearch 進行設定。一般來說，在預設情況下，Elasticsearch 會從 $ ES_HOME /config/ elasticsearch.yml 檔案載入其設定內容。我們還可以在命令列上指定設定，此時需要使用 "-e" 語法。在命令列設定 Elasticsearch 參數時，啟動指令如下：

```
./bin/elasticsearch -d -Ecluster.name=my_cluster -Enode.name=node_1
```

在 Elasticsearch 啟動後，我們可以在瀏覽器的網址列輸入 http://localhost:9200/ 來驗證 Elasticsearch 的啟動情況。按 "Enter" 鍵後，瀏覽器的頁面會顯示以下內容：

```
{
  "name" : "LAPTOP-1S8BALK3",
  "cluster_name" : "elasticsearch",
  "cluster_uuid" : "6UkGmpPrSwyXlzPHgSaivg",
  "version" : {
    "number" : "7.2.0",
    "build_flavor" : "default",
    "build_type" : "zip",
    "build_hash" : "508c38a",
    "build_date" : "2019-06-20T15:54:18.811730Z",
    "build_snapshot" : false,
    "lucene_version" : "8.0.0",
    "minimum_wire_compatibility_version" : "6.8.0",
    "minimum_index_compatibility_version" : "6.0.0-beta1"
  },
  "tagline" : "You Know, for Search"
}
```

此外，我們還可以設定 Elasticsearch 是否自動建立 x-pack 索引。x-pack 將嘗試在 Elasticsearch 中自動建立多個索引。在預設情況下，Elasticsearch 允許自動建立索引，且不需要其他步驟。

如果需要在 Elasticsearch 中禁用自動建立索引，則必須在 Elasticsearch.yml 中設定 action.auto_create_index，以允許 x-pack 建立以下索引：

```
action.auto_create_index: .monitoring*,.watches,.triggered_watches,.watcher-
history*,.ml*
```

2 在 Linux 系統下安裝 Elasticsearch

有 tar.gz 檔案格式的安裝套件可以安裝在任何 Linux 發行版本和 macOS 上。

在 Linux 系統下，取得 Elasticsearch V7.2.0 版本安裝套件的指令如下所示。

方法 1：

```
wget https://artifacts.elastic.co/downloads/elasticsearch/elasticsearch-7.2.0-
linux-x86_64.tar.gz
```

方法 2：

```
wget https://artifacts.elastic.co/downloads/elasticsearch/elasticsearch-7.2.0-
linux-x86_64.tar.gz. sha512
shasum -a 512 -c elasticsearch-7.2.0-linux-x86_64.tar.gz.sha512
```

其中，方法 2 中的 shasum 指令用於比較已下載的 tar.gz 安裝套件的 sha 和已發佈的校正碼，如果 shasum 指令能夠執行並傳回 OK，則證明所下載的安裝套件是正確的。

取得安裝套件後，可以使用以下指令解壓縮安裝套件：

```
tar -xzf elasticsearch-7.2.0-linux-x86_64.tar.gz
```

該指令執行完畢後，會在目前的目錄顯示 elasticsearch-7.2.0 資料夾，隨後執行以下指令切換到 elasticsearch-7.2.0 目錄下：

```
cd elasticsearch-7.2.0/
```

elasticsearch-7.2.0 目錄被稱為 $ES_HOME，該目錄下的檔案結構與 Windows 下的目錄結構相同。

安裝完 elasticsearch-7.2.0 後，在 elasticsearch-7.2.0 目錄中，我們可以執行以下指令啟動 Elasticsearch：

```
./bin/elasticsearch
```

在預設情況下，Elasticsearch 在前台執行，並將其記錄檔列印到標準輸出（stdout）。在 Elasticsearch 執行過程中，如需停止服務，則可以透過按組合鍵 Ctrl+C 停止服務。

在 Elasticsearch 啟動後，需要檢查 Elasticsearch 是否能夠執行。我們可以透過向本機主機上的通訊埠 9200 發送 HTTP 請求來測試本機 Elasticsearch 節點是否正在執行，發送請求如下：

```
curl http://localhost:9200/
```

我們會看到以下輸出結果：

```
{
  "name" : "Cp8oag6",
  "cluster_name" : "elasticsearch",
  "cluster_uuid" : "AT69_T_DTp-1qgIJlatQqA",
  "version" : {
    "number" : "7.2.0",
    "build_flavor" : "default",
    "build_type" : "tar",
    "build_hash" : "f27399d",
    "build_date" : "2016-03-30T09:51:41.449Z",
    "build_snapshot" : false,
    "lucene_version" : "8.0.0",
    "minimum_wire_compatibility_version" : "1.2.3",
    "minimum_index_compatibility_version" : "1.2.3"
  },
  "tagline" : "You Know, for Search"
}
```

在 Elasticsearch 執行過程中，如果需要將 Elasticsearch 作為守護處理程序執行，則需要在命令列上指定指令參數 "- d"，並使用 "- p" 選項將 Elasticsearch 的處理程序 ID 記錄在檔案中，啟動指令如下：

```
. / bin / elasticsearch - d - p pid
```

此時 Elasticsearch 的記錄檔訊息可以在 $es_home / logs / 目錄中找到。

在關閉 Elasticsearch 時，可以根據 PID 檔案中記錄的處理程序 ID 執行 pkill 指令，實際指令如下：

```
pkill -F pid
```

在啟動 Elasticsearch 過程中，我們還可以透過命令列對 Elasticsearch 進行設定。一般來說，在預設情況下，Elasticsearch 會從 $es_home/config/elasticsearch.yml 檔案載入其設定內容。我們還可以在命令列上指定設定，此時需要使用 "-e" 語法。當命令列設定 Elasticsearch 參數時，啟動指令如下：

```
./bin/elasticsearch -d -Ecluster.name=my_cluster -Enode.name=node_1
```

3.2.3 Elasticsearch 的設定

與近年來很多流行的架構和中介軟體一樣，Elasticsearch 的設定同樣遵循「約定大於設定」的設計原則。Elasticsearch 具有極好的預設值設定，使用者僅需要很少的設定即可使用 Elasticsearch。使用者既可以使用叢集更新設定 API 在正在執行的叢集上更改大多數設定，也可以透過設定檔對 Elasticsearch 進行設定。

一般來說，設定檔應包含特定節點的設定，如 node.name 和 paths 路徑等資訊，還會包含節點為了能夠加入 Elasticsearch 叢集而需要做出的設定，如 cluster.name 和 network.host 等。

1 設定檔位置資訊

在 Elasticsearch 中有三個設定檔，分別是 elasticsearch.yml、jvm.options 和 log4j2.properties，這些檔案位於 config 目錄下，如圖 3-8 所示。

elasticsearch.keystore	2019/7/29 19:15	KEYSTORE 文件	1 KB
elasticsearch.yml	2019/6/20 15:50	YML 文件	3 KB
jvm.options	2019/6/20 15:50	OPTIONS 文件	4 KB
log4j2.properties	2019/6/20 15:56	PROPERTIES 文件	17 KB
role_mapping.yml	2019/6/20 15:56	YML 文件	1 KB
roles.yml	2019/6/20 15:56	YML 文件	1 KB
users	2019/6/20 15:56	文件	0 KB
users_roles	2019/6/20 15:56	文件	0 KB

圖 3-8

其 中，elasticsearch.yml 用 於 設 定 Elasticsearch，jvm.options 用 於 設 定 Elasticsearch 依賴的 JVM 資訊，log4j2.properties 用於設定 Elasticsearch 記錄檔記錄中的各個屬性。

註：上述檔案位於 config 目錄下，這是預設位置。預設位置取決於我們安裝 Elasticsearch 時是否以下載的 tar.gz 套件或 zip 套件為基礎，如果是，則設定目錄預設位置為 $es_home/config。如果使用者想自訂設定目錄的位置，則可以透過 es_path_conf 環境變數進行更改，如下所示：

```
ES_PATH_CONF=/path/to/my/config ./bin/elasticsearch
```

或透過命令列或 shell 概要檔案匯出 es-path-conf 環境變數進行更改。

---

2 設定檔的格式

Elasticsearch 的設定檔格式為 yaml。下面展示一些更改資料和記錄檔目錄路徑的範例：

```
path:
    data: /var/lib/elasticsearch
    logs: /var/log/elasticsearch
```

除上述層級方式配製外，也可將層級路徑參數整合為一條參數路徑設定，如下所示：

```
  path.data: /var/lib/elasticsearch
path.logs: /var/log/elasticsearch
```

如果需要在設定檔中參考環境變數的值，則可以在設定檔中使用 ${...} 符號。參考的環境變數會取代環境變數原有的值，如下所示：

```
node.name:     ${HOSTNAME}
network.host: ${ES_NETWORK_HOST}
```

3 設定 JVM 選項

在 Elasticsearch 中，使用者很少需要更改 Java 虛擬機器（JVM）選項。一般來說，最可能的更改是設定堆大小。在預設情況下，Elasticsearch 設定 JVM 使用最小堆積空間和最大堆積空間的大小均為 1GB。

設定 JVM 選項（包含系統內容和 jvm 標示）的首選方法是透過 jvm.options 設定檔設定。此檔案的預設位置為 config/jvm.options。

在 Elasticsearch 中，我們透過 xms（最小堆大小）和 xmx（最大堆大小）這兩個參數設定 jvm.options 設定檔指定的整個堆大小，一般應將這兩個參數設定為相等。

在 jvm.options 設定檔中，包含了以下特殊語法行來分隔 JVM 參數列表。

（1）忽略由空白組成的行。

（2）以 "#" 開頭的行被視為註釋並被忽略，如下所示：

```
# this is a comment
```

（3）以 "-" 開頭的行被視為獨立於本機 JVM 版本編號的 JVM 選項，如下所示：

```
-Xmx2g
```

（4）以數字開頭，且後面為 ":" 的行被視為一個 JVM 選項，該選項僅在本機 JVM 的版本編號相互比對時適用，如下所示：

```
8:-Xmx2g
```

（5）以數字開頭，且後面為 "-" 的行被視為一個 JVM 選項，僅當本機 JVM 的版本編號大於或等於該數字版本編號時才適用，如下所示：

```
8-:-Xmx2g
```

（6）以數字開頭，且後面為 "-"，再後面為數字的行被視為一個 JVM 選項，僅當本機 JVM 的版本編號在這兩個數字版本編號的範圍內時才適用，如下所示：

```
8-9:-Xmx2g
```

（7）所有其他行都被拒絕解析。

此外，使用者還可以透過 ES_JAVA_OPTS 環境變數來設定 Java 虛擬機器選項，如下所示：

```
export ES_JAVA_OPTS="$ES_JAVA_OPTS -Djava.io.tmpdir=/path/to/temp/dir"
./bin/elasticsearch
```

4 安全設定

在 Elasticsearch 中，有些設定資訊比較敏感且需要保密，此時單純依賴檔案系統許可權便不足以保護這些資訊，因此需要設定安全維度的資訊。Elasticsearch 提供了一個金鑰函數庫和對應的金鑰函數庫工具來管理金鑰函數庫中的設定。這裡的所有指令都適用於 Elasticsearch 使用者。

需要指出的是，對金鑰函數庫所做的所有修改，都必須在重新啟動 Elasticsearch 之後才會生效。

此外，在目前 Elasticsearch 金鑰函數庫中只提供模糊處理，以後會增加密碼保護。

安全設定就像 elasticsearch.yml 設定檔中的正常設定一樣，需要在叢集中的每個節點上指定。目前，所有安全設定都是特定於節點的設定，每個節點上必須有相同的值。

安全設定的正常操作有建立金鑰函數庫、檢視金鑰函數庫中的設定清單、增加字串設定、增加檔案設定、刪除金鑰設定和可重新載入的安全設定等，下面一一介紹。

☑ 建立金鑰函數庫

想要建立 elasticsearch.keystore，需要使用 create 指令，如下所示：

```
bin/elasticsearch-keystore create
```

指令執行後，將建立 2 個檔案，檔案名稱分別為 elasticsearch.keystore 和 elasticsearch.yml。

☑ 檢視金鑰函數庫中的設定列表

使用 list 指令可以檢視金鑰函數庫中的設定列表，如下所示：

```
bin/elasticsearch-keystore list
```

☑ 增加字串設定

如果需要設定敏感的字串，如雲外掛程式的身份驗證憑據，則可以使用 add 指令增加，如下所示：

```
bin/elasticsearch-keystore add the.setting.name.to.set
```

指令執行後將提示輸入設定值。

使用者可以使用 --stdin 標示在視窗 stdin 中輸出待設定的目標值，如下所示：

```
cat /file/containing/setting/value | bin/elasticsearch-keystore add --stdin
the.setting. name.to.set
```

☑ 增加檔案設定

使用者可以使用增加檔案指令增加敏感資訊檔案，如雲外掛程式的身份驗證金鑰檔案。設定時需確保將檔案路徑作為參數包含在設定名稱之後，如下所示：

```
bin/elasticsearch-keystore add-file the.setting.name.to.set /path/example-
file.json
```

☑ 刪除金鑰設定

如果需要從金鑰函數庫中刪除設定，則使用 remove 指令，如下所示：

```
bin/elasticsearch-keystore remove the.setting.name.to.remove
```

☑ 可重新載入的安全設定

就像 elasticsearch.yml 中的設定值一樣，對金鑰函數庫內容的更改不會自動應用於正在執行的 Elasticsearch 節點，因此需要重新啟動節點才能重新讀取設定。

對於某些安全設定，我們可以標記為可重新載入，這樣設定後，就可以在正在執行的節點上重新讀取和應用了。

需要指出的是，所有安全設定的值（不論是否可重新載入），在所有叢集節點上必須相同。更改所需的安全設定後，使用 bin/elasticsearch keystore add 指令，呼叫：

```
POST _nodes/reload_secure_settings
```

該 API 介面將解密並重新讀取每個叢集節點上的整個金鑰函數庫，但只限於可多載的安全設定，對其他設定的更改將在下次重新啟動之後生效。

該 API 介面呼叫傳回後，重新載入就完成了，這表示依賴於這些設定的所有內部資料結構都已更改，一切設定資訊看起來好像從一開始就有了新的值。

當更改多個可重新載入的安全設定時，使用者需要在每個叢集節點上都修改所有設定，然後發出重新載入安全設定呼叫，而非在每次修改後就重新載入。

5 記錄檔記錄設定

在 Elasticsearch 中，使用 log4j2 來記錄記錄檔。使用者可以使用 log4j2. properties 檔案設定 log4j2。

Elasticsearch 公開了三個屬性資訊，分別是 $sys:es.logs.base_path、$sys:es. logs.cluster_name 和 $sys:es.logs.node_name，使用者可以在設定檔中參考這些屬性來確定記錄檔的位置。

屬性 $sys:es.logs.base_path 將解析為記錄檔目錄位址，$sys:es.logs.cluster_ name 將解析為叢集名稱（在預設設定中，用作記錄檔名稱的字首），$sys:es. logs.node_name_ 將解析為節點名稱（如果顯性地設定了節點名稱）。

舉例來說，假設使用者的記錄檔目錄（path.logs）是 /var/log/elasticsearch，叢集命名為 production，那麼 $sys:es.logs.base_path_ 將解析為 /var/log/elasticsearch，$sys:es.logs.base_path/sys:file. Separator/ $sys:es.logs.cluster_name.log 將解析為 /var/log/elasticsearch/production.log。

下面我們結合 log4j2.properties 檔案的主要設定資訊來介紹各個屬性的含義。log4j2.properties 檔案的設定資訊如下所示：

```
######## Server JSON ##########################
appender.rolling.type = RollingFile      編號 1
appender.rolling.name = rolling
appender.rolling.fileName = ${sys:es.logs.base_path}${sys:file.separator}
${sys:es.logs.cluster_name}_server.json  編號 2
appender.rolling.layout.type = ESJsonLayout    編號 3
appender.rolling.layout.type_name = server      編號 4
appender.rolling.filePattern = ${sys:es.logs.base_path}${sys:file.separator}
${sys:es.logs.cluster_name}-%d{yyyy-MM-dd}-%i.json.gz 編號 5
appender.rolling.policies.type = Policies
appender.rolling.policies.time.type = TimeBasedTriggeringPolicy   編號 6
appender.rolling.policies.time.interval = 1   編號 7
appender.rolling.policies.time.modulate = true    編號 8
appender.rolling.policies.size.type = SizeBasedTriggeringPolicy   編號 9
appender.rolling.policies.size.size = 256MB   編號 10
appender.rolling.strategy.type = DefaultRolloverStrategy
appender.rolling.strategy.fileIndex = nomax
appender.rolling.strategy.action.type = Delete  編號 11
appender.rolling.strategy.action.basepath = ${sys:es.logs.base_path}
appender.rolling.strategy.action.condition.type = IfFileName   編號 12
appender.rolling.strategy.action.condition.glob = ${sys:es.logs.cluster_name}
-*   編號 13
appender.rolling.strategy.action.condition.nested_condition.type =
IfAccumulatedFileSize   編號 14
appender.rolling.strategy.action.condition.nested_condition.exceeds = 2GB
編號 15
###############################################
```

其中，上述被編號的設定屬性含義如下所示。

編號 1：設定 RollingFile 的 appender 屬性。

編號 2：記錄檔資訊將輸出到 /var/log/elasticsearch/production.json 中。

編號 3：使用 JSON 格式輸出。

編號 4：type_name 是填充 ESJsonLayout 的類型欄位的標示，該欄位可以讓我
們在解析不同類型的記錄檔時更加簡單。

編號 5：將記錄檔捲動輸出到 /var/log/elasticsearch/production-yyyy-MM-dd-i.
json 檔案。記錄檔會被壓縮處理，i 呈遞增狀態。

編號 6：使用以時間戳記為基礎的新增記錄檔捲動策略。

編號 7：按天捲動新增記錄檔。

編號 8：在日期時間上對齊標準，而非按每 24 小時來新增一次捲動記錄檔。

編號 9：按記錄檔大小的策略來捲動新增記錄檔。

編號 10：每產生 256MB 的記錄檔，就捲動新增記錄檔一次。

編號 11：每次新增捲動記錄檔時執行刪除記錄檔動作。

編號 12：僅當檔案比對時才刪除記錄檔。

編號 13：該設定僅用於刪除記錄檔。

編號 14：只有當記錄檔目錄下累積了較多記錄檔時才刪除。

編號 15：壓縮記錄檔的條件是記錄檔大小達到 2 GB。

在 log4j2.properties 檔案中，我們還可以設定記錄檔記錄等級。設定記錄檔記
錄等級有四種方法，每種方法都有適合使用的場景。這四種設定方法分別是透
過命令列設定、透過 elasticsearch.yml 檔案設定、透過叢集設定和透過 log4j2.
properties 設定。

（1）透過命令列設定。

```
-e<name of logging hierarchy>=<level>（如 -e logger.org.elasticsearch.
transport=trace）。
```

適用場景：
當在單一節點上臨時偵錯一個問題（如在後啟動時或在開發過程中）時，這
是最適合的方法。

（2）透過 elasticsearch.yml 檔案設定。
所需要的設定屬性如下所示：

```
<name of logging hierarchy>:<level>
```

如 logger.org.elasticsearch.transport:trace。

適用場景：

當臨時偵錯一個問題，但沒有透過命令列啟動 Elasticsearch；或希望在更持久的基礎上調整記錄檔等級時，這是最適合的方法。

（3）透過叢集設定。

在叢集中設定記錄檔等級的方法如下所示：

```
PUT /_cluster/settings
{
  "transient": {
    "<name of logging hierarchy>": "<level>"
  }
}
```

範例如下所示：

```
PUT /_cluster/settings
{
  "transient": {
    "logger.org.elasticsearch.transport": "trace"
  }
}
```

適用場景：

當需要動態調整活動執行的叢集上的記錄檔等級時，這是最適合的方法。

（4）透過 log4j2.properties 設定。

在 log4j2.properties 中需要設定的屬性如下所示：

```
logger.<unique_identifier>.name = <name of logging hierarchy>
logger.<unique_identifier>.level = <level>
```

範例如下所示：

```
logger.transport.name = org.elasticsearch.transport
logger.transport.level = trace
```

適用場景：

當需要對記錄檔程式進行細粒度的控制時（如將記錄檔程式發送到另一個檔

案，或以不同的方式管理記錄檔程式），這是最適合的方法。

☑ deprecation 記錄檔

除正常記錄檔記錄外，Elasticsearch 還允許使用者啟用不推薦操作的記錄檔記錄。如果使用者需要遷移某些功能，則可以提前確定這部分屬性的設定。

在預設情況下，啟動警告等級記錄檔後，所有禁用記錄檔均可輸出到主控台和記錄檔中。實際設定如下所示：

```
logger.deprecation.level = warn
```

該設定生效後，將在記錄檔目錄中建立每日捲動 deprecation 記錄檔。使用者需要定期檢查此檔案，尤其是準備升級到新的主要版本時。

預設記錄檔記錄設定已將取消 deprecation 記錄檔的捲動策略設定為在 1GB 後捲動和壓縮，並最多保留五個記錄檔（四個捲動記錄檔和一個活動記錄檔）。

使用者可以在 config/log4j2.properties 檔案中透過將取消 deprecation 記錄檔等級設定為 error 來禁用它。

6 JSON 記錄檔格式

為了便於分析 Elasticsearch 的記錄檔，記錄檔預設以 JSON 格式列印。這是由 log4j 版面配置屬性 appender.rolling.layout.type=esjsonlayout 設定的。此版面配置需要設定一個 type_name 屬性，用於在分析時區分記錄檔流，實際設定如下所示：

```
appender.rolling.layout.type = ESJsonLayout
appender.rolling.layout.type_name = server
```

在設定生效後，記錄檔的每一行就是一個 JSON 格式的字串。

如果使用自訂版面配置，則需要用其他版面配置取代 appender.rolling.layout.type 行的設定，範例如下：

```
appender.rolling.type = RollingFile
appender.rolling.name = rolling
```

```
appender.rolling.fileName = ${sys:es.logs.base_path}${sys:file.separator}
${sys:es.logs.cluster_name}_server.log
appender.rolling.layout.type = PatternLayout
appender.rolling.layout.pattern = [%d{ISO8601}][%-5p][%-25c{1.}][%node _name]
%marker %.-10000m%n
appender.rolling.filePattern = ${sys:es.logs.base_path}${sys:file.separator}
${sys:es.logs.cluster_name}-%d{yyyy-MM-dd}-%i.log.gz
```

3.3 Elasticsearch 的核心概念

想要學好、用好 Elasticsearch，首先要了解其核心概念、名詞和屬性。這就像要看懂地圖，首先要知道地圖裡常用的標記符號一樣。

Elasticsearch 的核心概念有 Node 、Cluster、Shards、Replicas、Index、Type、Document、Settings、Mapping 和 Analyzer，其含義分別如下所示。

（1）Node：即節點。節點是組成 Elasticsearch 叢集的基本服務單元，叢集中的每個執行中的 Elasticsearch 伺服器都可稱之為節點。

（2）Cluster：即叢集。Elasticsearch 的叢集是由具有相同 cluster.name（預設值為 elasticsearch）的或多個 Elasticsearch 節點組成，各個節點協同工作，共用資料。同一個叢集內節點的名字不能重複，但叢集名稱一定要相同。

在實際使用 Elasticsearch 叢集時，一般需要給叢集起一個合適的名字來替代 cluster.name 的預設值。自訂叢集名稱的好處是，可以防止一個新啟動的節點加入相同網路中的另一個名稱相同的叢集中。

在 Elasticsearch 叢集中，節點的狀態有 Green、Yellow 和 Red 三種，分別如下所述。

① Green：綠色，表示節點執行狀態為健康狀態。所有的主分片和備份分片都可以正常執行，叢集 100% 健康。

② Yellow：黃色，表示節點的執行狀態為預警狀態。所有的主分片都可以正

常執行，但至少有一個備份分片不能正常執行。此時叢集依然可以正常執行，但叢集的高可用性在某種程度上已被弱化。

③ Red：紅色，表示叢集無法正常使用。此時，叢集中至少有一個分片的主分片及它的全部備份分片都無法正常執行。雖然叢集的查詢操作還可以進行，但是也只能傳回部分資料（其他正常分片的資料可以傳回），而分配到這個有問題分片的寫入請求將顯示出錯，最後導致資料遺失。

（3）Shards：即分片。當索引的資料量太大時，受限於單一節點的記憶體、磁碟處理能力等，節點無法足夠快地回應用戶端的請求，此時需要將一個索引上的資料進行水平拆分。拆分出來的每個資料部分稱之為一個分片。一般來說，每個分片都會放到不同的伺服器上。

進行分片操作之後，索引的規模變大，效能也隨之提升。

Elasticsearch 依賴 Lucene，Elasticsearch 中的每個分片其實都是 Lucene 中的索引檔案，因此每個分片必須有一個主分片和零到多個備份分片。

當軟體開發人員在一個設定有多分片的索引中寫入資料時，是透過路由來確定實際寫入哪個分片中，因此在建立索引時需要指定分片的數量，而且分片的數量一旦確定就不能更改。

當軟體開發人員在查詢索引時，需要在索引對應的多個分片進行查詢。Elasticsearch 會把查詢發送給每個相關的分片，並整理各個分片的查詢結果。對上層的應用程式而言，分片是透明的，即應用程式並不知道分片的存在。

在 Elasticsearch 中，預設為一個索引建立 5 個主分片，並分別為每個主分片建立一個備份。

（4）Replicas：即備份。備份指的是對主分片的備份，這種備份是精確複製模式。每個主分片可以有零個或多個備份，主分片和備份分片都可以對外提供資料查詢服務。當建置索引進行寫入操作時，首先在主分片完成資料的索引，然後資料會從主分片分發到備份分片進行索引。

當主分片不可用時，Elasticsearch 會在備份分片中選列出一個分片作為主分片，進一步避免資料遺失。

一方面，備份分片既可以提升 Elasticsearch 系統的高可用效能，又可以提升搜索時的平行處理效能；另一方面，備份分片也是一把雙面刃，即如果備份分片數量設定得太多，則在寫入操作時會增加資料同步的負擔。

（5）Index：即索引。在 Elasticsearch 中，索引由一個和多個分片組成。在使用索引時，需要透過索引名稱在叢集內進行唯一標識。

（6）Type：即類別。類別指的是索引內部的邏輯分區，透過 Type 的名字在索引內進行唯一標識。在查詢時如果沒有該值，則表示需要在整個索引中查詢。

（7）Document：即文件。索引中的每一筆資料叫作一個文件，與關聯式資料庫的使用方法類似，一筆文件資料透過 _id 在 Type 內進行唯一標識。

（8）Settings：Settings 是對叢集中索引的定義資訊，例如一個索引預設的分片數、備份數等。

（9）Mapping：Mapping 表示中儲存了定義索引中欄位（Field）的儲存類型、斷詞方式、是否儲存等資訊，有點類似關聯式資料庫（如 MySQL）中的表結構資訊。

在 Elasticsearch 中，Mapping 可以動態識別。如果沒有特殊需求，則不需要手動建立 Mapping，因為 Elasticsearch 會根據資料格式自動識別它的類型。當需要對某些欄位增加特殊屬性時，例如定義使用其他斷詞器、是否斷詞、是否儲存等，就需要手動設定 Mapping。一個索引的 Mapping 一旦建立，若已經儲存資料就不能修改。

（10）Analyzer：Analyzer 表示的是欄位斷詞方式的定義。一個 Analyzer 通常由一個 Tokenizer 和零到多個 Filter 組成。在 Elasticsearch 中，預設的標準 Analyzer 包含一個標準的 Tokenizer 和三個 Filter，即 Standard Token Filter、Lower Case Token Filter 和 Stop Token Filter。

3.4 Elasticsearch 的架構設計

Elasticsearch 的架構設計圖如圖 3-11 所示。

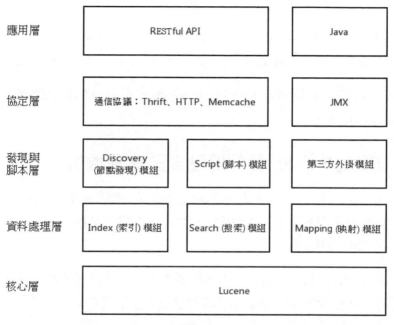

圖 3-11

如圖 3-11 所示，我們將 Elasticsearch 的架構自底向上分為五層，分別是核心層、資料處理層、發現與指令稿層、協定層和應用層。

其中，核心層是指 Lucene 架構——Elasticsearch 是以 Lucene 架構為基礎。

資料處理層主要是指在 Elasticsearch 中對資料的加工處理方式，常見的主要有 Index (索引) 模組、Search（搜索）模組和 Mapping（對映）模組。

發現與指令稿層主要是 Discovery（節點發現）模組、Script（指令稿）模組和協力廠商外掛程式模組。Discovery 模組是 Elasticsearch 自動發現節點的機制。Script 模組支援指令稿的執行，指令稿的應用使得我們能很方便的對查詢出來的資料進行加工處理，目前 Elasticsearch 支援 JavaScript、Python 等多種語言。協力廠商外掛程式模組表示 Elasticsearch 支援安裝很多協力廠商的外掛程式，如 elasticsearch-ik 斷詞外掛程式、elasticsearch-sql 外掛程式等。

協定層是 Elasticsearch 中的資料互動協定。目前 Elasticsearch 支援 Thrift、Memcached 和 HTTP 三種協定，預設的是 HTTP。

應用層指的是 Elasticsearch 的 API 支援模式。Elasticsearch 的特色之一就是 RESTFul 風格的 API，這種 API 介面風格也是目前十分流行的風格之一。

另外，圖 3-11 中的 JMX 指的是在 Elasticsearch 中對 Java 的管理架構，用來管理 Elasticsearch 應用。

3.4.1 Elasticsearch 的節點自動發現機制

在 Elasticsearch 內部，透過在叢集中設定一個相同的叢集名稱（即 cluster.name），就能將不同的節點連接到同一個叢集。這是怎麼實現的呢？本節就來揭曉節點自動發現機制。

Elasticsearch 內嵌自動發現功能，主要提供了 4 種可供選擇的發現機制。其中一種是預設實現，其他都是透過外掛程式實現，如下所示。

（1）Azure discovery 外掛程式方式：廣播模式。
（2）EC2 discovery 外掛程式方式：廣播模式。
（3）Google Compute Engine（GCE）discovery 外掛程式方式：廣播模式。
（4）Zen Discovery，預設實現方式，支援廣播模式和單一傳播模式。

Zen Discovery 是 Elasticsearch 內建的預設發現模組。發現模組用於發現叢集中的節點及選舉主節點（又稱 master 節點）。Zen Discovery 提供單一傳播模式和以檔案為基礎的發現，並且可以擴充為透過外掛程式支援其他形式的發現機制。

在設定前，我們需要了解廣播模式和單一傳播模式的設定參數。主要設定參數如下所示：

```
discovery.zen.ping.multicast.enabled: true
discovery.zen.fd.ping_timeout: 100s
discovery.zen.ping.timeout: 100s
discovery.zen.minimum_master_nodes: 2
discovery.zen.ping.unicast.hosts: ["172.31.X.Y"]
discovery.zen.ping.multicast.enabled
```

- discovery.zen.ping.multicast.enabled 表示關閉廣播模式的自動發現機制，主要是為了防止其他機器上的節點自動連入。

- discovery.zen.fd.ping_timeout 和 discovery.zen.ping.timeout 表示設定了節點與節點之間連接 ping 指令執行的逾時長。

- discovery.zen.minimum_master_nodes 表示叢集中選舉主節點時至少需要有多少個節點參與。

- discovery.zen.ping.unicast.hosts 表示在單一傳播模式下，節點應該自動發現哪些節點列表。action.auto_create_index: false 表示關閉自動建立索引。

1 單一傳播模式

Elasticsearch 支援廣播模式和單一傳播模式自動兩種節點發現機制，不過廣播模式已經不被大多數作業系統所支援，加之其安全性不高，所以一般我們會主動關閉廣播模式。關閉廣播模式的設定如下所示：

```
discovery.zen.ping.multicast.enabled: false   #關閉廣播
```

在 Elasticsearch 中，發現機制預設被設定為使用單一傳播模式，以防止節點無意中加入叢集。Elasticsearch 支援同一個主機啟動多個節點，因此只有在同一台機器上執行的節點才會自動組成叢集。當叢集的節點執行在不同的機器上時，在單一傳播模式下，我們需要為 Elasticsearch 設定一些它應該去嘗試連接的節點清單，設定方式如下所示：

```
#填寫叢集中的 IP 位址清單
discovery.zen.ping.unicast.hosts: ["192.168.X1.Y1:9300","192.168.X2.Y2:9300"]
```

因此，單一傳播模式下的設定資訊整理如下：

```
discovery.zen.ping.multicast.enabled: false
discovery.zen.fd.ping_timeout: 100s
```

```
discovery.zen.ping.timeout: 100s
discovery.zen.minimum_master_nodes: 2
discovery.zen.ping.unicast.hosts: ["192.168.X1.Y1:9300","192.168.X2.Y2:9300"]
```

設定後,叢集建置及主節點選舉過程如下:

節點啟動後先執行 ping 指令(這裡提及的 ping 指令不是 Linux 環境用的 ping 指令,而是 Elasticsearch 的 RPC 指令),如果 discovery.zen.ping.unicast.hosts 有設定,則 ping 設定中的 host;否則嘗試 ping localhost 的幾個通訊埠。

ping 指令的傳回結果會包含該節點的基本資訊及該節點認為的主節點。

在選舉開始時,主節點先從各節點認為的 master 中選。選舉規則比較簡單,即按照 ID 的字典序排序,取第一個。

如果各節點都沒有認為的 master,則從所有節點中選擇,規則同上。

需要注意的是,這裡有個叢集中節點整理最小值限制條件,即 discovery.zen.minimum_ master_nodes。如果節點數達不到最小值的限制,則循環上述過程,直到節點數超過最小限制值,才可以開始選舉。

最後選列出一個主節點,如果只有一個本機節點,則主節點就是它自己。

如果目前節點是主節點,則開始等待節點數達到 minimum_master_nodes,再提供服務。如果目前節點不是主節點,則嘗試加入主節點所在叢集。

2 廣播模式

在廣播模式下,我們僅需在每個節點設定好叢集名稱和節點名稱即可。互相通訊的節點會根據 Elasticsearch 自訂的服務發現協定,按照廣播的方式尋找網路上設定在同樣叢集內的節點。

3.4.2 節點類型

在 Elasticsearch 中,每個節點可以有多個角色,節點既可以是候選主節點,也可以是資料節點。

節點的角色設定在設定檔 /config/elasticsearch.yml 中設定即可，設定參數如下所示。在 Elasticsearch 中，預設都為 true。

```
node.master: true   // 是否為候選主節點
node.data: true     // 是否為資料節點
```

其中，資料節點負責資料的儲存相關的操作，如對資料進行增、刪、改、查和聚合等。正因為如此，資料節點常常對伺服器的設定要求比較高，特別是對 CPU、記憶體和 I/O 的需求很大。此外，資料節點整理通常隨著叢集的擴大而彈性增加，以便保持 Elasticsearch 服務的高性能和高可用。

候選主節點是被選舉為主節點的節點，在叢集中，只有候選主節點才有選舉權和被選舉權，其他節點不參與選舉工作。

一旦候選主節點被選舉為主節點，則主節點就要負責建立索引、刪除索引、追蹤叢集中節點的狀態，以及追蹤哪些節點是叢集的一部分，並決定將哪些分片分配給相關的節點等。

3.4.3 分片和路由

在 Elasticsearch 中，若要進行分片和備份設定，則需要儘早設定。因為當在一個多分片的索引中寫入資料時，需要透過路由來確定實際寫入哪一個分片中，所以在建立索引時需要指定分片的數量，並且分片的數量一旦確定就不能修改。

分片的數量和備份數量都可以透過建立索引時的 Settings 來設定，Elasticsearch 預設為一個索引建立 5 個主分片，並分別為每個分片建立一個備份。設定的參數如下所示：

```
index.number_of_shards: 5
index.number_of_replicas: 1
```

對文件的新增、索引和刪除請求等寫入操作，必須在主分片上面完成之後才能被複製到相關的備份分片。Elasticsearch 為了加快寫入的速度，寫入過程通常是平行處理。為了解決在平行處理寫的過程中出現的資料衝突的問題，

Elasticsearch 透過樂觀鎖進行控制，每個文件都有一個 version（版本編號），當文件被修改時版本編號遞增。

那分片如何使用呢？

當我們向 Elasticsearch 寫入資料時，Elasticsearch 根據文件識別符號 ID 將文件分配到多個分片。當查詢資料時，Elasticsearch 會查詢所有的分片並整理結果。對使用者而言，這個過程是透明的，使用者並不知道資料到底存在哪個分片。

為了避免在查詢時部分分片查詢失敗影響結果的準確性，Elasticsearch 引用了路由功能，即資料在寫入時，透過路由將資料寫入指定分片；在查詢時，可以透過相同的路由指明在哪個分片將資料查出來。在預設情況下，索引資料的分片演算法如下所示：

```
shard_num = hash(_routing) % num_primary_shards
```

其中，routing 欄位的設定值預設是 id 欄位或是 parent 欄位。routing 欄位在 Hash 分片之後再與有分片的數量取模，最後獲得這筆資料應該被分配在哪一個分片。

這樣做的目的是透過 Hash 分片來保障在每個分片資料量的均勻分佈，避免各個分片的儲存負載不均衡。在做資料檢索時，Elasticsearch 預設會搜索所有分片的資料，最後在主節點上整理各個分片資料並進行排序處理後，傳回最後的結果資料。

3.4.4 資料寫入過程

資料寫入操作是在 Elasticsearch 的記憶體中執行，資料會被分配到特定的分片和備份上，但最後資料需要儲存到磁碟上持久化。

在 Elasticsearch 中，資料的儲存路徑在設定檔 ../config/elasticsearch.yml 中進行設定，實際設定如下：

```
path.data: /path/to/data    // 索引資料
path.logs: /path/to/logs    // 記錄檔記錄
```

註：建議不要使用預設值，主要是考慮到當 Elasticsearch 升級時資料的安全性問題，防止因升級 Elasticsearch 而導致資料部分甚至全部遺失。

1 分段儲存

索引資料在磁碟上是以分段形式儲存。

「段」是 Elasticsearch 從 Lucene 中繼承的概念。在索引中，索引檔案被拆分為多個子檔案，其中每個子檔案就叫作段，每個段都是一個倒排索引的小單元。

段具有不變性，一旦索引的資料被寫入硬碟，就不能再修改。

為什麼要引用分段呢？

可以試想一下，如果我們全部的文件集合僅建置在一個很大的倒排索引檔案中，且資料量還在不斷增加，當進行修改時，我們需要全部更新目前的倒排索引檔案。這會使得資料更新時效性很差、且耗費大量資源，顯然並不是我們所樂見。

其實在 Lucene 中，分段的儲存模式可以避免在讀寫操作時使用鎖，進一步大幅提升 Elasticsearch 的讀寫效能。這有點類似 CurrentHashMap 中「分段鎖」的概念，二者有異曲同工之妙，都是為了減少鎖的使用，加強平行處理。

當分段被寫入磁碟後會產生一個提交點，提交點表示一個用來記錄所有段資訊的檔案已經產生。因此，一個段一旦擁有了提交點，就表示從此該段僅有讀的許可權，永遠失去了寫的許可權。

當段在記憶體中時，此時分段擁有寫入的許可權，資料還會不斷寫入，而不具備讀取資料的許可權，表示這部分資料不能被 Elasticsearch 使用者檢索到。

那麼，既然索引檔案分段儲存並且不可修改，那麼新增、更新和刪除如何處理呢？

其實新增非常容易處理。既然資料是新的，那麼只需在目前文件新增一個段即可。

刪除資料時，由於分段不可修改的特性，Elasticsearch 不會把文件從舊的段中移除，而是新增一個 .del 檔案，.del 檔案中會記錄這些被刪除文件的段資訊。被標記刪除的文件仍然可以被查詢比對，但它會在最後結果被傳回前透過 .del 檔案將其從結果集中移除。

當更新資料時，由於分段不可修改的特性，Elasticsearch 無法透過修改舊的段來反映文件的更新，於是，更新操作變成了兩個操作的結合，即先刪除、後新增。Elasticsearch 會將舊的文件從 .del 檔案中標記刪除，然後將文件的新版本索引到一個新的段中。在查詢資料時，兩個版本的文件都會被一個查詢比對到，但被刪除的舊版本文件在結果集傳回前就會被移除。

綜上所述，段作為不可修改具有一定的優勢，段的優勢主要表現在：不需要鎖，進一步提升 Elasticsearch 的讀寫效能。

分段不變性的主要缺點是儲存空間佔用量大──當刪除舊資料時，舊資料不會被馬上刪除，而是在 .del 檔案中被標記為刪除。而舊資料只能等到段更新時才能被移除，這樣就會導致儲存空間的浪費。倘若頻繁更新資料，則每次更新都是新增新的資料到新分段，並標記舊的分段中的資料，儲存空間的浪費會更多。

在刪除和更新資料時，儲存空間會浪費；在檢索資料時，依然有侷限──在查詢獲得的結果集中會包含所有的結果集，因此主節點需要排除被標記刪除的舊資料，隨之帶來的是查詢的負擔。

2 延遲寫策略

在 Elasticsearch 中，索引寫入磁碟的過程是非同步的。

因此，為了提升寫的效能，Elasticsearch 並沒有每新增一筆資料就增加一個段到磁碟上，而是採用延遲寫策略。延遲寫策略的執行過程如下。

每當有新的資料寫入時，就將其先寫入 JVM 的記憶體中。在記憶體和磁碟之間是檔案系統快取，檔案快取空間使用的是作業系統的空間。當達到預設的時間或記憶體的資料達到一定量時，會觸發一次更新（Refresh）操作。更新操作將記憶體中的資料產生到一個新的分段上並快取到檔案快取系統，稍後再被更新到磁碟中並產生提交點。

需要指出的是，由於新的資料會繼續寫入記憶體，而記憶體中的資料並不是以段的形式儲存，因此不能提供檢索功能。只有當資料經由記憶體更新到檔案快取系統，並產生新的段後，新的段才能供搜索使用，而不需要等到被更新到磁碟才可以搜索。

在 Elasticsearch 中，寫入和開啟一個新段的過程叫作更新。在預設情況下，每個分片會每秒自動更新一次，這就是 Elasticsearch 能做到近即時搜索的原因，雖然文件的變化並不是立即對搜索可見的，但會在一秒之內變為可見。

當然，除了自動更新之外，軟體開發人員也可以手動觸發更新。

我們還可以在建立索引時，在 Settings 中透過設定 refresh_interval 的值，來調整索引的更新頻率。在設定值時需要注意後面帶上時間單位，否則預設是毫秒。當 refresh_interval=-1 時，表示關閉索引的自動更新。

雖然延遲寫策略可以減少資料往磁碟上寫的次數，提升 Elasticsearch 的整體寫入能力，但檔案快取系統的引用同時也帶來了資料遺失的風險，如機房斷電等。

為此，Elasticsearch 引用交易記錄檔（Translog）機制。交易記錄檔用於記錄所有還沒有持久化到磁碟的資料。

於是，在增加了交易記錄檔機制後，資料寫入索引的流程如下所示。

（1）新文件被索引之後，先被寫入記憶體中。為了防止資料遺失，Elasticsearch 會追加一份資料到交易記錄檔中。

（2）新的文件持續在被寫入記憶體時，同時也會記錄到交易記錄檔中。當然，此時的新資料還不能被檢索和查詢。

（3） 當達到預設的更新時間，或記憶體中的資料達到一定量後，Elasticsearch 會觸發一次更新，將記憶體中的資料以一個新段形式更新到檔案快取系統中並清空記憶體。這時新段雖未被提交到磁碟，但已經可以對外提供文件的檢索功能且不被修改。

（4） 隨著新文件索引不斷被寫入，當記錄檔資料大小超過某個值（如 512MB），或超過一定時間（如 30 min）時，Elasticsearch 會觸發一次 Flush。

此時，記憶體中的資料被寫入一個新段，同時被寫入檔案快取系統，檔案快取系統中的資料透過 Fsync 更新到磁碟中，產生提交點。而記錄檔被刪除，建立一個空的新記錄檔。

3 段合併

在 Elasticsearch 自動更新流程中，每秒都會建立一個新的段。這自然會導致短時間內段的數量暴增，而當段數量太多時會帶來較大的資源消耗，如對檔案控制代碼、記憶體和 CPU 的消耗。而在內容搜索階段，由於搜索請求要檢查到每個段，然後合併查詢結果，因此段越多，搜索速度越慢。

為此，Elasticsearch 引用段合併機制。段合併機制在後台定期進行，小的段會被合併到大的段，然後這些大的段再被合併到更大的段。

在段合併過程中，Elasticsearch 會將那些舊的已刪除文件從檔案系統中清除。被刪除的文件不會被拷貝到新的大段中，當然，在合併的過程中不會中斷索引和搜索。

段合併是自動進行索引和搜索，在合併處理程序中，會選擇一小部分大小相似的段，在後台將它們合併到更大的段中，這些段既可以是未提交的，也可以是已提交的。

在合併結束後，舊的段會被刪除，新的段會被 Flush 到磁碟，同時寫入一個包含新段且排除舊的和較小的段的新提交點。開啟新的段之後，可以用來搜索。

由於段合併的計算量較大，對磁碟 I/O 的消耗也較大，因此段合併會影響正常的資料寫入速率，然而 Elasticsearch 不會放任自流，讓段合併影響搜索效能。

Elasticsearch 在預設情況下會對合併流程進行資源限制，這就是搜索服務仍然有足夠的資源可以執行的原因。

3.5 基礎知識連結

1 樂觀鎖

Elasticsearch 引用了樂觀鎖機制來解決平行處理寫過程中資料衝突的問題，其實樂觀鎖在多個維度均有應用。

在資料庫中，我們用樂觀鎖來控制表結構，減少長交易中資料庫加鎖的負擔，達到資料表「讀多寫少」場景下的高性能；

在 Java 中，Java 引用了 CAS（Compare And Swap）樂觀鎖實現機制實現多執行緒同步的原子指令，如 AtomicInteger。

▨ 命名的藝術

本章重點介紹了 Elasticsearch 的核心概念，這些概念的英文命名方法很值得我們學習參考，如 Shard 英文原意為碎片，這個詞很具體地解釋了倒排索引分解的結果，我們透過這個單字就能見名知意。

其實，命名的學問不僅在 Elasticsearch 中用得很巧，在 Java 中也隨處可見。如研發人員經常使用的 "<>" 運算符號，英文原意為 Diamond Operator。這個命名很有想像力，"<>" 很像一個菱形，而菱形的英文單字是 Diamond。

2 設定檔格式

前面我們介紹了設定檔格式 YML。該檔案格式是由 Clark Evans、Ingy döt Net 和 Oren Ben-Kiki 在 2001 年第一次發表。

YAML 是 "YAML Ain't a Markup Language"（YAML 不是一種置標語言）的字首縮寫。有意思的是，在開發這種語言時，YAML 的初衷本是 "Yet Another

Markup Language"（仍是一種置標語言）。後來為了強調 YAML 語言以資料作為中心，而非以置標語言為重點，因而採用返璞詞來重新命名。

設定檔先後經歷了 ini 格式、JSON 格式、XML 格式、Properties 格式和 HOCON 格式。其中，HOCON（Human - Optimized Config Object Notation）格式由 Lightbend 公司開發，它被用於 Sponge，以及利用 SpongeAPI 的獨立外掛程式以儲存重要的資料，HOCON 檔案通常以 .conf 作為副檔名。

在設定檔的格式變遷中，我們能看到設定的方式都在追求語法簡單、能繼承、支援註釋等特性。

如表 3-1 所示，Elasticsearch 中的索引（Index）如果對標關聯式資料庫中的資料庫（DataBase）的話，則表（Table）與類型（Type）對應——一個資料庫下面可以有多張表（Table），就像 1 個索引（Index）下面有多種類型（Type）一樣。

表 3-1

Elasticsearch	關聯式資料庫
索引（Indices）	庫（Databases）
類型（Types）	表（Tables）
文件（Documents）	行（Rows）
欄位（Fields）	列（Columns）
對映（Mapping）	schema
Put/Post/Delete/Update /Get	增刪改查

行（ROW）與文件（Document）對應——一個資料庫表（Table）下的資料由多行（ROW）組成，就像 1 個類型 Type 由多個文件（Document）組成一樣。

列（column）與欄位（Field）對應——資料庫表（Table）中一行資料由多列（column）組成，就像 1 個文件（Document）由多個欄位（Field）組成一樣。

關聯式資料庫中的 schema 與 Elasticsearch 中的對映（Mapping）對應——在關聯式資料庫中，schema 定義了表、表中欄位、表和欄位之間的關係，就像在 Elasticsearch 中，Mapping 定義了索引下 Type 的欄位處理規則，即索引的建立、索引的類型、是否儲存原始索引 JSON 文件、是否壓縮原始 JSON 文件、是否需要斷詞處理、如何進行斷詞處理等。

關聯式資料庫中的增（Insert）刪（Delete）改（Update）查（Select）操作可與 Elasticsearch 中的增（Put/Post）刪（Delete）改（Update）查（GET）一一對應。

其實，Elasticsearch 中的分片與關聯式資料庫中常用的分函數庫分表方法有異曲同工之妙！

3 備份

備份技術是分散式系統中常見的一種資料組織形式，在日常工作中，「備份」技術也十分常見。例如各級領導都需要指定和培養「二責」人選，當自己出差或請假時，「二責」可以組織團隊中的工作。又如在團隊中，團隊成員之間的工作常常需要多人間相互備份，防止某位成員有事或離職時，相關工作不能繼續展開。

在分散式系統中，備份是如何由來，為什麼這麼有必要性呢？

備份（Replica 或稱 Copy）一般指在分散式系統中為資料或服務提供的容錯。這種容錯設計是加強分散式系統容錯率、加強可用性的常用方法。

在服務備份方面，一般指的是在不同伺服器中部署同一份程式。如 Tomcat/Jetty 叢集部署服務，叢集中任意一台伺服器都是叢集中其他伺服器的備份。

在資料備份方面，一般指的是在不同的節點上持久化同一份資料。當某節點中儲存的資料遺失時，系統就可以從備份中讀到資料。

可以說，資料備份是分散式系統解決資料遺失或例外的唯一方法，因此備份協定也成為貫穿整個分散式系統的理論核心。

☑ 備份的資料一致性

分散式系統透過備份控制協定，讓使用者透過一定的方式即可讀取分散式系統內部各個備份的資料，這些資料在一定的限制條件下是相同的，即備份資料一致性（Consistency）。備份資料一致性是針對分散式系統中的各個節點而言，不是針對某節點的某個備份而言。

在分散式系統中，一致性分為強一致性（Strong Consistency）、弱一致性（Week Consistency），還有介於二者之間的階段一致性（Session Consistency）和最後一致性（Eventual Consistency）。

其中，強一致性最難實現。強一致性要求任何時刻使用者都可以讀到最近一次成功更新的備份資料。弱一致性與強一致性正好相反，資料更新後，使用者無法在一定時間內讀到最新的值，因此很少在實際中使用。

階段一致性指的是在一次階段內，使用者一旦讀到某個資料的某個版本的更新資料，則在這個階段中就不會再讀到比目前版本更老舊的資料。最後一致性指的是叢集中各個備份的資料最後能達到完全一致的狀態。

從備份的角度而言，強一致性最佳，但對於分散式系統而言，還要考慮其他方面，如分散式系統的整體效能（即系統的吞吐）、系統的可用性、系統的可擴充性等。這也是系統設計要全盤考慮的原因。

▨ 備份資料的分佈方式

備份的資料是如何分發合格的呢？這就有關資料的分佈方式。

一般來説，資料的分佈方式主要有雜湊方式、按資料範圍分佈、按資料量分佈和一致性雜湊方式（Consistent Hashing）等。

其中雜湊方式最為簡單，簡單是其最大的優勢，但缺點同樣明顯。一方面，可擴充性不高——一旦儲存規模需要擴大，則所有資料都需要重新按雜湊值分發；另一方面，雜湊方式容易導致儲存空間的資料分佈不均勻。

按資料範圍分佈也比較常見，一般來說，是將資料按特徵值的範圍劃分為不同的區間，使得叢集中不同的伺服器處理不同區間的資料。這種方式可以避免雜湊值帶來的儲存空間資料分佈不均勻的情況。

按資料量分佈和按資料範圍分佈核心想法比較接近，一般是將資料看作一個順序增長的，並將資料集按照某一較為固定的大小劃分為許多資料區塊，把不同的資料區塊分佈到不同的伺服器上。

而一致性雜湊是在專案實作中使用較為廣泛的資料分佈方式。一致性雜湊的基本想法是使用一個雜湊函數計算資料的雜湊值，而雜湊函數的輸出值會作為一個封閉的環，我們會根據雜湊值將節點隨機分佈到這個環上，每個節點負責處理從自己開始順時鐘至下一個節點的全部雜湊值域上的資料。

有了資料分佈的方法，那麼資料以何種形態進行分佈呢？一般來說有兩種，一種是以伺服器為核心，另一種是以資料為核心。

以機器為核心時，機器之間互為備份，備份機器之間的資料完全相同。以機器為核心的策略適用於上述各種資料分佈方式，最主要的優點就是簡單，容易落地；而缺點也很明顯，一旦資料出問題，在資料恢復時就需要恢復多台伺服器中的資料，效率很低；而且增加伺服器後，會帶來可擴充性低的問題。

以資料為核心時，一般將資料拆分為許多個資料段，以資料段為單位去分發。一般來說，每個資料段的大小儘量相等，而且限制資料量大小的上線。在不同的系統中，資料段有很多不同的稱謂，如在 Lucene 和 Elasticsearch 中稱之為 segment，在 Kafka 中稱之為 chunk 和 partition 等。

以資料為核心並不適合所有資料分佈方式，一般會採用雜湊方式或一致性雜湊方式。

將資料拆分為資料段表示備份的管理將以資料段為單位進行展開，因此備份與機器不再強相關，每台機器都可以負責一定資料段的備份。這帶來的好處是當某台伺服器中的資料有問題時，我們可以從叢集中的任何其他伺服器恢復資料，因此資料的恢復效率很高。

▨ 備份分發策略

備份分發策略指的是主節點和備份節點之間備份資料同步的方法。一般來說分為兩大類：中心化方式和去中心化方式。

中心化方式的基本上線想法是由一個中心節點協調備份資料的更新、維護備份之間的一致性。資料的更新可以是主節點主動向備份節點發送，也可以是備份節點向主節點發送。中心化方式的優點是設計想法較為簡單，而缺點也很明顯，資料的同步及系統的可用性都有「單點依賴」的風險，即依賴於中心化節點。一旦中心化節點發生例外，則資料同步和系統的可用性都會受到影響。

在去中心化方式中則沒有中心節點，所有的節點都是 P2P 形式，地位對等，節點之間透過平等協商達到一致，因此去中心化節點不會因為某個節點的例外而導致系統的可用性受到影響。但有得必有失，去中心化方式的最大缺點在於各個節點達成共識的過程較長，需要反覆進行訊息通訊來確認內容，實現較為複雜。

3.6 小結

本章主要介紹了 Elasticsearch 的基本情況及其安裝、設定，另外，還介紹了 Elasticsearch 中的核心概念及其架構設計。

在基礎知識連結部分，先後以樂觀鎖、命名的藝術、設定檔格式的變遷、備份技術、Elasticsearch 和關聯式資料庫核心概念的類比為核心，進行了連結基礎知識的介紹。

初級用戶端實戰

本 章主要介紹在 Elasticsearch 中初級用戶端的 API 使用方法。

想要使用 Elasticsearch 服務，則要先取得一個 Elasticsearch 用戶端。取得 Elasticsearch 用戶端的方法很簡單，最常見的就是建立一個可以連接到叢集的傳輸用戶端物件。

在 Elasticsearch 中，用戶端有初級用戶端和進階用戶端兩種。它們均使用 Elasticsearch 提供了 RESTful 風格的 API，因此，本書中的用戶端 API 使用也以 RESTful 風格的 API 為主。在使用 RESTful API 時，一般透過 9200 通訊埠與 Elasticsearch 進行通訊。

初級用戶端是 Elasticsearch 提供給使用者的官方版初級用戶端。初級用戶端允許透過 HTTP 與 Elasticsearch 叢集進行通訊，它將請求封裝發給 Elasticsearch 叢集，將 Elasticsearch 叢集的回應封裝傳回給使用者。初級用戶端與所有 Elasticsearch 版本都相容。

進階用戶端是用於彈性搜索的進階用戶端，以初級用戶端為基礎。進階用戶端公開了 API 特定的方法，並負責處理未編組的請求和回應。

4.1 初級用戶端初始化

在介紹如何使用初級用戶端之前，我們先要了解初級用戶端的主要功能，其主要功能包含：

（1）跨所有可用節點的負載平衡。

（2）在節點故障和特定回應程式時的容錯移轉。

（3）失敗連接的懲罰機制。判斷一個失敗節點是否重試，取決於用戶端連接時連續失敗的次數；失敗的嘗試次數越多，用戶端再次嘗試同一節點之前等待的時間越長。

（4）持久連接。

（5）請求和回應的追蹤記錄檔記錄。

（6）自動發現叢集節點，該功能可選。

在介紹完初級用戶端的主要功能後，下面介紹初級用戶端的使用。

❶ 取得用戶端元件

首先，我們需要從 Maven 儲存函數庫中取得初級用戶端的 jar 套件。目前，初級用戶端託管在 Maven Central 中央倉庫中，其所需的 Java 最低版本為 1.7。

需要指出的是，初級用戶端保持了與 Elasticsearch 相同的發佈週期。而用戶端的版本和用戶端可以通訊的 Elasticsearch 的版本之間沒有關係，即初級用戶端可以與所有 Elasticsearch 版本相容。

從 Maven Central 中央倉庫，可以查到目前初級用戶端的 Maven 依賴和 Gradle 依賴分別如下。

Maven 依賴：

```
<dependency>
    <groupId>org.elasticsearch.client</groupId>
    <artifactId>elasticsearch-rest-client</artifactId>
    <version>7.2.0</version>
</dependency>
```

```
Gradle 依賴:
dependencies {
    compile 'org.elasticsearch.client:elasticsearch-rest-client:7.2.0'
}
```

而 elasticsearch-rest-client 也有依賴項，這是因為是初級用戶端內部需要使用
Apache HTTP 非同步用戶端發送 HTTP 請求，因此依賴於以下元件，即非同
步 HTTP 用戶端及其本身的可傳遞依賴項，實際如下所示：

```
org.apache.httpcomponents:httpasyncclient
org.apache.httpcomponents:httpcore-nio
org.apache.httpcomponents:httpclient
org.apache.httpcomponents:httpcore
commons-codec:commons-codec
commons-logging:commons-logging
```

此時，專案中的 POM 檔案（以 Maven 依賴為例）如下所示：

```xml
<?xml version="1.0" encoding="UTF-8"?>
<project xmlns="http://maven.apache.org/POM/4.0.0" xmlns:xsi="http://
www.w3.org/2001/XMLSchema-instance"
xsi:schemaLocation="http://maven.apache.org/POM/4.0.0
http://maven.apache.org/xsd/maven-4.0.0.xsd">
  <modelVersion>4.0.0</modelVersion>
  <groupId>com.niudong</groupId>
  <artifactId>esdemo</artifactId>
  <version>0.0.1-SNAPSHOT</version>
  <packaging>jar</packaging>
  <name>esdemo</name>
  <description>ElasticSearch Demo project for Spring Boot</description>
  <parent>
    <groupId>org.springframework.boot</groupId>
    <artifactId>spring-boot-starter-parent</artifactId>
    <version>2.1.6.RELEASE</version>
    <relativePath /> <!-- lookup parent from repository -->
  </parent>
  <properties>
    <project.build.sourceEncoding>UTF-8</project.build.sourceEncoding>
```

```xml
    <project.reporting.outputEncoding>UTF-8</project.reporting.
outputEncoding>
    <java.version>1.8</java.version>
</properties>
<dependencies>
    <dependency>
        <groupId>org.springframework.boot</groupId>
        <artifactId>spring-boot-starter-web</artifactId>
    </dependency>
    <dependency>
        <groupId>org.springframework.boot</groupId>
        <artifactId>spring-boot-starter-test</artifactId>
        <scope>test</scope>
    </dependency>
    <!-- 增加 Elasticsearch 的依賴 -->
    <dependency>
        <groupId>org.elasticsearch.client</groupId>
        <artifactId>elasticsearch-rest-client</artifactId>
        <version>7.2.0</version>
    </dependency>
    <!-- 增加 Elasticsearch 的依賴 begin-->
    <dependency>
        <groupId>org.apache.httpcomponents</groupId>
        <artifactId>httpasyncclient</artifactId>
        <version>4.1.4</version>
    </dependency>
    <dependency>
        <groupId>org.apache.httpcomponents</groupId>
        <artifactId>httpcore-nio</artifactId>
        <version>4.4.11</version>
    </dependency>
    <dependency>
        <groupId>org.apache.httpcomponents</groupId>
        <artifactId>httpclient</artifactId>
        <version>4.5.9</version>
    </dependency>
    <dependency>
```

```
        <groupId>org.apache.httpcomponents</groupId>
        <artifactId>httpcore</artifactId>
        <version>4.4.11</version>
    </dependency>
    <dependency>
        <groupId>commons-codec</groupId>
        <artifactId>commons-codec</artifactId>
        <version>1.12</version>
    </dependency>
    <dependency>
        <groupId>commons-logging</groupId>
        <artifactId>commons-logging</artifactId>
        <version>1.2</version>
    </dependency>
    <!-- 增加 Elasticsearch 的依賴 end -->
</dependencies>
<build>
    <plugins>
        <plugin>
            <groupId>org.springframework.boot</groupId>
            <artifactId>spring-boot-maven-pluqin</artifactId>
        </plugin>
    </plugins>
</build>
</project>
```

2 用戶端初始化

使用者可以透過對應的 RestClientBuilder 類別來建置 RestClient 實例,該類別是透過 RestClient builder(HttpHost…)靜態方法建立。

RestClient 在初始化時,唯一需要的參數是用戶端將與之通訊的或多個主機作為 HttpHost 實例提供的,如下所示:

```
RestClient restClient = RestClient.builder(
    new HttpHost("localhost", 9200, "http"),
    new HttpHost("localhost", 9201, "http")).build();
```

對於 RestClient 類別而言，RestClient 類別是執行緒安全的。在理想情況下，它與使用它的應用程式具有相同的生命週期。因此，當不再需要時，應該關閉它，以便釋放它使用的所有資源及底層 HTTP 客戶端實例及其執行緒，這一點很重要。關閉的方法如下所示：

```
restClient.close();
```

下面透過一段程式展示初級用戶端初始化的使用。程式分為 3 層，分別是 Controller 層、Service 層和 ServiceImpl 實現層。

其中，Controller 層的程式如下所示：

```
package com.niudong.esdemo.controller;
import org.springframework.beans.factory.annotation.Autowired;
import org.springframework.web.bind.annotation.RequestMapping;
import org.springframework.web.bind.annotation.RestController;
import com.niudong.esdemo.service.MeetElasticSearchService;
@RestController
@RequestMapping("/springboot/es")
public class MeetElasticSearchController {
  @Autowired
  private MeetElasticSearchService meetElasticSearchService;

  @RequestMapping("/init")
  public String initElasticSearch() {
    meetElasticSearchService.initEs();
    return "Init ElasticSearch Over!";
  }
}
```

Service 層的程式如下所示：

```
package com.niudong.esdemo.service;
public interface MeetElasticSearchService {
  public void initEs();
  public void closeEs();
}
```

ServiceImpl 實現層的程式如下所示：

```
package com.niudong.esdemo.service.impl;
import javax.annotation.PostConstruct;
import org.apache.commons.logging.Log;
import org.apache.commons.logging.LogFactory;
import org.apache.http.HttpHost;
import org.elasticsearch.client.RestClient;
import org.springframework.stereotype.Service;
import com.niudong.esdemo.service.MeetElasticSearchService;
@Service
public class MeetElasticSearchServiceImpl implements MeetElasticSearchService {
  private static Log log = LogFactory.getLog (MeetElasticSearchServiceImpl.
      class);
  private RestClient restClient;
  @PostConstruct
  public void initEs() {
    restClient = RestClient
        .builder(new HttpHost("localhost", 9200, "http"), new HttpHost
            ("localhost", 9201, "http"))
        .build();
    log.info("ElasticSearch init in service.");
  }
  public void closeEs() {
    try {
      restClient.close();
    } catch (Exception e) {
      e.printStackTrace();
    }
  }
}
```

在執行程式之前，首先啟動 Elasticsearch 服務，啟動後，輸出內容如下所示。

```
2019-07-30 13:51:57.742  INFO 52088 --- [nio-8080-exec-2] c.n.e.s.i.MeetElasticSearchServiceImpl    : ElasticSearch init
in service.
```

然後編譯專案，在專案根目錄下輸入以下指令：

```
mvn clean package
```

接著透過下面的指令啟動專案服務：

```
java -jar ./target/esdemo-0.0.1-SNAPSHOT.jar
```

在專案服務啟動後，在瀏覽器中呼叫以下介面檢視 Elasticsearch 用戶端的連接情況：

```
http://localhost:8080/springboot/es/init
```

如圖 4-1 所示。

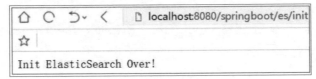

圖 4-1

可以看到，瀏覽器頁面輸出了 "Init ElasticSearch Over!"，即 Elasticsearch 用戶端已經成功連接。

RestClientBuilder 允許在建置 RestClient 時選擇性地設定以下設定參數。

（1）請求標頭設定方法。

```
// 帶請求標頭
public void initEsWithHeader() {
RestClientBuilder builder = RestClient.builder(new HttpHost("localhost",
    9200, "http"));
// 設定每個請求需要發送的預設請求標頭，以防止在每個請求中指定它們
Header[] defaultHeaders = new Header[] {new BasicHeader("header",
    "value")};
builder.setDefaultHeaders(defaultHeaders);
}
```

（2）設定監聽器。

```
// 帶失敗監聽
public void initEsWithFail() {
RestClientBuilder builder = RestClient.builder(new HttpHost("localhost",
    9200, "http"));
```

```
// 設定一個監聽器，該監聽器在每次節點失敗時都會收到通知，在啟用偵測時在內部使用
builder.setFailureListener(new RestClient.FailureListener() {
  @Override
  public void onFailure(Node node) {
  }
});
}
```

（3）設定節點選擇器。

```
// 節點選擇器
public void initEsWithNodeSelector() {
RestClientBuilder builder = RestClient.builder(new HttpHost("localhost",
    9200, "http"));
builder.setNodeSelector(NodeSelector.SKIP_DEDICATED_MASTERS);
}
```

（4）設定逾時。

```
// 設定逾時
public void initEsWithTimeout() {
  RestClientBuilder builder = RestClient.builder(new HttpHost("localhost",
      9200, "http"));
  builder.setRequestConfigCallback(new RestClientBuilder.
RequestConfigCallback() {
    @Override
    public RequestConfig.Builder customizeRequestConfig(
        RequestConfig.Builder requestConfigBuilder) {
      return requestConfigBuilder.setSocketTimeout(10000);
    }
  });
}
```

4.2 提交請求

在建立好用戶端之後，執行請求之前還需建置請求物件。使用者可以透過呼叫用戶端的 performRequest 和 performRequestAsync 方法來發送請求。

其中，performRequest 是同步請求方法，它將阻塞呼叫執行緒，並在請求成功時傳回回應，或在請求失敗時引發例外。

而 performRequestAsync 是非同步方法，它接收一個 ResponseListener 物件作為參數。如果請求成功，則該參數使用回應進行呼叫；如果請求失敗，則使用例外進行呼叫。

1 建置請求物件 Request

請求物件 Request 的請求方式與 The HTTP 的請求方式相同，如 GET、POST、HEAD 等，程式如下所示：

```
/**
 * 本部分用於介紹如何建置對 Elasticsearch 服務的請求
 */
public Request buildRequest() {
  Request request = new Request("GET", // 與 The HTTP 的請求方式相，如 GET、
// POST、HEAD 等
      "/");
  return request;
}
```

2 請求的執行

在建置請求（Request）後，即可執行請求。請求有同步執行和非同步執行兩種方式。下面將分別展示如何使用 performRequest 方法和 performRequestAsync 方法發送請求。

◪ 同步方式

當以同步方式執行 Request 時，用戶端會等待 Elasticsearch 伺服器傳回的查詢結果 Response。在收到 Response 後，用戶端繼續執行相關的邏輯程式。以同步方式執行的程式如下所示，我們仍然建置三層程式，分別是 Controller 層、Service 層和 ServiceImpl 實現層。

其中，在 Controller 層的 MeetElasticSearchController 類別中增加以下程式：

```
@RequestMapping("/buildRequest")
  public String executeRequestForElasticSearch() {
    return meetElasticSearchService.executeRequest();
  }
```

隨後，在 Service 層的 MeetElasticSearchService 類別中增加以下程式：

```
/**
  * 本部分用於介紹如何建置對 Elasticsearch 服務的請求
  */
  public String  executeRequest();
```

在 ServiceImpl 實現層的 MeetElasticSearchServiceImpl 類別中增加以下程式：

```
/**
  * 本部分用於介紹如何建置對 Elasticsearch 服務的請求
  */
  public String  executeRequest() {
    Request request = new Request("GET",
// 與 The HTTP 的請求方式相同，如 GET、POST、HEAD 等
      "/");
    // 在伺服器上請求
    try {
      Response response = restClient.performRequest(request);
      return response.toString();
    } catch (Exception e) {
      e.printStackTrace();
    }
    try {
      restClient.close();
    } catch (Exception e) {
      e.printStackTrace();
    }

    return "Get result failed!";
  }
```

隨後編譯專案,在專案根目錄下輸入以下指令:

```
mvn clean package
```

透過以下指令啟動專案服務:

```
java -jar ./target/esdemo-0.0.1-SNAPSHOT.jar
```

在專案服務啟動後,在瀏覽器中透過呼叫以下介面檢視 Elasticsearch 用戶端的連接情況:

```
http://localhost:8080/springboot/es/buildRequest
```

在瀏覽器中可以看到請求發送成功,顯示的 Response 資訊如下所示。

```
Response{requestLine=GET / HTTP/1.1, host=http://localhost:9200, response=HTTP/1.1 200 OK}
```

✍ 非同步方式

當以非同步方式執行請求時,初級用戶端不必同步等待請求結果的傳回,可以直接向介面呼叫方傳回非同步介面執行成功的結果。

為了處理非同步傳回的回應資訊或處理在請求執行過程中引發的例外資訊,使用者需要指定監聽器。以非同步方式呼叫的程式如下所示。我們在 ServiceImpl 實現層的 MeetElasticSearchServiceImpl 類別中增加以下程式:

```
// 以非同步方式在伺服器上執行請求
  public String buildRequestAsync() {
    Request request = new Request("GET", // 與 The HTTP 的請求方式相同,如 GET、
// POST、HEAD 等
        "/");
    // 在伺服器上請求
    restClient.performRequestAsync(request, new ResponseListener() {
      @Override
      public void onSuccess(Response response) {
      }
      @Override
      public void onFailure(Exception exception) {
      }
    });
```

```
  try {
    restClient.close();
  } catch (Exception e) {
    e.printStackTrace();
  }
  return "Get result failed!";
}
```

在非同步請求處理後，如果請求執行成功，則呼叫 ResponseListener 類別中的 onResponse 方法進行相關邏輯的處理；如果請求執行失敗，則呼叫 ResponseListener 類別中的 onFailure 方法進行相關邏輯的處理。

3 可選參數設定

不論同步請求，還是非同步請求，我們都可以在請求中增加參數，增加方法如下所示：

```
request.addParameter("pretty", "true");
```

可以將請求的主體設定為任意 HttpEntity，設定方法如下所示：

```
request.setEntity(new NStringEntity(
        "{\"json\":\"text\"}",
        ContentType.APPLICATION_JSON));
```

還可以將其設定為一個字串，在 Elasticsearch 中，預設使用 application/json 的內容格式，設定方法如下所示：

```
request.setJsonEntity("{\"json\":\"text\"}");
```

此外，Request 還有一些可選的請求建置選項，透過 RequestOptions 來實現。

需要說明的是，在 RequestOptions 類別中儲存的請求，可以在同一應用程式的多個請求之間共用。因此使用者可以建立一個單實例，然後在所有請求之間共用。實際方法如下所示：

```
private static final RequestOptions COMMON_OPTIONS;
static {
    RequestOptions.Builder builder = RequestOptions.DEFAULT.toBuilder();
```

```
    builder.addHeader("Authorization", "Bearer " + TOKEN);
    builder.setHttpAsyncResponseConsumerFactory(
        new HttpAsyncResponseConsumerFactory
            .HeapBufferedResponseConsumerFactory(30 * 1024 * 1024 * 1024));
    COMMON_OPTIONS = builder.build();
}
```

上述程式中的 TOKEN 表示增加所有請求所需的任何標頭。

而 AddHeader 用於授權或在 Elasticsearch 前使用代理所需的標頭資訊。在使用時，不需要設定 Content-Type 標頭，因為用戶端將自動在請求的 HttpEntity 中設定 Content-Type 標頭。

在建立好 RequestOptions 單實例 COMMON_OPTIONS 後，我們就可以在發出請求時使用它了，使用方法如下所示：

```
request.setOptions(COMMON_OPTIONS);
```

Elasticsearch 允許使用者根據每個請求訂製這些選項，如增加一個額外的標題：

```
RequestOptions.Builder options = COMMON_OPTIONS.toBuilder();
options.addHeader("title","any other things");
request.setOptions(options);
```

RequestOptions 在 ServiceImpl 實現層的 MeetElasticSearchServiceImpl 類別中的程式如下所示：

```
// 設定全域單實例 RequestOptions
private static final RequestOptions COMMON_OPTIONS;
static {
    RequestOptions.Builder builder = RequestOptions.DEFAULT.toBuilder();
    builder.addHeader("Authorization", "Bearer " + "my-token");
    builder.setHttpAsyncResponseConsumerFactory(
        new HttpAsyncResponseConsumerFactory
            .HeapBufferedResponseConsumerFactory(30 * 1024 * 1024 * 1024));
    COMMON_OPTIONS = builder.build();
}
```

```java
/**
 * 本部分用於介紹如何建置對 Elasticsearch 服務的請求
 */
public String buildRequestWithRequestOptions() {
  Request request = new Request("GET",
// 與 The HTTP 的請求方式相同，如 GET、POST、HEAD 等
      "/");
  // 在伺服器上請求
  try {
    Response response = restClient.performRequest(request);

    RequestOptions.Builder options = COMMON_OPTIONS.toBuilder();
    options.addHeader("title", "u r my dear!");
    request.setOptions(COMMON_OPTIONS);

    return response.toString();
  } catch (Exception e) {
    e.printStackTrace();
  }
  try {
    restClient.close();
  } catch (Exception e) {
    e.printStackTrace();
  }
  return "Get result failed!";
}
```

4 多個平行非同步作業

除單一操作的執行外，Elasticsearch 的用戶端還可以並存執行許多操作。下面
透過 ServiceImpl 實現層 MeetElasticSearchServiceImpl 類別中的範例，展示如
何對多文件進行平行索引。程式如下：

```java
// 平行處理處理文件資料
public void multiDocumentProcess(HttpEntity[] documents) {
  final CountDownLatch latch = new CountDownLatch(documents.length);
```

```
for (int i = 0; i < documents.length; i++) {
  Request request = new Request("PUT", "/posts/doc/" + i);
  // 假設 documents 儲存在 HttpEntity 陣列中
  request.setEntity(documents[i]);
  restClient.performRequestAsync(request, new ResponseListener() {
    @Override
    public void onSuccess(Response response) {
      latch.countDown();
    }
    @Override
    public void onFailure(Exception exception) {
      latch.countDown();
    }
  });
}
try {
  latch.await();
} catch (Exception e) {
  e.printStackTrace();
}
}
```

4.3 對請求結果的解析

4.2 節介紹了用戶端中請求物件的建置和請求方式，本節介紹對獲得的回應結果 Response 的解析。

請求物件有兩種請求方式，分別是同步請求和非同步請求，因此對於請求的回應結果 Response 的解析也分為兩種。

同步請求獲得的回應物件是由 performRequest 方法傳回的；而非同步請求獲得的回應物件是透過 ResponseListener 類別下 onSuccess(Response) 方法中的參數所接收。回應物件中包裝 HTTP 用戶端傳回的回應物件，並公開一些附加資訊。

下面透過程式，學習對請求結果的解析。以同步請求方式為例，對請求結果的解析程式如下所示。

範例程式共分為三層，分別是 Controller 層、Service 層和 ServiceImpl 實現層。

首先，在 Controller 層的 MeetElasticSearchController 類別中增加以下程式：

```
@RequestMapping("/parseEsResponse")
  public String parseElasticSearchResponse() {
    meetElasticSearchService.parseElasticSearchResponse();
    return "Parse ElasticSearch Response Is  Over!";
  }
```

然後在 Service 層的 MeetElasticSearchService 類別中增加以下程式：

```
/**
   * 本部分用於介紹如何解析 Elasticsearch 服務的傳回結果
   */
  public void parseElasticSearchResponse();
```

接著在 ServiceImpl 實現層的 MeetElasticSearchServiceImpl 類別中增加以下程式：

```
/**
   * 本部分用於介紹如何解析 Elasticsearch 服務的傳回結果
   */
  public void parseElasticSearchResponse() {
    try {
      Response response = restClient.performRequest(new Request("GET", "/"));

      // 已執行請求的資訊
      RequestLine requestLine = response.getRequestLine();
      //Host 傳回的資訊
      HttpHost host = response.getHost();
      // 回應狀態行，從中可以解析狀態碼
      int statusCode = response.getStatusLine().getStatusCode();
      // 回應標頭，也可以透過 getheader（string）按名稱取得
      Header[] headers = response.getHeaders();
      String responseBody = EntityUtils.toString(response.getEntity());
```

```
    log.info("parse ElasticSearch Response,responseBody is :" +
        responseBody);
} catch (Exception e) {
    e.printStackTrace();
}
}
```

隨後編譯專案，在專案根目錄下輸入以下指令：

```
mvn clean package
```

透過以下指令啟動專案服務：

```
java -jar ./target/esdemo-0.0.1-SNAPSHOT.jar
```

當專案服務啟動後，在瀏覽器中呼叫以下介面檢視 Elasticsearch 用戶端的連接
情況：

```
http://localhost:8080/springboot/es/parseEsResponse
```

在伺服器主控台中列印 responseBody 的內容，如圖 4-2 所示。

```
2019-07-30 20:19:18.213  INFO 22020 --- [nio-8080-exec-1] c.n.e.s.i.MeetElasticSearchServiceImpl   : parse ElasticSearc
Response,responseBody is :{
  "name" : "LAPTOP-1S8BALK3",
  "cluster_name" : "elasticsearch",
  "cluster_uuid" : "6UkGmpPrSwyXlzPHgSaivg",
  "version" : {
    "number" : "7.2.0",
    "build_flavor" : "default",
    "build_type" : "zip",
    "build_hash" : "508c38a",
    "build_date" : "2019-06-20T15:54:18.811730Z",
    "build_snapshot" : false,
    "lucene_version" : "8.0.0",
    "minimum_wire_compatibility_version" : "6.8.0",
    "minimum_index_compatibility_version" : "6.0.0-beta1"
  },
  "tagline" : "You Know, for Search"
}
```

圖 4-2

在瀏覽器頁面輸出介面請求成功的內容，如下所示：

```
Parse ElasticSearch Response Is Over!
```

4.4 常見通用設定

除上述用戶端 API 外，用戶端還支援一些常見通用設定，如逾時設定、執行緒數設定、節點選擇器設定和設定偵測器等。

▌**1** 逾時設定

我們可以在建置 RestClient 時提供 requestconfigCallback 的實例來完成逾時設定。該介面有一個方法，它接收 org.apache.http.client.config.requestconfig. builder 的實例作為參數，並且具有相同的傳回類型。

使用者可以修改請求設定產生器 org.apache.http.client.config.requestconfig. builder 的實例，然後傳回。

在下面的範例中，增加了連接逾時（預設為 1s）和通訊端逾時（預設為 30s），程式如下所示：

```
.setConnectTimeout(5000)
.setSocketTimeout(60000);
```

在實際使用時，詳見 MeetElasticSearchServiceImpl 中的程式，部分程式如下所示：

```
// 設定逾時
  public void initEsWithTimeout() {
    RestClientBuilder builder = RestClient.builder(new HttpHost("localhost",
        9200, "http"));
    builder.setRequestConfigCallback(new RestClientBuilder.
RequestConfigCallback() {
      @Override
      public RequestConfig.Builder customizeRequestConfig(
          RequestConfig.Builder requestConfigBuilder) {
        return requestConfigBuilder.setSocketTimeout(10000).setSocketTimeout
          (60000);
      }
    });
  }
```

2 執行緒數設定

Apache HTTP 非同步用戶端預設啟動一個排程程式執行緒,連線管理員使用的多個工作執行緒。一般執行緒數與本機檢測到的處理器數量相同,執行緒數主要取決於 Runtime.getRuntime(). availableProcessors() 傳回的結果。

Elasticsearch 允許使用者修改執行緒數,修改程式如下所示,詳見 MeetElasticSearchServiceImpl 類別:

```
/**
 * 本部分用於介紹使用 Elasticsearch 用戶端的通用設定
 */
 public void setThreadNumber(int number) {
   RestClientBuilder builder = RestClient.builder(new HttpHost("localhost",
     9200))
      .setHttpClientConfigCallback(new HttpClientConfigCallback() {
       @Override
       public HttpAsyncClientBuilder customizeHttpClient(
         HttpAsyncClientBuilder httpClientBuilder) {
        return httpClientBuilder.setDefaultIOReactorConfig(
          IOReactorConfig.custom().setIoThreadCount(number).build());
       }
     });
 }
```

3 節點選擇器設定

在預設情況下,用戶端會以輪詢的方式將每個請求發送到設定的各個節點中。

Elasticsearch 允許使用者自由選擇需要連接的節點。一般透過初始化用戶端來設定節點選擇器,以便篩選節點。

該功能在啟用偵測器時很有用,以防止 HTTP 請求只命中專用的主節點。

設定後,對於每個請求,用戶端都透過節點選擇器來篩選備選節點。

程式如下所示,詳見 MeetElasticSearchServiceImpl 類別:

```
// 設定節點選擇器
  public void setNodeSelector() {
    RestClientBuilder builder = RestClient.builder(new HttpHost("localhost",
        9200, "http"));
    builder.setNodeSelector(new NodeSelector() {
      @Override
      public void select(Iterable<Node> nodes) {
        boolean foundOne = false;
        for (Node node : nodes) {
          String rackId = node.getAttributes().get("rack_id").get(0);
          if ("targetId".equals(rackId)) {
            foundOne = true;
            break;
          }
        }

        if (foundOne) {
          Iterator<Node> nodesIt = nodes.iterator();
          while (nodesIt.hasNext()) {
            Node node = nodesIt.next();
            String rackId = node.getAttributes().get("rack_id").get(0);
            if ("targetId".equals(rackId) == false) {
              nodesIt.remove();
            }
          }
        }

      }
    });
  }
```

4 設定偵測器

偵測器允許自動發現執行中的 Elasticsearch 叢集中的節點，並將其設定為現有
的 RestClient 實例。

在預設情況下，偵測器使用 nodes info API 檢索屬於叢集的節點，並使用 jackson 解析獲得的 JSON 回應。

目前，偵測器與 Elasticsearch 2.X 及更新版本相容。

在使用偵測器之前需增加相關的依賴，程式如下所示：

```
// Maven 版本依賴
<!-- 增加 Elasticsearch 的偵測器 begin -->
    <dependency>
        <groupId>org.elasticsearch.client</groupId>
        <artifactId>elasticsearch-rest-client-sniffer</artifactId>
        <version>7.2.1</version>
    </dependency>
    <!-- 增加 Elasticsearch 的偵測器 end -->
// Gradle 依賴
dependencies {
    compile 'org.elasticsearch.client:elasticsearch-rest-client-sniffer:7.2.1'
}
```

在建立好 RestClient 實例（如初始化中程式所示）後，就可以將偵測器與其進行連結了。偵測器利用 RestClient 提供的定期機制（在預設情況下定期時間為 5min），從叢集中取得目前節點的列表，並透過呼叫 RestClient 類別中的 setNodes 方法來更新它們。

偵測器的使用程式詳見 ServiceImpl 實現層的 MeetElasticSearchServiceImpl 類別，部分程式如下所示：

```
// 設定偵測器
  public void setSniffer() {
    RestClient restClient = RestClient.builder(new HttpHost("localhost", 9200,
"http")).build();
    Sniffer sniffer = Sniffer.builder(restClient)
    // 偵測器預設每 5min 更新一次節點。可以透過 setSniffIntervalMillis（以毫秒為單
    // 位）自訂此間隔
        .setSniffIntervalMillis(60000).build();
    // 使用後，結束用戶端和偵測器
    try {
```

```
      sniffer.close();
      restClient.close();
   } catch (Exception e) {
      e.printStackTrace();
   }
}
```

當然，除在用戶端啟動時設定偵測器外，還可以在失敗時啟用偵測器。這表示在每次失敗後，節點清單都會立即更新，而不是在接下來的普通偵測循環中更新。

在這種情況下，首先需要建立一個 SniffOnFailureListener，然後在建立 RestClient 時設定。在建立偵測器後，同一個 SniffOnFailureListener 實例會相互連結，以便在每次失敗時都通知該實例，並使用偵測器執行偵測動作。

偵測器 SniffOnFailureListener 的使用程式詳見 ServiceImpl 實現層的 MeetElasticSearchServiceImpl 類別，部分程式如下所示：

```
// 設定偵測器
public void setSnifferWhenFail(int failTime) {
   SniffOnFailureListener sniffOnFailureListener = new SniffOnFailure
        Listener();
   RestClient restClient = RestClient.builder(new HttpHost("localhost",
        9200))
        .setFailureListener(sniffOnFailureListener).build();
   Sniffer sniffer = Sniffer.builder(restClient).
        setSniffAfterFailureDelayMillis (failTime).build();
   sniffOnFailureListener.setSniffer(sniffer);
   // 使用後，結束用戶端和偵測器
   try {
      sniffer.close();
      restClient.close();
   } catch (Exception e) {
      e.printStackTrace();
   }
}
```

由於 Elasticsearch 節點資訊 API 不會傳回連接到節點時要使用的協定,而是只傳回它們的 host:port,因此在預設情況下會使用 HTTP。如果需要使用 HTTPS,則必須手動建立並提供 ElasticSearchNodesNiffer 實例,相關程式如下所示:

```
// 設定以 HTTPS 為基礎的偵測器
  public void setSnifferWithHTTPS(int failTime) {
    RestClient restClient = RestClient.builder(new HttpHost("localhost", 9200,
        "http")).build();
    NodesSniffer nodesSniffer = new ElasticsearchNodesSniffer(restClient,
        ElasticsearchNodesSniffer.DEFAULT_SNIFF_REQUEST_TIMEOUT,
        ElasticsearchNodesSniffer.Scheme.HTTPS);
    Sniffer sniffer = Sniffer.builder(restClient).setNodesSniffer(nodesSniffer).
        build();
    // 使用後,結束用戶端和偵測器
    try {
      sniffer.close();
      restClient.close();
    } catch (Exception e) {
      e.printStackTrace();
    }
  }
```

4.5 進階用戶端初始化

目前,官方計畫在 Elasticsearch 7.0 版本中關閉 TransportClient,並且在 8.0 版本中完全刪除 TransportClient。作為替代品,我們應該使用進階用戶端。進階用戶端可以執行 HTTP 請求,而不是序列化 Java 請求。

進階用戶端基於初級用戶端來實現。

進階用戶端的主要目標是公開特定的 API 方法,這些 API 方法將接收請求作為參數並傳回回應結果,以便由用戶端本身處理請求和回應結果。

與初級用戶端一樣，進階用戶端也有同步、非同步兩種 API 呼叫方式。其中，以同步呼叫方式呼叫後直接傳回回應物件，而非同步呼叫方式則需要設定一個監聽器參數才能呼叫，該參數在收到回應或錯誤後會獲得對應的結果通知。

進階用戶端需要 Java 1.8 及其以上的 JDK 環境，而且依賴於 Elasticsearch core 專案，它接收與 TransportClient 相同的請求參數，並傳回相同的回應物件。這其實也是為了加強相容性而做的設計。

進階用戶端的版本與 Elasticsearch 版本同步。進階用戶端能夠與執行著相同主版本和更新版本上的任何 Elasticsearch 節點進行有效通訊。進階用戶端無須與它通訊的 Elasticsearch 節點處於同一個小版本，這是因為向前相容設計的緣故。這也表示進階用戶端支援與 Elasticsearch 的較新版本進行有效通訊。

舉例來說，6.0 版本的用戶端可以與任何 6.X 版本的 Elasticsearch 節點進行通訊，而 6.1 版本的用戶端可以確保與 6.1 版本、6.2 版本和任何更新版本的 6.X 節點進行通訊，但在與低於 6.0 版本的 Elasticsearch 節點進行通訊時可能會出現不相容問題。

下面介紹進階用戶端的使用，首先介紹進階用戶端的初始化方法。

在建置進階用戶端之前，我們需要在專案中引用必要的依賴設定項目。在 Maven 包裝和管理依賴時，在 POM 檔案中新增的依賴內容如下所示：

```
<!-- 增加 Elasticsearch 的進階用戶端 begin -->
    <dependency>
        <groupId>org.elasticsearch.client</groupId>
        <artifactId>elasticsearch-rest-high-level-client</artifactId>
        <version>7.2.1</version>
    </dependency>
    <!-- 增加 Elasticsearch 的進階用戶端 end -->
```

如果以 Gradle 來管理依賴，則在 Gradle 為基礎的設定檔 build.gradle 中增加以下內容：

```
dependencies {
    compile 'org.elasticsearch.client:elasticsearch-rest-high-level-
client:7.2.1'
}
```

jar 套件 elasticsearch-rest-high-level-client 又依賴以下 jar 套件，實際如下所示：

```
org.elasticsearch.client:elasticsearch-rest-client
org.elasticsearch:elasticsearch
```

因此，在 POM 檔案中需補充上述二者的依賴。由於在初級用戶端的依賴建置過程中已經增加了 elasticsearch-rest-client，所以僅需增加對 Elasticsearch 的依賴即可。此時在 POM 檔案中新增內容如下所示：

```
<!-- 增加 Elasticsearch 的進階用戶端的依賴 begin -->
    <dependency>
        <groupId>org.elasticsearch</groupId>
        <artifactId>elasticsearch</artifactId>
        <version>7.2.1</version>
    </dependency>
    <!-- 增加 Elasticsearch 的進階用戶端的依賴 end -->
```

當依賴增加完畢後，即可進入進階用戶端的初始化。進階用戶端的初始化是基於 RestHighLevelClient 實現的，而 RestHighLevelClient 是以初級用戶端產生器建置為基礎。

進階用戶端在內部建立執行請求的初級用戶端，該初級用戶端會維護一個連接池並啟動一些執行緒，因此當對進階用戶端的介面呼叫完成時，應該關閉它，因為它將同步關閉內部初級用戶端，以釋放這些資源。

在 Service 層中增加 MeetHighElasticSearchService 類別，程式如下所示：

```
package com.niudong.esdemo.service;
public interface MeetHighElasticSearchService {
  /**
   * 本部分用於介紹如何與 Elasticsearch 建置連接和關閉連接
   */
```

```
  public void initEs();
  public void closeEs();
}
```

在 ServiceImpl 實現層中增加 MeetHighElasticSearchServiceImpl 類別，程式如
下所示：

```
package com.niudong.esdemo.service.impl;
import javax.annotation.PostConstruct;
import org.apache.commons.logging.Log;
import org.apache.commons.logging.LogFactory;
import org.apache.http.HttpHost;
import org.elasticsearch.client.RestClient;
import org.elasticsearch.client.RestHighLevelClient;
import org.springframework.stereotype.Service;
import com.niudong.esdemo.service.MeetHighElasticSearchService;
@Service
public class MeetHighElasticSearchServiceImpl implements MeetHighElasticSearch
    Service {
  private static Log log = LogFactory.getLog
(MeetHighElasticSearchServiceImpl.class);
  private RestHighLevelClient restClient;
  /**
   * 本部分用於介紹如何與 ElasticSearch 建置連接和關閉連接
   */
  // 初始化連接
  @PostConstruct
  public void initEs() {
    restClient = new RestHighLevelClient(RestClient.builder(new HttpHost
        ("localhost", 9200, "http"),
      new HttpHost("localhost", 9201, "http")));

    log.info("ElasticSearch init in service.");
  }
  // 關閉連接
  public void closeEs() {
    try {
      restClient.close();
```

```
    } catch (Exception e) {
      e.printStackTrace();
    }
  }
}
```

4.6 建立請求物件模式

在進階用戶端中,請求物件 Request、請求結果的解析與初級用戶端中的用法相同;與之類似的還有 RequestOptions 的使用及用戶端的常見設定,不再贅述。

4.7 基礎知識連結

初級用戶端的同步和非同步的思想遍及各種中介軟體和資料儲存平台。如 MySQL 叢集、Redis 叢集、HBase 叢集中各個節點的資料同步方式,甚至區塊鏈中各個節點的資料同步方式,都有同步和非同步兩種。訊息通知機制中也常用到同步和非同步兩種方式。

那麼,什麼是同步和非同步呢?它們的差別和關係又是什麼呢?

和同步、非同步相近的一組詞是阻塞和非阻塞。

我們可以從執行緒的維度了解同步和非同步、阻塞和非阻塞。

同步和非同步,顧名思義是相對的概念。對同一個執行緒而言,會有阻塞和非阻塞之分;對不同執行緒而言,則會有同步和非同步之分。

換言之,對同一個執行緒來說,一個時間拍一個快照,這時執行緒不是處於阻塞狀態,就是處於非阻塞狀態。因為阻塞狀態一般是目前執行緒同步等待其他執行緒的處理結果;而非阻塞狀態一般是目前執行緒非同步於需配合的其他執行緒的結果處理過程。

一般來說，同步狀態是目前執行緒發起請求呼叫後，在被呼叫的執行緒訊息過程中，目前執行緒必須等被呼叫的執行緒完才能傳回結果；如果被呼叫的執行緒還未處理完，則目前執行緒就不能傳回結果，目前執行緒必須主動等待所需的結果。

非同步狀態是目前執行緒發起請求呼叫後，被呼叫的執行緒直接傳回，但是並沒有傳回給目前執行緒對應的結果，而是等被呼叫的執行緒處理完訊息之後，透過狀態、通知或回呼函數來通知呼叫執行緒，目前執行緒處於被動接收結果的狀態。

從上述說明不難發現，同步和非同步最主要的差別在於工作或介面呼叫完成時訊息通知的方式。

一般來說，阻塞狀態是被呼叫的執行緒在傳回處理結果前，目前執行緒會被暫停，不釋放 CPU 執行權，目前執行緒也不能做其他事情，只能等待，直到被呼叫的執行緒傳回處理結果後，才能接著向下執行。

非阻塞狀態是目前執行緒在沒有取得被呼叫的執行緒的處理結果前，不是一直等待，而是繼續向下執行。如果此時是同步狀態，則目前執行緒可以透過輪詢的方式檢查被呼叫執行緒的處理結果是否傳回；如果此時是非同步狀態，則只有在被呼叫的執行緒後才會通知目前執行緒回呼。

從上述說明不難發現，阻塞和非阻塞最主要的區別在於目前執行緒發起工作或介面呼叫後是否能繼續執行。

同步和非同步、阻塞和非阻塞還可以兩兩組合，進一步產生 4 種組合，即同步阻塞、同步非阻塞、非同步阻塞和非同步非阻塞。

前端人員熟悉的 Node.js、運行維護人員和前後端人員都熟悉的 Nginx 就是非同步非阻塞的典型代表。

那麼什麼是同步阻塞、同步非阻塞和非同步阻塞、非同步非阻塞呢？如表 4-1 所示。

表 4-1

名稱	說明
同步阻塞	目前執行緒在得不到呼叫結果前不傳回，目前執行緒進入阻塞態等待
同步非阻塞	目前執行緒在得不到呼叫結果前不傳回，但目前執行緒不阻塞，一直在 CPU 中執行
非同步阻塞	目前執行緒呼叫其他執行緒，目前執行緒自己並不阻塞，但其他執行緒會阻塞來等待結果
非同步非阻塞	目前執行緒呼叫其他執行緒，其他執行緒一直在執行，直到得出結果

下面我們透過一個網購的實例來實際解釋。

假設現在筆者在電子商務平台購物，此時會有以下幾種選擇：

（1）如果筆者付款後，什麼事情也不做，僅是坐等送貨員來送貨上門——這是同步阻塞。

（2）如果筆者付款後，什麼事情也不做，僅是坐等送貨員來送貨上門；送貨員到樓下之後會打電話和筆者確認是否在家——這是非同步阻塞。

（3）如果筆者付款後，就去健身了，在休息的間隙，不時地看看 App 中的送貨狀態——這就是同步非阻塞。

（4）如果筆者付款後，就去健身了；送貨員到樓下之後會打電話和筆者確認是否在家——這就是非同步非阻塞。

4.8　小結

本章主要介紹了在 Elasticsearch 中初級客戶端相關 API 的使用，例如建置用戶端初始化、建置請求物件、解析傳回結果及用戶端常用的設定屬性等。

由於用戶端在發出請求時分為同步和非同步兩種方式，因此在基礎知識連結部分，詳細介紹了同步和非同步、阻塞和非阻塞的差別和關聯。

進階用戶端文件實戰一

力學如力耕
勤惰爾自知

在 Elasticsearch 中，進階用戶端支援以下文件相關的 API：

（1）Single document APIs——單文件操作 API。

（2）Index API——文件索引 API。

（3）Get API——文件取得 API。

（4）Exists API——文件存在性判斷 API。

（5）Delete API——文件刪除 API。

（6）Update API——文件更新 API。

（7）Term Vectors API——詞向量 API。

（8）Bulk API——批次處理 API。

（9）Multi-Get API——多文件取得 API。

（10）ReIndex API——重新索引 API。

（11）Update By Query API——查詢更新 API。

（12）Delete By Query API——查詢刪除 API。

（13）Multi Term Vectors API——多詞條向量 API。

其中，前 7 項較為簡單，後 6 項較為複雜，因此我們將用兩章介紹這些 API 的使用。

5.1 文件

在 Elasticsearch 中，使用 JavaScript 物件符號（JavaScript Object Notation），作為文件序列化格式，也就是我們常說的 JSON。

JSON 物件由鍵值對組成，鍵（key）是欄位或屬性名稱，值（value）可以是多種類型，如字串、數字、布林值、一個物件、值陣列、日期的字串、地理位置物件等。

目前，JSON 格式已經被大多數語言所接受並支援，而且已經成為 NoSQL 領域的標準格式。它簡潔且容易閱讀，常見的 JSON 格式如下所示：

```
{
    "name":  "niudong",
    "age":  2,
    "sex":  "man",
    "date":  "2019-06-01",
    "home":  "beijing",
    "company": [
        {
            "company_type": "Electronic Commerce",
            "company_name":  "Alibaba"
        },
        {
            "company_type": "Education",
            "company_name":  "TAL"
        }
    ]
}
```

在 Elasticsearch 中，儲存並索引的 JSON 資料被稱為文件，文件會以唯一 ID 進行標識並儲存於 Elasticsearch 中。文件不僅有本身的鍵值對資料資訊，還包含了中繼資料，即關於文件的資訊。文件一般包含三個必需的中繼資料資訊：

（1）index：文件儲存的資料結構，即索引。

（2）type：文件代表的物件類型。

（3）ID：文件的唯一標識。

index 表示的是索引。這有點類似於開發人員常用的關聯式資料庫中的「資料庫」的概念，索引表示的是 Elasticsearch 儲存和索引連結資料的資料結構。需要指出的是，文件資料最後是被儲存和索引在分片之中，索引多個分片儲存的邏輯空間。

type 表示的是文件代表的物件類型。在 Java 語言堆疊中，開發人員對物件導向程式設計很熟悉。在物件導向程式設計時，每個物件都是一個類別的實例，這個類別定義了物件的多個屬性或與物件連結的資料和方法。

就像我們用「資料庫」類比「索引 _index」，「_type 物件類型」也可以類比「表結構」，這不難了解。在關聯式資料庫中，研發人員經常將相同類的物件儲存在同一張表中，因為它們具有相同的結構──在資料庫中稱之為表結構。

同樣地，在 Elasticsearch 中，每個物件類型都有自己的對映結構，所有類型下的文件都被儲存在同一個索引下。

在實際使用時，type 的命名有一定的標準，名字中的字母可以是大寫，也可以是小寫，但不能包含底線或逗點。

id 表示的是文件的唯一標識，如 ID，文件標識會以字串形式出現。當在 Elasticsearch 中建立一個文件時，使用者既可以自訂 id，也可以讓 Elasticsearch 自動產生。

id、index 和 type 三個元素組合使用時，可以在 Elasticsearch 中唯一標識一個文件。

5.2 文件索引

Elasticsearch 針對的是文件，它可以儲存整個文件。

但 Elasticsearch 對文件的操作不僅限於儲存，Elasticsearch 還會索引每個文件的內容使之可以被搜索。在 Elasticsearch 中，使用者可以對文件資料進行索引、搜索、排序和過濾等操作，而這也是 Elasticsearch 能夠執行複雜的全文檢索搜尋的原因之一。

下面一一介紹在 Elasticsearch 中對文件操作的各種 API。首先介紹文件索引 API 的使用。

文件索引 API 允許使用者將一個類型化的 JSON 文件索引到一個特定的索引中，並且使它可以被搜索到。

在索引 JSON 文件前，首先要建置 JSON 文件。

1 建置 JSON 文件

在 Elasticsearch 中，建置一個 JSON 文件可以用以下幾種方式：

（1）手動使用本機 byte[] 或使用 String 來建置 JSON 文件。

（2）使用 Map，Elasticsearch 會自動把它轉換成與其相等的 JSON 文件。

（3）使用如 Jackson 這樣的協力廠商類別庫來序列化開發人員建置的 Java Bean，以便建置 JSON 文件。

（4）使用內建的幫助類別 XContentFactory.jsonBuilder 來建置 JSON 文件。

在 Elasticsearch 內部，每種類型都被轉換成 byte[]，即位元組陣列。因此，使用者可以直接使用已經是這種形式的物件；而 JsonBuilder 是高度最佳化過的 JSON 產生器，它可以直接建置一個 byte[]。

文件透過索引 API 被索引後，表示文件資料可以被儲存和搜索。文件透過其 index、type 和 ID 確定其唯一儲存所在。對於 ID，使用者可以自己提供一個 ID，或使用索引 API 自動產生。

下面透過程式展示 JSON 文件的建置方法。

在 ServiceImpl 實現層的 MeetHighElasticSearchServiceImpl 類別中增加以下程式：

```
/**
 * 本部分用於介紹 Elasticsearch 索引 API 的使用
```

```java
    */
// 基於 String 建置 IndexRequest
public void buildIndexRequestWithString(String indexName, String document) {
  // 索引名稱
  IndexRequest request = new IndexRequest(indexName);
  // 文件 ID
  request.id(document);
  // String 類型的文件
  String jsonString = "{" + "\"user\":\"niudong\"," + "\"postDate\":\"2019-
      07-30\","
    + "\"message\":\"Hello Elasticsearch\"" + "}";
  request.source(jsonString, XContentType.JSON);
}
// 基於 Map 建置 IndexRequest
public void buildIndexRequestWithMap(String indexName, String document) {
  Map<String, Object> jsonMap = new HashMap<>();
  jsonMap.put("user", "niudong");
  jsonMap.put("postDate", new Date());
  jsonMap.put("message", "Hello Elasticsearch");
  // 以 Map 形式提供文件來源，Elasticsearch 會自動將 Map 形式轉為 JSON 格式
  IndexRequest indexRequest = new IndexRequest(indexName).id(document).
      source(jsonMap);
}
// 基於 XContentBuilder 建置 IndexRequest
public void buildIndexRequestWithXContentBuilder(String indexName, String
      document) {
  try {
    XContentBuilder builder = XContentFactory.jsonBuilder();
    builder.startObject();
    {
      builder.field("user", "niudong");
      builder.timeField("postDate", new Date());
      builder.field("message", "Hello Elasticsearch");
    }
    builder.endObject();
    // 以 XContentBuilder 物件提供文件來源，Elasticsearch 內建的幫助器自動將其產生
    // JSON 格式內容
```

```
      IndexRequest indexRequest = new IndexRequest(indexName).id(document).
          source(builder);
  } catch (Exception e) {
      e.printStackTrace();
  }
}
// 基於鍵值對建置 IndexRequest
public void buildIndexRequestWithKV(String indexName, String document) {
  // 以鍵值對提供文件來源，Elasticsearch 自動將其轉為 JSON 格式
  IndexRequest indexRequest = new IndexRequest(indexName).id(document).
      source("user", "niudong",
    "postDate", new Date(), "message", "Hello Elasticsearch");
}
```

如上述程式所示，展示了四種建置 JSON 文件的方法，並展示了在
IndexRequest（索引請求）中必選參數的設定方法。

在 IndexRequest 中，除三個必選參數外，還有一些可選參數。可選參數包含
路由、逾時、版本、版本類型和索引管線名稱等，程式如下所示：

```
// 設定 IndexRequest 的其他參數
public void buildIndexRequestWithParam(String indexName, String document) {
  // 將以鍵值對提供的文件來源轉為 JSON 格式
  IndexRequest request = new IndexRequest(indexName).id(document).
      source("user", "niudong",
    "postDate", new Date(), "message", "Hello Elasticsearch");
  request.routing("routing");// 路由值

  // 設定逾時
  request.timeout(TimeValue.timeValueSeconds(1));
  request.timeout("1s");

  // 設定逾時策略
  request.setRefreshPolicy(WriteRequest.RefreshPolicy.WAIT_UNTIL);
  request.setRefreshPolicy("wait_for");

  // 設定版本
```

```
    request.version(2);

    // 設定版本類型
    request.versionType(VersionType.EXTERNAL);

    // 設定操作類型
    request.opType(DocWriteRequest.OpType.CREATE);
    request.opType("create");

    // 在索引文件之前要執行的接收管線的名稱
    request.setPipeline("pipeline");
}
```

❷ 執行索引文件請求

在 JSON 文件建置後，即可執行索引文件請求。在 Elasticsearch 中，索引文件有兩種方式，即同步方式和非同步方式。

✍ 同步方式

當以同步方式執行 IndexRequest 時，用戶端必須等待傳回結果 IndexResponse。在收到 IndexResponse 之後，用戶端方可繼續執行程式。以同步方式執行的程式如下所示：

```
IndexResponse indexResponse = client.index(request, RequestOptions.DEFAULT);
```

當然，在以同步方式執行過程中可能會出現解析 IndexResponse 回應失敗、請求逾時或沒有從伺服器傳回回應的情況，進一步引發 IOException。因此，在程式中需要進行對應的處理。

另外，在 Elasticsearch 伺服器傳回 4XX 或 5XX 錯誤程式的情況下，進階用戶端會嘗試解析 IndexResponse 來回應正文中的錯誤詳細資訊，並拋出一個通用的例外 ElasticSearchException。因此，在程式中需要進行對應的處理。

在 ServiceImpl 實現層的 MeetHighElasticSearchServiceImpl 類別中增加以下程式：

```
// 索引文件
  public void indexDocuments(String indexName, String document) {
    // 將以鍵值對提供的文件來源轉為 JSON 格式
    IndexRequest indexRequest = new IndexRequest(indexName).id(document).
        source("user", "niudong",
      "postDate", new Date(), "message", "Hello Elasticsearch");
    try {
      restClient.index(indexRequest, RequestOptions.DEFAULT);
    } catch (Exception e) {
      e.printStackTrace();
    }
  }
```

☑ 非同步方式

當以非同步方式執行 IndexRequest 時，進階用戶端不必同步等待請求結果的
傳回，可以直接向介面呼叫方傳回非同步介面執行成功的結果。

為了處理非同步傳回的回應資訊或處理在請求執行過程中引發的例外資訊，
使用者需要指定監聽器。以非同步方式呼叫的程式如下所示：

```
client.indexAsync(request, RequestOptions.DEFAULT, listener);
```

其中，request 是要執行的 IndexRequest，listener 是執行完成時使用的
ActionListener。

在非同步請求處理後，如果請求執行成功，則呼叫 ActionListener 類別
中的 onResponse 方法進行相關邏輯的處理；如果請求執行失敗，則呼叫
ActionListener 類別中的 onFailure 方法進行相關邏輯的處理。

當然，在非同步呼叫過程中可能會出現例外或處理失敗的情況，因此使用者
需要在程式中做對應的處理。

在 ServiceImpl 實現層的 MeetHighElasticSearchServiceImpl 類別中增加以下程
式：

```
// 以非同步方式索引文件
  public void indexDocumentsAsync(String indexName, String document) {
```

```
// 將以鍵值對提供的文件來源轉為 JSON 格式
IndexRequest indexRequest = new IndexRequest(indexName).id(document).source
    ("user", "niudong",
        "ostdate", new Date(),"message", "Hello Elasticsearch");
ActionListener listener = new ActionListener<IndexResponse>() {
    @Override
    public void onResponse(IndexResponse indexResponse) {
    }
    @Override
    public void onFailure(Exception e) {
    }
};
try {
    restClient.indexAsync(indexRequest, RequestOptions.DEFAULT, listener);
} catch (Exception e) {
    e.printStackTrace();
}
  }
```

③ 對回應結果的解析

在文件索引的請求發出後，不論同步方式，還是非同步方式，用戶端均會收到對應的結果 IndexResponse。使用者可以透過解析 IndexResponse 來判斷文件索引是成功還是失敗。

程式分為三層，分別是 Controller 層、Service 層和 ServiceImpl 實現層。其中，Controller 層中 MeetHighElasticSearchController 類別的程式如下所示：

```
package com.niudong.esdemo.controller;
import org.elasticsearch.common.Strings;
import org.springframework.beans.factory.annotation.Autowired;
import org.springframework.web.bind.annotation.RequestMapping;
import org.springframework.web.bind.annotation.RestController;
import com.niudong.esdemo.service.MeetHighElasticSearchService;
@RestController
@RequestMapping("/springboot/es/high")
public class MeetHighElasticSearchController {
```

```
@Autowired
private MeetHighElasticSearchService meetHighElasticSearchService;

@RequestMapping("/index/put")
public String putIndexInHighElasticSearch(String indexName, String document) {
    if(Strings.isEmpty(indexName) || Strings.isEmpty(document)) {
        return "Parameters are error!";
    }
    meetHighElasticSearchService.indexDocuments(indexName, document);
    return "Index High ElasticSearch Client Successed!";
}
}
```

Service 層中 MeetHighElasticSearchService 類別的程式如下所示：

```
package com.niudong.esdemo.service;
public interface MeetHighElasticSearchService {
    /**
     * 本部分用於介紹如何與 Elasticsearch 建置連接和關閉連接
     */
    public void initEs();
    public void closeEs();
    // 索引文件
    public void indexDocuments(String indexName, String document);
}
```

ServiceImpl 層中 MeetHighElasticSearch Service Impl 類別的程式如下所示：

```
package com.niudong.esdemo.service.impl;
import java.util.Date;
import java.util.HashMap;
import java.util.Map;
import javax.annotation.PostConstruct;
import org.apache.commons.logging.Log;
import org.apache.commons.logging.LogFactory;
import org.apache.http.HttpHost;
import org.elasticsearch.action.ActionListener;
import org.elasticsearch.action.DocWriteRequest;
import org.elasticsearch.action.index.IndexRequest;
```

```java
import org.elasticsearch.action.index.IndexResponse;
import org.elasticsearch.action.support.WriteRequest;
import org.elasticsearch.action.support.replication.ReplicationResponse;
import org.elasticsearch.client.RequestOptions;
import org.elasticsearch.client.RestClient;
import org.elasticsearch.client.RestHighLevelClient;
import org.elasticsearch.common.unit.TimeValue;
import org.elasticsearch.common.xcontent.XContentBuilder;
import org.elasticsearch.common.xcontent.XContentFactory;
import org.elasticsearch.common.xcontent.XContentType;
import org.elasticsearch.index.VersionType;
import org.springframework.stereotype.Service;
import org.elasticsearch.action.DocWriteResponse;
import com.niudong.esdemo.service.MeetHighElasticSearchService;
@Service
public class MeetHighElasticSearchServiceImpl implements
MeetHighElasticSearchService {
  private static Log log = LogFactory.getLog
(MeetHighElasticSearchServiceImpl.class);
  private RestHighLevelClient restClient;
  /**
   * 本部分介紹如何與 Elasticsearch 建置連接和關閉連接
   */
  // 初始化連接
  @PostConstruct
  public void initEs() {
    restClient = new RestHighLevelClient(RestClient.builder(new HttpHost
        ("localhost", 9200, "http"),
      new HttpHost("localhost", 9201, "http")));
    log.info("ElasticSearch init in service.");
  }
  // 關閉連接
  public void closeEs() {
    try {
      restClient.close();
    } catch (Exception e) {
      e.printStackTrace();
```

```
  }
}
/**
 * 本部分介紹 Elasticsearch 索引 API 的使用
 */
// 基於 String 建置 IndexRequest
public void buildIndexRequestWithString(String indexName, String document) {
  // 索引名稱
  IndexRequest request = new IndexRequest(indexName);
  // 文件 ID
  request.id(document);
  // String 類型的文件
  String jsonString = "{" + "\"user\":\"niudong\"," + "\"postDate\":\"2019-
      07-30\","
    + "\"message\":\"Hello Elasticsearch\"" + "}";
  request.source(jsonString, XContentType.JSON);
}
// 基於 Map 建置 IndexRequest
public void buildIndexRequestWithMap(String indexName, String document) {
  Map<String, Object> jsonMap = new HashMap<>();
  jsonMap.put("user", "niudong");
  jsonMap.put("postDate", new Date());
  jsonMap.put("message", "Hello Elasticsearch");
  // 將以 Map 形式提供的文件來源轉為 JSON 格式
  IndexRequest indexRequest = new IndexRequest(indexName).id(document).
      source(jsonMap);
}
// 基於 XContentBuilder 建置 IndexRequest
public void buildIndexRequestWithXContentBuilder(String indexName, String
    document) {
  try {
    XContentBuilder builder = XContentFactory.jsonBuilder();
    builder.startObject();
    {
      builder.field("user", "niudong");
      builder.timeField("postDate", new Date());
      builder.field("message", "Hello Elasticsearch");
```

```
      }
      builder.endObject();
      // 將 XContentBuilder 物件提供的文件來源轉為 JSON 格式
      IndexRequest indexRequest = new IndexRequest(indexName).id(document).
          source(builder);
    } catch (Exception e) {
      e.printStackTrace();
    }
}
// 基於鍵值對建置 IndexRequest
public void buildIndexRequestWithKV(String indexName, String document) {
    // 將以鍵值對形式提供的文件來源轉為 JSON 格式
    IndexRequest indexRequest = new IndexRequest(indexName).id(document).
        source("user", "niudong",
        "postDate", new Date(), "message", "Hello Elasticsearch");
}
// 建置 IndexRequest 的其他參數設定
public void buildIndexRequestWithParam(String indexName, String document) {
    // 將以鍵值對形式提供的文件來源轉為 JSON 格式
    IndexRequest request = new IndexRequest(indexName).id(document).source
        ("user", "niudong",
        "postDate", new Date(), "message", "Hello Elasticsearch");
    request.routing("routing");// 路由值
    // 設定逾時
    request.timeout(TimeValue.timeValueSeconds(1));
    request.timeout("1s");
    // 設定逾時策略
    request.setRefreshPolicy(WriteRequest.RefreshPolicy.WAIT_UNTIL);
    request.setRefreshPolicy("wait_for");
    // 設定版本
    request.version(2);
    // 設定版本類型
    request.versionType(VersionType.EXTERNAL);
    // 設定操作類型
    request.opType(DocWriteRequest.OpType.CREATE);
    request.opType("create");
    // 在索引文件之前要執行的接收管線的名稱
```

```
    request.setPipeline("pipeline");
  }
  // 索引文件
  public void indexDocuments(String indexName, String document) {
    // 將以鍵值對形式提供的文件來源轉為 JSON 格式
    IndexRequest indexRequest = new IndexRequest(indexName).id(document).
        source("user", "niudong",
        "postDate", new Date(), "message",
```

"Hello Elasticsearch! 北京時間 8 月 1 日淩晨 2 點，美聯儲公佈 7 月議息會議結果。一如市場預期，美聯儲本次降息 25 個基點，將聯邦基金利率的目標範圍調至 2.00%~2.25%。此次是 2007—2008 年間美國為應對金融危機啟動降息週期後，美聯儲十年多以來第一次降息。美聯儲公佈利率決議後，美股下跌，美金上漲，人民幣匯率下跌 ");

```
    try {
      IndexResponse indexResponse = restClient.index(indexRequest,
RequestOptions.DEFAULT);
      // 解析索引結果
      processIndexResponse(indexResponse);
    } catch (Exception e) {
      e.printStackTrace();
    }
}
// 關閉 Elastisearch 連接
 closeEs();
  }
  // 解析索引結果
  private void processIndexResponse(IndexResponse indexResponse) {
    String index = indexResponse.getIndex();
    String id = indexResponse.getId();
    log.info("index is " + index + ", id is " + id);
    if (indexResponse.getResult() == DocWriteResponse.Result.CREATED) {
      // 文件建立時
      log.info("Document is created!");
    } else if (indexResponse.getResult() == DocWriteResponse.Result.UPDATED) {
      // 文件更新時
      log.info("Document has updated!");
    }
    ReplicationResponse.ShardInfo shardInfo = indexResponse.getShardInfo();
    if (shardInfo.getTotal() != shardInfo.getSuccessful()) {
```

```
      // 處理成功，shards 小於總 shards 的情況
      log.info("Successed shards are not enough!");
    }
    if (shardInfo.getFailed() > 0) {
      for (ReplicationResponse.ShardInfo.Failure failure : shardInfo.
          getFailures()) {
        String reason = failure.reason();
        log.info("Fail reason is " + reason);
      }
    }
  }
  // 非同步索引文件
  public void indexDocumentsAsync(String indexName, String document) {
    // 將以鍵值對形式提供的文件來源轉為 JSON 格式
    IndexRequest indexRequest = new IndexRequest(indexName).id(document).
        source("user", "niudong",
      "postDate", new Date(), "message", "Hello Elasticsearch");
    ActionListener listener = new ActionListener<IndexResponse>() {
      @Override
      public void onResponse(IndexResponse indexResponse) {
      }
      @Override
      public void onFailure(Exception e) {
      }
    };
    try {
      restClient.indexAsync(indexRequest, RequestOptions.DEFAULT, listener);
    } catch (Exception e) {
      e.printStackTrace();
    }
  }
// 關閉 Elasticsearch 連接
    closeEs();
}
```

隨後編輯專案，在專案根目錄下輸入以下指令：

```
mvn clean package
```

透過以下指令啟動專案服務：

```
java -jar ./target/esdemo-0.0.1-SNAPSHOT.jar
```

在專案服務啟動後，在瀏覽器中呼叫以下介面檢視文件索引的情況：

```
http://localhost:8080/springboot/es/high/index/put?indexName=
ultraman&document=1
```

呼叫介面後，伺服器輸出內容如下所示。

```
2019-08-01 09:44:31.451  INFO 63876 --- [nio-8080-exec-7] n.e.s.i.MeetHighElasticSearchServiceImpl : index is ultraman,
id is 1
2019-08-01 09:44:31.453  INFO 63876 --- [nio-8080-exec-7] n.e.s.i.MeetHighElasticSearchServiceImpl : Document is created
!
```

更換索引的字串內容後第二次呼叫，結果如下所示。

```
2019-08-01 09:45:18.094  INFO 40436 --- [nio-8080-exec-1] n.e.s.i.MeetHighElasticSearchServiceImpl : index is ultraman,
id is 1
2019-08-01 09:45:18.096  INFO 40436 --- [nio-8080-exec-1] n.e.s.i.MeetHighElasticSearchServiceImpl : Document has update
d!
```

兩次呼叫後瀏覽器頁面均傳回：

```
Index High ElasticSearch Client Successed!
```

需要指出的是，在兩次索引中，雖然索引內容不同，但由於傳入的文件 ID 均為 1，因此，從第二次介面呼叫中可以看到，在索引後，文件 ID 依然是 1，和預期相同。

5.3 文件索引查詢

Elasticsearch 提供了文件索引查詢 API，即 Get API。

▉ 建置文件索引查詢請求

在執行文件索引查詢請求前，需要建置文件索引查詢請求，即 GetRequest。GetRequest 有兩個必選參數，即索引名稱和文件 ID。建置 GetRequest 的程式如下所示：

```
// 建置 GetRequest
  public void buildGetRequest(String indexName, String document) {
    GetRequest getRequest = new GetRequest(indexName, document);
  }
```

在建置 GetRequest 的過程中，可以選擇其他可選參數進行對應的設定。可選
參數有禁用來源檢索、為特定欄位設定來源包含關係、為特定欄位設定來源
排除關係、為特定儲存欄位設定檢索、設定路由值、設定偏好值、設定在檢
索文件之前執行更新、設定版本編號、設定版本類型等，程式如下所示：

```
// 建置 GetRequest
  public void buildGetRequest(String indexName, String document) {
    GetRequest getRequest = new GetRequest(indexName, document);
    // 可選設定參數
    // 禁用來源檢索，在預設情況下啟用
    getRequest.fetchSourceContext(FetchSourceContext.DO_NOT_FETCH_SOURCE);
    // 為特定欄位設定來源包含
    String[] includes = new String[] {"message", "*Date"};
    String[] excludes = Strings.EMPTY_ARRAY;
    FetchSourceContext fetchSourceContext = new FetchSourceContext(true,
        includes, excludes);
    getRequest.fetchSourceContext(fetchSourceContext);
    // 為特定欄位設定來源排除
    includes = Strings.EMPTY_ARRAY;
    excludes = new String[] {"message"};
    fetchSourceContext = new FetchSourceContext(true, includes, excludes);
    getRequest.fetchSourceContext(fetchSourceContext);
getRequest.storedFields("message"); // 為特定儲存欄位設定檢索
// 要求在對映中單獨儲存欄位
    try {
      GetResponse getResponse = restClient.get(getRequest,
RequestOptions.DEFAULT);
      String message = getResponse.getField("message").getValue();
// 檢索訊息儲存欄位（要求該欄位單獨儲存在對映中）
      log.info("message is " + message);
    } catch (Exception e) {
      e.printStackTrace();
```

```
  }
  // 路由值
  getRequest.routing("routing");
  // 偏好值
  getRequest.preference("preference");
  // 將即時標示設定為假（預設為真）
  getRequest.realtime(false);
  // 在檢索文件之前執行更新（預設為 false）
  getRequest.refresh(true);
  // 設定版本編號
  getRequest.version(2);
  // 設定版本類型
  getRequest.versionType(VersionType.EXTERNAL);
}
```

2 執行文件索引查詢請求

在 GetRequest 建置後，即可執行文件索引查詢請求。與文件索引請求類似，
文件索引查詢請求也有同步和非同步兩種執行方式。

同步方式

當以同步方式執行 GetRequest 時，用戶端會等待 Elasticsearch 伺服器傳回的
查詢結果 GetResponse。在收到 GetResponse 後，用戶端會繼續執行相關的邏
輯程式。以同步方式執行的程式如下所示：

```
// 以同步方式執行 GetRequest
  public void getIndexDocuments(String indexName, String document) {
    GetRequest getRequest = new GetRequest(indexName, document);
    try {
      GetResponse getResponse = restClient.get(getRequest,
RequestOptions.DEFAULT);
    } catch (Exception e) {
      e.printStackTrace();
}
// 關閉 Elasticsearch 連接
closeEs();
  }
```

非同步方式

當以非同步方式執行 GetRequest 時，進階用戶端不必同步等待請求結果的傳回，可以直接向介面呼叫方傳回非同步介面執行成功的結果。為了處理非同步傳回的回應資訊或處理在請求執行過程中引發的例外資訊，使用者需要指定監聽器 ActionListener。

如果請求執行成功，則呼叫 ActionListener 中的 onResponse 方法進行對應邏輯的處理。如果請求執行失敗，則呼叫 ActionListener 中的 onFailure 方法進行對應邏輯的處理。

當然，在非同步請求執行過程中可能會出現例外，例外的處理與同步方式執行情況相同。

以非同步方式執行的程式如下所示：

```java
// 以非同步方式執行 GetRequest
public void getIndexDocumentsAsync(String indexName, String document) {
  GetRequest getRequest = new GetRequest(indexName, document);
  ActionListener<GetResponse> listener = new ActionListener<GetResponse>() {
    @Override
    public void onResponse(GetResponse getResponse) {
      String id = getResponse.getId();
      String index = getResponse.getIndex();
      log.info("id is " + id + ", index is " + index);
    }
    @Override
    public void onFailure(Exception e) {
    }
  };
  try {
    restClient.getAsync(getRequest, RequestOptions.DEFAULT, listener);
  } catch (Exception e) {
    e.printStackTrace();
  }
}
```

❸ 傳回結果的解析

不論同步方式，還是非同步方式，在 GetRequest 執行後都會收到請求回應結果 GetResponse。使用者可以透過解析 GetResponse 檢視文件索引的結果。

解析 GetResponse 的程式共分為三層，分別是 Controller 層、Service 層和 ServiceImpl 實現層。

在 Controller 層的 MeetHighElasticSearchController 類別中增加以下程式：

```
// 以同步方式執行 GetRequest
@RequestMapping("/index/get")
public String getIndexInHighElasticSearch(String indexName, String
    document) {
  if (Strings.isEmpty(indexName) || Strings.isEmpty(document)) {
    return "Parameters are error!";
  }
  meetHighElasticSearchService.getIndexDocuments(indexName, document);
  return "Get Index High ElasticSearch Client Successed!";
}
```

在 Service 層的 MeetHighElasticSearchService 類別中增加以下程式：

```
// 以同步方式執行 GetRequest
public void getIndexDocuments(String indexName, String document);
```

在 ServiceImpl 實現層的 MeetHighElasticSearchServiceImpl 類別中增加以下程式：

```
// 以同步方式執行 GetRequest
public void getIndexDocuments(String indexName, String document) {
  GetRequest getRequest = new GetRequest(indexName, document);
  try {
    GetResponse getResponse = restClient.get(getRequest,
RequestOptions.DEFAULT);
    // 處理 GetResponse
    processGetResponse(getResponse);
  } catch (Exception e) {
    e.printStackTrace();
```

```
  }
// 關閉 Elasticsearch 連接
closeEs();
  }
  // 處理 GetResponse
  private void processGetResponse(GetResponse getResponse) {
    String index = getResponse.getIndex();
    String id = getResponse.getId();
    log.info("id is " + id + ", index is " + index);
    if (getResponse.isExists()) {
      long version = getResponse.getVersion();
// 以字串形式檢索文件
      String sourceAsString = getResponse.getSourceAsString();
// 以 Map<String, Object> 形式檢索文件
      Map<String, Object> sourceAsMap = getResponse.getSourceAsMap();
// 以 byte[] 形式檢索文件
      byte[] sourceAsBytes = getResponse.getSourceAsBytes();
      log.info("version is " + version + ", sourceAsString is " +
          sourceAsString);
    } else {
// 當找不到文件時在此處處理。注意，儘管傳回的回應具有 404 狀態碼，但傳回的是有效的
// getResponse，而非引發例外。這樣的回應不包含任何來源文件，並且其 isexists 方法返
// 回 false
    }
  }
```

隨後編輯專案，在專案根目錄下輸入以下指令：

```
mvn clean package
```

透過以下指令啟動專案服務：

```
java -jar ./target/esdemo-0.0.1-SNAPSHOT.jar
```

在專案服務啟動後，在瀏覽器中呼叫以下介面檢視文件索引的查詢結果：

```
http://localhost:8080/springboot/es/high/index/get?indexName=
ultraman&document=1
```

伺服器執行結果如圖 5-1 所示。

```
2019-08-01 11:06:50.566  INFO 52128 --- [nio-8080-exec-2] n.e.s.i.MeetHighElasticSearchServiceImpl : id is 1, index is u
ltraman
2019-08-01 11:06:50.674  INFO 52128 --- [nio-8080-exec-2] n.e.s.i.MeetHighElasticSearchServiceImpl : version is 2, sourc
eAsString is {"user":"niudong","postDate":"2019-08-01T01:45:17.741Z","message":"Hello Elasticsearch!北京时间8月1日凌晨2
点。美联储公布7月议息会议结果。一如市场预期，美联储本次降息25个基点，将联邦基金利率的目标范围调至2.00%-2.25%。此次是2007
-2008年间美国为应对金融危机启动降息周期后，美联储十年多以来首次降息。美联储公布利率决议后，美股下跌，美元上涨，人民币汇
率下跌"}
```

圖 5-1（編按：本圖為簡體中文介面）

比較之前的索引內容，可以看到二者的結果相同。

而瀏覽器端則顯示以下內容，表示介面呼叫成功：

```
Get Index High ElasticSearch Client Successed!
```

5.4 文件存在性驗證

Elasticsearch 還提供了驗證文件是否存在於某索引中的介面 API，即 Exists
API。當呼叫該介面時，如果被驗證的文件存在，則 Exists API 會傳回 true，
否則傳回 false。

這個介面特別適合只檢查某文件是否存在的場景。

1 建置文件存在性驗證請求

Exists API 依賴於 GetRequest 的實例，這點很像文件索引查詢 API，即 Get
API。同理，在 Exists API 中也會支援 GetRequest 的所有可選參數。

由於在 Exists API 的傳回結果中只有布林型的值，即 true 或 false，因此建議
使用者關閉分析來源和任何儲存欄位，使請求的建置維持輕量級。

Exists API 的使用程式如下所示：

```
// 以同步方式驗證索引文件是否存在
  public void checkExistIndexDocuments(String indexName, String document) {
    GetRequest getRequest = new GetRequest(indexName, document);
    // 禁用分析來源
    getRequest.fetchSourceContext(new FetchSourceContext(false));
    // 禁用分析儲存欄位
    getRequest.storedFields("_none_");

  }
```

2 執行文件存在性驗證請求

在 GetRequest 建置後，即可執行文件存在性驗證請求。與文件索引請求和文件索引查詢請求類似，文件存在性驗證請求也有同步和非同步兩種執行方式。

同步方式

當以同步方式執行文件存在性驗證請求時，用戶端會等待 Elasticsearch 伺服器傳回的查詢結果。在收到查詢結果後，用戶端會繼續執行相關的邏輯程式。以同步方式執行的程式如下所示：

```
// 以同步方式驗證索引文件是否存在
public void checkExistIndexDocuments(String indexName, String document) {
  GetRequest getRequest = new GetRequest(indexName, document);
  // 禁用分析來源
  getRequest.fetchSourceContext(new FetchSourceContext(false));
  // 禁用分析儲存欄位
  getRequest.storedFields("_none_");
  try {
    boolean exists = restClient.exists(getRequest, RequestOptions.DEFAULT);
    log.info("索引 " + indexName + " 下的 " + document + " 文件的存在性是 " +
        exists);
  } catch (Exception e) {
    e.printStackTrace();
  }
  // 關閉 Elasticsearch 連接
  closeEs();
}
```

非同步方式

當以非同步方式執行 GetRequest 時，進階用戶端不必同步等待請求結果的傳回，可以直接向介面呼叫方傳回非同步介面執行成功的結果。

為了處理非同步傳回的回應資訊或處理在請求執行過程中引發的例外資訊，使用者需要指定監聽器。以非同步方式執行的核心程式如下所示：

```
client.existsAsync(getRequest, RequestOptions.DEFAULT, listener);
```

其中，listener 為監聽器。

以非同步方式執行後不會阻塞，而是立即傳回。在請求執行後，如果請求執行成功，則呼叫 ActionListener 類別中的 onResponse 方法進行相關邏輯的處理；如果請求執行失敗，則呼叫 ActionListener 類別中的 onFailure 方法進行相關邏輯的處理。

當然，在非同步請求執行過程中可能會出現例外，例外的處理與同步方式執行情況相同。

以非同步方式執行的全部程式如下所示：

```java
// 以非同步方式驗證索引文件是否存在
public void checkExistIndexDocumentsAsync(String indexName, String
    document) {
  GetRequest getRequest = new GetRequest(indexName, document);
  // 禁用分析來源
  getRequest.fetchSourceContext(new FetchSourceContext(false));
  // 禁用分析儲存欄位
  getRequest.storedFields("_none_");
  // 定義監聽器
  ActionListener<Boolean> listener = new ActionListener<Boolean>() {
    @Override
    public void onResponse(Boolean exists) {
      log.info("索引" + indexName + "下的" + document + "文件的存在性是" +
          exists);
    }
    @Override
    public void onFailure(Exception e) {
    }
  };
  try {
    restClient.existsAsync(getRequest, RequestOptions.DEFAULT, listener);
  } catch (Exception e) {
    e.printStackTrace();
  }
  // 關閉 Elasticsearch 連接
```

```
    closeEs();
  }
```

隨後編輯專案,在專案根目錄下輸入以下指令:

```
mvn clean package
```

透過以下指令啟動專案服務:

```
java -jar ./target/esdemo-0.0.1-SNAPSHOT.jar
```

在專案服務啟動後,在瀏覽器中呼叫以下介面檢視特定文件在特定索引下的存在情況:

```
http://localhost:8080/springboot/es/high/index/check?indexName=
ultraman&document=1
```

伺服器輸出的查詢結果下所示:

```
2019-08-01 11:47:31.568  INFO 31100 --- [io-8080-exec-10] n.e.s.i.MeetHighElasticSearchServiceImpl : 索引ultraman下的1文
档的存在性是true
```

顯然上述結果與預期相同,而此時瀏覽器端會顯示介面呼叫成功,內容如下所示:

```
Check Index High ElasticSearch Client Successed!
```

5.5 刪除文件索引

Elasticsearch 提供了在索引中刪除文件介面的 API。

1 建置刪除文件索引請求

在執行刪除文件索引請求前,需要建置刪除文件索引請求,即 DeleteRequest。在 DeleteRequest 中有兩個必選參數,即索引名稱和文件 ID。

DeleteRequest 的初始化程式如下所示:

```
// 建置 DeleteRequest
  public void buildDeleteRequestIndexDocuments (String indexName, String
```

```
    document) {
  DeleteRequest request = new DeleteRequest(indexName, document);
  }
}
```

在 DeleteRequest 的建置過程中,可以設定一些可選參數。可選參數有路由設定參數、逾時設定參數、更新策略設定參數、版本設定參數和版本類型設定參數等。各個參數的設定方法如下所示:

```
// 建置 DeleteRequest
  public void buildDeleteRequestIndexDocuments (String indexName, String
      document) {
  DeleteRequest request = new DeleteRequest(indexName, document);
  // 設定路由
  request.routing("routing");
  // 設定逾時
  request.timeout(TimeValue.timeValueMinutes(2));
  request.timeout("2m");
  // 設定更新策略
  request.setRefreshPolicy(WriteRequest.RefreshPolicy.WAIT_UNTIL);
  request.setRefreshPolicy("wait_for");
  // 設定版本
  request.version(2);
  // 設定版本類型
  request.versionType(VersionType.EXTERNAL);
  }
```

2 執行刪除文件索引請求

在 DeleteRequest 建置後,即可執行刪除文件索引請求。與文件索引請求類似,刪除文件索引請求也有同步和非同步兩種執行方式。

◪ 同步方式

當以同步方式執行刪除文件索引請求時,用戶端會等待 Elasticsearch 伺服器傳回的查詢結果 DeleteResponse。在收到 DeleteResponse 後,用戶端繼續執行相關的邏輯程式。以同步方式執行的程式如下所示:

```
// 以同步方式刪除文件索引請求
  public void deleteIndexDocuments(String indexName, String document) {
    DeleteRequest request = new DeleteRequest(indexName, document);
    try {
      DeleteResponse deleteResponse = restClient.delete(request,
RequestOptions.DEFAULT);
    } catch (Exception e) {
      e.printStackTrace();
    }
    // 關閉 Elasticsearch 連接
    closeEs();
  }
```

非同步方式

當以非同步方式執行刪除文件索引請求時，進階用戶端不必同步等待請求結果的傳回，可以直接向介面呼叫方傳回非同步介面執行成功的結果。

在刪除文件索引請求執行後，如果請求執行成功，則呼叫 ActionListener 類別中的 onResponse 方法進行相關邏輯的處理；如果請求執行失敗，則呼叫 ActionListener 類別中的 onFailure 方法進行相關邏輯的處理。當然，在非同步請求執行過程中可能會出現例外，例外的處理與同步方式執行情況相同。

為了處理非同步傳回的回應資訊或處理在請求執行過程中引發的例外資訊，使用者需要指定監聽器。

以非同步方式執行的程式如下所示：

```
// 以非同步方式刪除文件索引請求
  public void deleteIndexDocumentsAsync(String indexName, String document) {
    DeleteRequest request = new DeleteRequest(indexName, document);
    ActionListener listener = new ActionListener<DeleteResponse>() {
      @Override
      public void onResponse(DeleteResponse deleteResponse) {
        String id = deleteResponse.getId();
        String index = deleteResponse.getIndex();
        long version = deleteResponse.getVersion();
```

```
    log.info("delete id is " + id + ", index is " + index + ",version is
        " + version);
    }
    @Override
    public void onFailure(Exception e) {
    }
};
try {
    restClient.deleteAsync(request, RequestOptions.DEFAULT, listener);
} catch (Exception e) {
    e.printStackTrace();
}
// 關閉 Elasticsearch 連接
closeEs();
}
```

3 對回應結果的解析

不論同步方式，還是非同步方式，在請求執行後都會傳回刪除結果 DeleteResponse。使用者可以透過解析 DeleteResponse 來檢視文件在索引中的刪除資訊。

範例程式分為三層，分別是 Controller 層、Service 層和 ServiceImpl 實現層。

在 Controller 層的 MeetHighElasticSearchController 類別中增加以下程式：

```
// 以同步方式刪除文件索引請求
@RequestMapping("/index/delete")
public String deleteIndexInHighElasticSearch(String indexName, String
    document) {
    if (Strings.isEmpty(indexName) || Strings.isEmpty(document)) {
        return "Parameters are error!";
    }
    meetHighElasticSearchService.deleteIndexDocuments(indexName, document);
    return "Delete Index High ElasticSearch Client Successed!";
}
```

在 Service 層的 MeetHighElasticSearchService 類別中增加以下程式：

```
// 以同步方式刪除文件索引請求
  public void deleteIndexDocuments(String indexName, String document);
```

在 ServiceImpl 實現層的 MeetHighElasticSearchServiceImpl 類別中增加以下程式：

```
// 以同步方式刪除文件索引請求
  public void deleteIndexDocuments(String indexName, String document) {
    DeleteRequest request = new DeleteRequest(indexName, document);
    try {
      DeleteResponse deleteResponse = restClient.delete(request,
RequestOptions.DEFAULT);

      // 處理 DeleteResponse
      processDeleteRequest(deleteResponse);
    } catch (Exception e) {
      e.printStackTrace();
    }
    // 關閉 Elasticsearch 連接
    closeEs();
  }
  private void processDeleteRequest(DeleteResponse deleteResponse) {
    String index = deleteResponse.getIndex();
    String id = deleteResponse.getId();
    long version = deleteResponse.getVersion();
    log.info("delete id is " + id + ", index is " + index + ",version is " +
        version);
    ReplicationResponse.ShardInfo shardInfo = deleteResponse.getShardInfo();
    if (shardInfo.getTotal() != shardInfo.getSuccessful()) {
      log.info("Success shards are not enough");
    }
    if (shardInfo.getFailed() > 0) {
      for (ReplicationResponse.ShardInfo.Failure failure : shardInfo.
        getFailures()) {
        String reason = failure.reason();
        log.info("Fail reason is " + reason);
      }
    }
  }
```

隨後編輯專案，在專案根目錄下輸入以下指令：

```
mvn clean package
```

透過以下指令啟動專案服務：

```
java -jar ./target/esdemo-0.0.1-SNAPSHOT.jar
```

在專案服務啟動後，在瀏覽器中呼叫以下介面檢視刪除 ultraman 索引下文件編號為 2 的結果：

```
http://localhost:8080/springboot/es/high/index/delete?indexName=
ultraman&document=2
```

在介面呼叫後，伺服器輸出內容如下所示：

```
2019-08-01 16:03:50.027  INFO 39848 --- [nio-8080-exec-1] n.e.s.i.MeetHighElasticSearchServiceImpl : delete id is 2, ind
ex is ultraman,version is 2
```

從 Elasticsearch 主控台的輸出內容可以看到，在文件刪除後，文件的版本編號增加了 1，變成了 2。此時瀏覽器端顯示的內容如下所示：

```
Delete Index High ElasticSearch Client Successed!
```

5.6 更新文件索引

Elasticsearch 提供了在索引中更新文件介面的 API。

1 建置更新文件索引請求

在執行更新索引文件請求前，需要建置更新文件索引請求，即 UpdateRequest。UpdateRequest 有兩個必選參數，即索引名稱和文件 ID。

UpdateRequest 的初始化程式如下所示：

```
// 建置 UpdateRequest
  public void buildUpdateRequestIndexDocuments(String indexName, String
      document) {
    UpdateRequest request = new UpdateRequest(indexName, document);
  }
```

在建置 UpdateRequest 的過程中,還需要設定一些可選參數。可選參數有路由設定參數和逾時設定參數,另外,還可以設定平行處理對同一文件更新時的重試次數。此外,使用者可以設定啟用來源檢索參數,為特定欄位設定來源包含關係,為特定欄位設定來源排除關係。各個參數的設定方法如下所示:

```java
// 建置 UpdateRequest
  public void buildUpdateRequestIndexDocuments(String indexName, String
      document) {
  UpdateRequest request = new UpdateRequest(indexName, document);
  // 設定路由
  request.routing("routing");
  // 設定逾時
  request.timeout(TimeValue.timeValueSeconds(1));
  request.timeout("1s");
  // 設定更新策略
  request.setRefreshPolicy(WriteRequest.RefreshPolicy.WAIT_UNTIL);
  request.setRefreshPolicy("wait_for");
  // 設定:如果更新的文件在更新時被另一個操作更改,則重試更新操作的次數
  request.retryOnConflict(3);
  // 啟用來源檢索,在預設情況下禁用
  request.fetchSource(true);
  // 為特定欄位設定來源包含關係
  String[] includes = new String[] {"updated", "r*"};
  String[] excludes = Strings.EMPTY_ARRAY;
  request.fetchSource(new FetchSourceContext(true, includes, excludes));
  // 為特定欄位設定來源排除關係
  includes = Strings.EMPTY_ARRAY;
  excludes = new String[] {"updated"};
  request.fetchSource(new FetchSourceContext(true, includes, excludes));
  }
```

此外,還可以使用其他方式建置 UpdateRequest,例如使用部分文件更新方式。當使用部分文件更新方式時,部分文件將與現有文件合併。

可以用以下幾種方式提供被更新的部分檔案的內容:

(1)以 JSON 格式提供部分文件來源。

（2）以 Map 形式提供部分文件來源，Elasticsearch 自動將其轉為 JSON 格式。

（3）以 XContentBuilder 物件形式提供文件來源，Elasticsearch 自動將其轉為 JSON 格式。

（4）以鍵值對形式提供部分文件來源，Elasticsearch 自動將其轉為 JSON 格式。

在 ServiceImpl 實現層的 MeetHighElasticSearchServiceImpl 類別下的 buildUpdateRequestIndexDocuments 方法中增加以下程式：

```
/*
 * 用其他方式建置 UpdateRequest
 */
    // 方式 2：以 JSON 格式建置文件
    request = new UpdateRequest(indexName, document);
    String jsonString = "{" + "\"updated\":\"2019-12-31\"," +
        "\"reason\":\"Year update！\"" + "}";
    request.doc(jsonString, XContentType.JSON);
    // 方式 3：以 Map 形式提供文件來源，Elasticsearch 自動將其轉為 JSON 格式
    Map<String, Object> jsonMap = new HashMap<>();
    jsonMap.put("updated", new Date());
    jsonMap.put("reason", "Year update!");
    request = new UpdateRequest(indexName, document).doc(jsonMap);
    // 方式 4：以 XContentBuilder 物件形式提供文件來源，Elasticsearch 內建的幫助器
    // 自動將其產生為 JSON 格式的內容
    try {
      XContentBuilder builder = XContentFactory.jsonBuilder();
      builder.startObject();
      {
        builder.timeField("updated", new Date());
        builder.field("reason", "Year update!");
      }
      builder.endObject();
      request = new UpdateRequest(indexName, document).doc(builder);
    } catch (Exception e) {
      e.printStackTrace();
    }
```

```
// 方式 5：以鍵值對形式提供文件來源，Elasticsearch 自動將其轉為 JSON 格式
request =
    new UpdateRequest(indexName, document).doc("updated", new Date(),
        "reason", "Year update！");
```

如果被更新的文件不存在，則可以使用 upsert 方法將某些內容定義為新文件。
upsert 方法的程式如下所示。

在 ServiceImpl 實 現 層 的 MeetHighElasticSearchServiceImpl 類 別 下 的
buildUpdateRequestIndexDocuments 方法中增加以下程式：

```
/*
 * 下面展示 upsert 的使用
 */
    jsonString = "{\"created\":\"2019-12-31\"}";
    request.upsert(jsonString, XContentType.JSON);
```

與部分文件更新類似，在使用 upsert 方法時，可以使用字串、Map 對映、
XContentBuilder 物件或鍵值對定義 upsert 文件的內容。

② 執行更新文件索引請求

在 UpdateRequest 建置後，即可執行更新文件索引請求。與文件索引請求類
似，更新文件索引請求也有同步和非同步兩種執行方式。

◎ 同步方式

當以同步方式執行更新文件索引請求時，用戶端會等待 Elasticsearch 伺服器傳
回的查詢結果 UpdateResponse。在收到 UpdateResponse 後，用戶端繼續執行
相關的邏輯程式。以同步方式執行的程式如下所示：

```
// 以同步方式更新文件索引請求
  public void updateIndexDocuments(String indexName, String document) {
    UpdateRequest request = new UpdateRequest(indexName, document);

    try {
      UpdateResponse updateResponse = restClient.update(request,
RequestOptions.DEFAULT);
```

```
  } catch (Exception e) {
    e.printStackTrace();
  }
  // 關閉 Elasticsearch 連接
  closeEs();
}
```

◨ 非同步方式

當以非同步方式執行 UpdateRequest 時,進階用戶端不必同步等待請求結果的
傳回,可以直接向介面呼叫方傳回非同步介面執行成功的結果。

在 UpdateRequest 執行後,如果請求執行成功,則呼叫 ActionListener 類別中
的 onResponse 方法執行相關邏輯;如果請求執行失敗,則呼叫 ActionListener
類別中的 onFailure 方法執行相關邏輯。

當然,在非同步請求執行過程中,可能會出現例外,例外的處理與同步方式
執行情況相同。

為了處理非同步傳回的回應資訊或處理在請求執行過程中引發的例外資訊,
使用者需要指定監聽器。以非同步方式執行的程式如下所示:

```
// 以非同步方式更新文件索引請求
public void updateIndexDocumentsAsync(String indexName, String document) {
  UpdateRequest request = new UpdateRequest(indexName, document);
  ActionListener listener = new ActionListener<UpdateResponse>() {
    @Override
    public void onResponse(UpdateResponse updateResponse) {
    }
    @Override
    public void onFailure(Exception e) {
    }
  };
  try {
    restClient.updateAsync(request, RequestOptions.DEFAULT, listener);
  } catch (Exception e) {
    e.printStackTrace();
```

```
    }
    // 關閉 Elasticsearch 連接
    closeEs();
}
```

3 解析更新文件索引請求的回應結果

不論同步方式，還是非同步方式，均會傳回更新文件索引請求的回應結果 UpdateResponse。使用者可以透過解析 UpdateResponse 檢視文件的更新情況。

以同步方式為例，UpdateResponse 的解析程式共分為三層，分別是 Controller 層、Service 層和 ServiceImpl 實現層。

在 Controller 層的 MeetHighElasticSearchController 類別中增加以下程式：

```
// 以同步方式更新文件索引請求
@RequestMapping("/index/update")
public String updateIndexInHighElasticSearch(String indexName, String
    document) {
  if (Strings.isEmpty(indexName) || Strings.isEmpty(document)) {
    return "Parameters are error!";
  }
  meetHighElasticSearchService.updateIndexDocuments(indexName, document);
  return "Update Index High ElasticSearch Client Successed!";
}
```

在 Service 層的 MeetHighElasticSearchService 類別中增加以下程式：

```
// 以同步方式更新文件索引請求
public void updateIndexDocuments(String indexName, String document);
```

在 ServiceImpl 實現層的 MeetHighElasticSearchServiceImpl 類別中增加以下程式：

```
// 以同步方式更新文件索引請求
public void updateIndexDocuments(String indexName, String document) {
  UpdateRequest request = new UpdateRequest(indexName, document);
  Map<String, Object> jsonMap = new HashMap<>();
  jsonMap.put("updated", new Date());
```

```java
jsonMap.put("reason", "Year update!");
jsonMap.put("content",
    "2015 年 12 月，美聯儲開啟新一輪加息週期，至 2018 年 12 月，美聯儲累計加息 9
    次，每次加息 25 個基點，其中 2018 年共加息四次，將聯邦基金利率的目標範圍調至
    2.25%-2.50% 區間。
    今年以來，美聯儲未有一次加息動作，聯邦基金利率的目標範圍維持不變。\r\n");
request = new UpdateRequest(indexName, document).doc(jsonMap);
try {
    UpdateResponse updateResponse = restClient.update(request,
RequestOptions.DEFAULT);
    // 處理 UpdateResponse
    processUpdateRequest(updateResponse);
} catch (Exception e) {
    e.printStackTrace();
}
// 關閉 Elasticsearch 連接
closeEs();
}
// 處理 UpdateResponse
private void processUpdateRequest(UpdateResponse updateResponse) {
    String index = updateResponse.getIndex();
    String id = updateResponse.getId();
    long version = updateResponse.getVersion();
    log.info("update id is " + id + ", index is " + index + ",version is " +
        version);
    if (updateResponse.getResult() == DocWriteResponse.Result.CREATED) {
        // 建立文件成功
    } else if (updateResponse.getResult() == DocWriteResponse.Result.UPDATED) {
        // 更新文件成功
        // 檢視更新的資料
        log.info(updateResponse.getResult().toString());
    } else if (updateResponse.getResult() == DocWriteResponse.Result.DELETED) {
        // 刪除文件成功
    } else if (updateResponse.getResult() == DocWriteResponse.Result.NOOP) {
        // 無文件操作
    }
}
```

隨後編輯專案，在專案根目錄下透過以下指令實現：

```
mvn clean package
```

透過以下指令啟動專案服務：

```
java -jar ./target/esdemo-0.0.1-SNAPSHOT.jar
```

在專案服務啟動後，在瀏覽器中透過呼叫以下介面檢視索引的更新情況：

```
http://localhost:8080/springboot/es/high/index/update?indexName=
ultraman&document=1
```

此時伺服器列印輸出的內容如下所示：

```
2019-08-01 17:05:11.831  INFO 69104 --- [nio-8080-exec-1] n.e.s.i.MeetHighElasticSearchServiceImpl : update id is 1, ind
ex is ultraman,version is 4
```

當再次執行文件查詢介面進行 GET 查詢時，伺服器列印輸出內容如下所示：

```
2019-08-01 17:06:32.180  INFO 67544 --- [nio-8080-exec-7] n.e.s.i.MeetHighElasticSearchServiceImpl : id is 1, index is u
ltraman
2019-08-01 17:06:32.451  INFO 67544 --- [nio-8080-exec-7] n.e.s.i.MeetHighElasticSearchServiceImpl : version is 4, sourc
eAsString is {"user":"niudong","postDate":"2019-08-01T01:45:17.741Z","message":"Hello Elasticsearch!北京时间8月1日凌晨2
点，美联储公布7月议息会议结果。一如市场预期，美联储本次降息25个基点，将联邦基金利率的目标范围调至2.00%-2.25%。此次是2007
-2008年间美国为应对金融危机启动降息周期后，美联储十年多以来首次降息。美联储公布利率决议后，美股下跌，美元上涨，人民币汇
率下跌","reason":"Year update!","updated":"2019-08-01T09:05:11.3662Z","content":"2015年12月，美联储开启新一轮加息周期，至
2018年12月，美联储累计加息9次，每次加息25个基点，其中2018年共加息四次，将联邦基金利率的目标范围调至2.25%-2.50%区间。今年
以来，美联储未有一次加息动作，联邦基金利率的目标范围维持不变。\r\n"}
```

（編按：本圖為簡體中文介面）

透過比較資料，可以發現更新的資料和之前的資料進行了合併，即對部分文件使用更新時，部分文件將與現有文件合併。此外，對版本編號 version 欄位也進行了加 1 處理。

其實，在 Elasticsearch 的索引中處理文件的增、刪、改請求時，文件的 version 會隨著文件的改變而加 1。Elasticsearch 透過使用這個 version 來保障所有修改都被正確排序。當一個舊版本出現在新版本之後，它會被簡單地忽略。

使用者可以利用 version 的這一優點確保資料不會因為修改衝突而遺失，因此使用者可以指定文件的 version 做想要的更改。當然，如果想要修改的版本編號不是最新的，則修改請求會失敗。

5.7 取得文件索引的詞向量

詞向量，在 Elasticsearch 中的英文名稱為 "Term Vectors"。那麼什麼是詞向量呢？我們可以通俗地了解為：詞向量是關於詞的一些統計資訊的統稱。

因此，詞向量相關的 API 介面主要用於傳回特定文件中詞語的資訊和統計資訊。需要指出的是，這裡說的「文件」既可以是儲存在索引中的文件，也可以人工提供。

◼ 建置詞向量請求

在執行詞向量請求前，使用者需要建置詞向量請求，即 TermVectorsRequest。

TermVectorsRequest 有三個必選參數，即索引名稱、檢索資訊的欄位和文件 ID。

建置 TermVectorsRequest 的程式如下所示：

```
// 建置 TermVectorsRequest
public void buildTermVectorsRequest(String indexName, String document,
    String field) {
  TermVectorsRequest request = new TermVectorsRequest(indexName, document);
  request.setFields(field);
}
```

人工文件需要基於 XContentBuilder 物件提供，XContentBuilder 物件是產生 JSON 內容的 Elasticsearch 內建幫助器，程式如下所示：

```
// 建置 TermVectorsRequest
  public void buildTermVectorsRequest(String indexName, String document,
      String field) {
    // 方式 1：索引中存在的文件
    TermVectorsRequest request = new TermVectorsRequest(indexName, document);
    request.setFields(field);
    // 方式 2：索引中不存在的文件，可以人工為文件產生詞向量
    try {
      XContentBuilder docBuilder = XContentFactory.jsonBuilder();
```

```
   docBuilder.startObject().field("user", "niudong").endObject();
   request = new TermVectorsRequest(indexName, docBuilder);
 } catch (Exception e) {
   e.printStackTrace();
 }
}
```

TermVectorsRequest 除索引名稱、檢索資訊的欄位和文件 ID 三個必選參數外，還有一些可選參數需要設定。可選參數有 FieldStatistics 欄位、TermStatistics 欄位、位置欄位、偏移量欄位、有效酬載欄位、FilterSettings 欄位、PerFieldAnalyzer 欄位、即時檢索欄位和路由欄位等資訊。

我們在 buildTermVectorsRequest 方法中增加 TermVectorsRequest 的可選參數，程式如下所示：

```
/*
 * 可選參數
 */
  // 當把 FieldStatistics 設定為 false（預設為 true）時，可忽略文件計數、文件頻率總
  // 和及總術語頻率總和
  request.setFieldStatistics(false);

  // 將 TermStatistics 設定為 true（預設為 false），以顯示術語總頻率和文件頻率
  request.setTermStatistics(true);

  // 將 "位置" 設定為 "假"（預設為 "真"），忽略位置的輸出
  request.setPositions(false);

  // 將 "偏移" 設定為 "假"（預設為 "真"），忽略偏移的輸出
  request.setOffsets(false);

  // 將 "有效酬載" 設定為 "假"（預設為 "真"），忽略有效酬載的輸出
  request.setPayloads(false);
  Map<String, Integer> filterSettings = new HashMap<>();
  filterSettings.put("max_num_terms", 3);
  filterSettings.put("min_term_freq", 1);
  filterSettings.put("max_term_freq", 10);
```

```
filterSettings.put("min_doc_freq", 1);
filterSettings.put("max_doc_freq", 100);
filterSettings.put("min_word_length", 1);
filterSettings.put("max_word_length", 10);
// 設定 FilterSettings，根據 TF-IDF 分數篩選可傳回的詞條
request.setFilterSettings(filterSettings);
Map<String, String> perFieldAnalyzer = new HashMap<>();
perFieldAnalyzer.put("user", "keyword");
// 設定 PerFieldAnalyzer，指定與欄位已有的分析器不同的分析器
request.setPerFieldAnalyzer(perFieldAnalyzer);
// 將 Realtime 設定為 false（預設為 true），以便在 Realtime 附近檢索術語向量
request.setRealtime(false);

// 設定路由
request.setRouting("routing");
```

2 執行詞向量請求

當 TermVectorsRequest 建置後，即可執行取得文件詞向量請求。與文件索引請求類似，取得文件詞向量請求也有同步和非同步兩種執行方式。

同步方式

當以同步方式執行取得文件詞向量請求時，用戶端會等待 Elasticsearch 伺服器傳回的查詢結果 TermVectorsResponse 。在收到 TermVectorsResponse 後，用戶端會繼續執行相關的邏輯程式。以同步方式執行的程式如下所示：

```
// 以同步方式執行取得文件詞向量請求
  public void exucateTermVectorsRequest(String indexName, String document,
      String field) {
    TermVectorsRequest request = new TermVectorsRequest(indexName, document);
    request.setFields(field);
    try {
      TermVectorsResponse response = restClient.termvectors(request,
RequestOptions.DEFAULT);
    } catch (Exception e) {
      e.printStackTrace();
    }
```

```
    // 關閉 Elasticsearch 連接
    closeEs();
}
```

✍ 非同步方式

當以非同步方式執行取得文件詞向量請求時，進階用戶端不必同步等待請求結果的傳回，可以直接向介面呼叫方傳回非同步介面執行成功的結果。

為了處理非同步傳回的回應資訊或處理在請求執行過程中引發的例外資訊，使用者需要指定監聽器。以非同步方式執行的程式如下所示：

```
client.termvectorsAsync(request, RequestOptions.DEFAULT, listener);
```

其中，listener 為監聽器。

當然，在非同步請求執行過程中可能會出現例外，例外的處理與同步方式執行情況相同。

在非同步請求執行後，如果請求執行成功，則呼叫 ActionListener 類別中的 onResponse 方法進行相關邏輯的處理；如果請求執行失敗，則呼叫 ActionListener 類別中的 onFailure 方法進行相關邏輯的處理。

以非同步方式呼叫的程式如下所示：

```
// 以非同步方式執行 TermVectorsRequest
public void exucateTermVectorsRequestAsync(String indexName, String
    document, String field) {
  TermVectorsRequest request = new TermVectorsRequest(indexName, document);
  request.setFields(field);
  ActionListener listener = new ActionListener<TermVectorsResponse>() {
    @Override
    public void onResponse(TermVectorsResponse termVectorsResponse) {
    }
    @Override
    public void onFailure(Exception e) {
    }
  };
  try {
```

```
    restClient.termvectorsAsync(request, RequestOptions.DEFAULT, listener);
  } catch (Exception e) {
    e.printStackTrace();
  }
  // 關閉 Elasticsearch 連接
  closeEs();
}
```

3 解析詞向量請求的回應結果

不論同步方式，還是非同步方式，Elasticsearch 伺服器都會傳回詞向量請求的回應結果 TermVectorsResponse。使用者可以透過解析 TermVectorsResponse 取得相關內容。以同步執行方式為例，取得 TermVectorsResponse 後的解析程式共分為三層，分別是 Controller 層、Service 層和 ServiceImpl 實現層。

在 Controller 層的 MeetHighElasticSearchController 類別中增加以下程式：

```
// 以同步方式執行 TermVectorsRequest
  @RequestMapping("/index/term")
  public String termVectorsInHighElasticSearch(String indexName,
String document, String field) {
    if (Strings.isEmpty(indexName) || Strings.isEmpty(document) ||
Strings.isEmpty (field)) {
      return "Parameters are error!";
    }
    meetHighElasticSearchService.exucateTermVectorsRequest(indexName,
document, field);
    return "Test TermVectorsRequest High ElasticSearch Client Successed!";
  }
```

在 Service 層的 MeetHighElasticSearchService 類別中增加以下程式：

```
// 以同步方式執行 TermVectorsRequest
  public void exucateTermVectorsRequest(String indexName, String document,
String field);
```

在 ServiceImpl 實現層的 MeetHighElasticSearchServiceImpl 類別中增加以下程式：

```
// 以同步方式執行 TermVectorsRequest
  public void exucateTermVectorsRequest(String indexName, String document,
      String field) {
    TermVectorsRequest request = new TermVectorsRequest(indexName, document);
    request.setFields(field);
    try {
      TermVectorsResponse response = restClient.termvectors(request,
RequestOptions.DEFAULT);
      // 處理 TermVectorsResponse
      processTermVectorsResponse(response);
    } catch (Exception e) {
      e.printStackTrace();
    }
    // 關閉 Elasticsearch 連接
    closeEs();
  }
  // 處理 TermVectorsResponse
  private void processTermVectorsResponse(TermVectorsResponse response) {
    String index = response.getIndex();
    String type = response.getType();
    String id = response.getId();
    // 指示是否找到文件
    boolean found = response.getFound();
    log.info("index is " + index + ",id is " + id + ", type is " + type + ",
        found is " + found);
    List<TermVector> list = response.getTermVectorsList();
    log.info("list is " + list.size());
    for (TermVector tv : list) {
      processTermVector(tv);
    }
  }
  // 處理 TermVector
  private void processTermVector(TermVector tv) {
    String fieldname = tv.getFieldName();
    int docCount = tv.getFieldStatistics().getDocCount();
    long sumTotalTermFreq = tv.getFieldStatistics().getSumTotalTermFreq();
    long sumDocFreq = tv.getFieldStatistics().getSumDocFreq();
```

```
   log.info("fieldname is " + fieldname + "; docCount is " + docCount + ";
      sumTotalTermFreq is "
    + sumTotalTermFreq + ";sumDocFreq is " + sumDocFreq);
  if (tv.getTerms() == null) {
    return;
  }
  List<TermVectorsResponse.TermVector.Term> terms = tv.getTerms();
  for (TermVectorsResponse.TermVector.Term term : terms) {
    String termStr = term.getTerm();
    int termFreq = term.getTermFreq();
    int docFreq = term.getDocFreq() == null ? 0 : term.getDocFreq();
    long totalTermFreq = term.getTotalTermFreq() == null ? 0 :
        term.getTotalTermFreq();
    float score = term.getScore() == null ? 0 : term.getScore();
    log.info("termStr is " + termStr + "; termFreq is " + termFreq + ";
        docFreq is " + docFreq
      + ";totalTermFreq is " + totalTermFreq + ";score is " + score);
    if (term.getTokens() != null) {
      List<TermVectorsResponse.TermVector.Token> tokens = term.getTokens();
      for (TermVectorsResponse.TermVector.Token token : tokens) {
        int position = token.getPosition() == null ? 0 : token.
getPosition();
        int startOffset = token.getStartOffset() == null ? 0 :
            token.getStartOffset();
        int endOffset = token.getEndOffset() == null ? 0 : token.
            getEndOffset ();
        String payload = token.getPayload();
        log.info("position is " + position + "; startOffset is " +
            startOffset + "; endOffset is "
          + endOffset + ";payload is " + payload);
      }
    }
  }
}
```

隨後編輯專案，在專案根目錄下輸入以下指令：

```
mvn clean package
```

透過以下指令啟動專案服務：

```
java -jar ./target/esdemo-0.0.1-SNAPSHOT.jar
```

在專案服務啟動後，在瀏覽器中呼叫以下介面檢視取得文件索引的詞向量的情況：

```
http://localhost:8080/springboot/es/high/index/term?indexName=
ultraman&document=1&field=content
```

在介面執行後，瀏覽器中如果顯示以下內容，則表示介面執行成功：

```
Test TermVectorsRequest High ElasticSearch Client Successed!
```

在伺服器中會輸出如下所示內容。

```
2019-08-01 19:08:50.811  INFO 71176 --- [nio-8080-exec-1] n.e.s.i.MeetHighElasticSearchServiceImpl : index is ultraman,i
d is 1, type is _doc, found is true
2019-08-01 19:08:50.813  INFO 71176 --- [nio-8080-exec-1] n.e.s.i.MeetHighElasticSearchServiceImpl : list is 1
2019-08-01 19:08:50.822  INFO 71176 --- [nio-8080-exec-1] n.e.s.i.MeetHighElasticSearchServiceImpl : fieldname is conten
t; docCount is 1; sumTotalTermFreq is 95;sumDocFreq is 57
2019-08-01 19:08:50.832  INFO 71176 --- [nio-8080-exec-1] n.e.s.i.MeetHighElasticSearchServiceImpl : termStr is 12; term
Freq is 2; docFreq is 0;totalTermFreq is 0;score is 0.0
2019-08-01 19:08:50.850  INFO 71176 --- [nio-8080-exec-1] n.e.s.i.MeetHighElasticSearchServiceImpl : position is 2; star
tOffset is 5; endOffset is 7;payload is null
2019-08-01 19:08:50.872  INFO 71176 --- [nio-8080-exec-1] n.e.s.i.MeetHighElasticSearchServiceImpl : position is 19; sta
rtOffset is 28; endOffset is 30;payload is null
2019-08-01 19:08:50.903  INFO 71176 --- [nio-8080-exec-1] n.e.s.i.MeetHighElasticSearchServiceImpl : termStr is 2.25; te
rmFreq is 1; docFreq is 0;totalTermFreq is 0;score is 0.0
2019-08-01 19:08:50.918  INFO 71176 --- [nio-8080-exec-1] n.e.s.i.MeetHighElasticSearchServiceImpl : position is 61; sta
rtOffset is 79; endOffset is 83;payload is null
2019-08-01 19:08:50.932  INFO 71176 --- [nio-8080-exec-1] n.e.s.i.MeetHighElasticSearchServiceImpl : termStr is 2.50; te
rmFreq is 1; docFreq is 0;totalTermFreq is 0;score is 0.0
2019-08-01 19:08:50.943  INFO 71176 --- [nio-8080-exec-1] n.e.s.i.MeetHighElasticSearchServiceImpl : position is 62; sta
rtOffset is 85; endOffset is 89;payload is null
2019-08-01 19:08:50.957  INFO 71176 --- [nio-8080-exec-1] n.e.s.i.MeetHighElasticSearchServiceImpl : termStr is 2015; te
rmFreq is 1; docFreq is 0;totalTermFreq is 0;score is 0.0
2019-08-01 19:08:50.971  INFO 71176 --- [nio-8080-exec-1] n.e.s.i.MeetHighElasticSearchServiceImpl : position is 0; star
tOffset is 0; endOffset is 4;payload is null
2019-08-01 19:08:50.994  INFO 71176 --- [nio-8080-exec-1] n.e.s.i.MeetHighElasticSearchServiceImpl : termStr is 2018; te
rmFreq is 1; docFreq is 0;totalTermFreq is 0;score is 0.0
2019-08-01 19:08:51.002  INFO 71176 --- [nio-8080-exec-1] n.e.s.i.MeetHighElasticSearchServiceImpl : position is 17; sta
rtOffset is 23; endOffset is 27;payload is null
2019-08-01 19:08:51.023  INFO 71176 --- [nio-8080-exec-1] n.e.s.i.MeetHighElasticSearchServiceImpl : position is 40; sta
rtOffset is 54; endOffset is 58;payload is null
2019-08-01 19:08:51.034  INFO 71176 --- [nio-8080-exec-1] n.e.s.i.MeetHighElasticSearchServiceImpl : termStr is 25; term
Freq is 1; docFreq is 0;totalTermFreq is 0;score is 0.0
2019-08-01 19:08:51.052  INFO 71176 --- [nio-8080-exec-1] n.e.s.i.MeetHighElasticSearchServiceImpl : position is 34; sta
rtOffset is 46; endOffset is 48;payload is null
2019-08-01 19:08:51.099  INFO 71176 --- [nio-8080-exec-1] n.e.s.i.MeetHighElasticSearchServiceImpl : termStr is 9; termF
req is 1; docFreq is 0;totalTermFreq is 0;score is 0.0
2019-08-01 19:08:51.115  INFO 71176 --- [nio-8080-exec-1] n.e.s.i.MeetHighElasticSearchServiceImpl : position is 28; sta
rtOffset is 39; endOffset is 40;payload is null
2019-08-01 19:08:51.140  INFO 71176 --- [nio-8080-exec-1] n.e.s.i.MeetHighElasticSearchServiceImpl : termStr is 一; term
Freq is 2; docFreq is 0;totalTermFreq is 0;score is 0.0
2019-08-01 19:08:51.178  INFO 71176 --- [nio-8080-exec-1] n.e.s.i.MeetHighElasticSearchServiceImpl : position is 10; sta
rtOffset is 15; endOffset is 16;payload is null
2019-08-01 19:08:51.186  INFO 71176 --- [nio-8080-exec-1] n.e.s.i.MeetHighElasticSearchServiceImpl : position is 74; sta
rtOffset is 103; endOffset is 104;payload is null
2019-08-01 19:08:51.218  INFO 71176 --- [nio-8080-exec-1] n.e.s.i.MeetHighElasticSearchServiceImpl : termStr is 不; term
Freq is 1; docFreq is 0;totalTermFreq is 0;score is 0.0
2019-08-01 19:08:51.228  INFO 71176 --- [nio-8080-exec-1] n.e.s.i.MeetHighElasticSearchServiceImpl : position is 93; sta
rtOffset is 123; endOffset is 124;payload is null
2019-08-01 19:08:51.265  INFO 71176 --- [nio-8080-exec-1] n.e.s.i.MeetHighElasticSearchServiceImpl : termStr is 個; term
Freq is 1; docFreq is 0;totalTermFreq is 0;score is 0.0
```

索引 ultraman 中編號為 1 的文件下的 content 中詞語的統計資訊。

5.8 文件處理過程解析

前面介紹了 Elasticsearch 中文件及對文件的一些操作的介面實戰，相信讀者已經對文件有一定的認知，這是一個很好的開始。基礎打好之後，就更容易學習和了解較為複雜的內容。下面介紹 Elasticsearch 內部如何處理文件。

5.8.1 文件的索引過程

首先需要明確一點，寫入磁碟的倒排索引具有不可變性。Elasticsearch 為什麼要這樣做，主要是基於以下幾個考量：

（1）讀寫操作輕量級，不需要鎖。如果 Elasticsearch 從來不需要更新一個索引，就不必擔心多個程式同時嘗試修改索引的情況。

（2）一旦索引被檔案系統的記憶體讀取，它就會一直在那裏，因為不會改變。此外，當檔案系統記憶體有足夠大的空間時，大部分的索引讀寫操作可以直接存取記憶體，而不需要磁碟，這顯然有助於提升 Elasticsearch 的效能。

（3）當寫入單一大的倒排索引時，Elasticsearch 可以壓縮資料，以減少磁碟 I/O 和需要儲存索引的記憶體大小。

當然，倒排索引的不可變性也是一把「雙面刃」，不可變的索引也有它的缺點。首先就是因為它的不可變性，使用者如果想要搜索一個新文件，只能重建整個索引。這不僅嚴重限制了一個索引所能裝下的資料，還限制了一個索引可以被更新的頻率。

那麼 Elasticsearch 是如何在保持倒排索引不可變好處的同時又能更新倒排索引呢？答案是，使用多個索引。

Elasticsearch 不是重新定義整個倒排索引，而是增加額外的索引反映最近的變化。每個倒排索引都可以按順序查詢，從最「老舊」的索引開始查詢，最後把結果聚合起來。

Elasticsearch 的底層依賴於 Lucene，Lucene 中的索引其實是 Elasticsearch 中的分片，Elasticsearch 中的索引是分片的集合。當 Elasticsearch 搜索索引時，它發送查詢請求給該索引下的所有分片，然後過濾這些結果，最後聚合成全域的結果。

為了避免混淆，Elasticsearch 引用了 per-segment search 的概念。一個段（segment）就是一個有完整功能的倒排索引。Lucene 中的索引指的是段的集合，再加上提交點（commit point，包含所有段的檔案）。新的文件在被寫入磁碟的段之前，需要先寫入記憶體區的索引。

一個 per-segment search 的工作流程如下所示：

（1）新的文件首先被寫入記憶體區的索引。

（2）記憶體中的索引不斷被提交，新段不斷產生。當新的提交點產生時就將這些新段的資料寫入磁碟，包含新段的名稱。
　　　寫入磁碟是與檔案同步寫入，也就是説，所有的寫入操作都需要等待檔案系統記憶體的資料同步到磁碟，確保它們可以被物理寫入。

（3）新段被開啟，於是它包含的文件就可以被檢索到。

（4）記憶體被清除，等待接收新的文件。

當一個請求被接收，所有段依次被查詢時，所有段上的 term 統計資訊會被聚合，確保每個 term 和文件的相關性被正確計算。透過這種方式，新的文件就能夠以較小的代價加入索引。

段具有不可變性，那麼 Elasticsearch 是如何刪除和更新文件資料的呢？

段的不可變特性，表示文件既不能從舊的段中移除，舊的段中的文件也不能被更新。於是 Elasticsearch 在每一個提交點都引用一個 .del 檔案，包含了段上已經被刪除的文件。

當一個文件被刪除時，它實際上只是在 .del 檔案中被標記為刪除。在進行文件查詢時，被刪除的文件依然可以被比對查詢，但是在最後傳回之前會從結果中刪除。

當一個文件被更新時，舊版本的文件會被標記為刪除，新版本的文件在新的段中被索引。當對文件進行查詢時，該文件的不同版本都會比對一個查詢請求，但是較舊的版本會從結果中被刪除。

被刪除的檔案越累積越多，每個段消耗的如檔案控制代碼、記憶體、CPU 等資源越來越大。如果每次搜索請求都需要依次檢查每個段，則段越多，查詢就越慢。這些勢必會影響 Elasticsearch 的效能，那麼 Elasticsearch 是如何處理的呢？ Elasticsearch 引用了段合併段。在段合併時，我們會展示被刪除的檔案是如何從檔案系統中被清除。

Elasticsearch 透過後台合併段的方式解決了上述問題，在段合併過程中，小段被合併成大段，大段再合併成更大的段。在合併段時，被刪除的文件不會被合併到大段中。

在索引過程中，refresh 會建立新的段，並開啟它。合併過程是在後台選擇一些小的段，把它們合併成大的段。在這個過程中不會中斷索引和搜索。當新段合併後，即可開啟搜索；而舊段會被刪除。

需要指出的是，合併大的段會消耗很多 I/O 和 CPU。為了不影響 Elasticsearch 的搜索效能，在預設情況下，Elasticsearch 會限制合併過程，這樣搜索才有足夠的資源進行。

除自動完成段合併外，Elasticsearch 還提供了 optimize API，以便根據需要強制合併段。optimize API 強制分片合併段以達到指定 max_num_segments 參數，這會減少段的數量（通常為 1），達到加強搜索效能的目的。

需要指出的是，不要在動態的索引上使用 optimize API。optimize API 的典型場景是記錄記錄檔，記錄檔是按照每天、每周、每月存入索引，舊的索引一般只讀取不修改。在這種場景下，使用者主動把每個索引的段降至 1 就很有效，因為搜索過程會用到更少的資源，效能更好。

5.8.2 文件在檔案系統中的處理過程

前面介紹了文件的索引過程，由於資料最後會持久化在磁碟的檔案系統中，那麼 Elasticsearch 是如何將文件儲存到檔案系統中的呢？本節將揭曉。

在 Elasticsearch 的設定檔 elasticsearch-7.2.0\config\elasticsearch.yml 中，有一個設定屬性 path.data，該屬性包含了 Elasticsearch 中儲存的資料的資料夾的路徑，下面就從本目錄開始講起。Elasticsearch 根目錄的內容如圖 5-2 所示。

圖 5-2

從圖 5-2 中可見，Elasticsearch 根目錄中預設有一個 data 目錄，data 目錄即為 Elasticsearch 中預設的 path.data 屬性的值。

在 Elasticsearch 中，是使用 Lucene 來處理分片等級的索引和查詢，因此 data 目錄中的檔案由 Elasticsearch 和 Lucene 寫入。

兩者的職責非常明確：Lucene 負責寫和維護 Lucene 索引檔案，而 Elasticsearch 在 Lucene 之上寫與功能相關的中繼資料，例如欄位對映、索引設定和其他叢集中繼資料等。

開啟 data 目錄，如圖 5-3 所示。

圖 5-3

nodes 資料夾用於儲存本機中的節點資訊。開啟 nodes 資料夾，如圖 5-4 所示。

圖 5-4

在目前的 Elasticsearch 中，因為只有一個主節點分片，所以只有一個編號為 0 的資料夾，開啟該資料夾，如圖 5-5 所示。

📁	_state	2019/9/11 10:02	文件夾
📁	indices	2019/9/6 16:17	文件夾
📄	node.lock	2019/7/29 19:15	LOCK 文件

圖 5-5

圖 5-5 中的 node.lock 檔案用於確保一次只能從一個資料目錄讀取、寫入一個 Elasticsearch 相關的資訊。_state 資料夾用於儲存目前節點的索引資訊，indices 資料夾用於儲存目前節點的索引資訊。

開啟 state 資料夾，如圖 5-6 所示。

📄	global-30.st	2019/9/11 10:02	ST 文件
📄	manifest-276.st	2019/9/11 10:02	ST 文件
📄	node-20.st	2019/9/11 10:02	ST 文件

圖 5-6

圖 5-6 中有一個名稱為 global-30.st 的檔案，其中 global 字首表示這是一個全域狀態檔案，而 .st 副檔名表示這是一個包含中繼資料的狀態檔案。這種二進位標頭檔案有關使用者叢集的全域中繼資料，數字 30 表示叢集中繼資料的版本。

開啟 indices 資料夾，如圖 5-7 所示。

📁	3MJyBU9yQ-m0xyjN1NQDCQ	2019/9/11 10:02	文件夾
📁	5v8K-jIqRqC38JXOkY--6A	2019/9/11 10:02	文件夾
📁	BJJw9kdcRLy5gohfoYNR8g	2019/9/11 10:02	文件夾
📁	fcSpMAb6RfmtruwrH5e1Bg	2019/9/11 10:02	文件夾
📁	fY1q2BI2SsaDdIRR4g_O8Q	2019/9/11 10:02	文件夾
📁	iVI4JyVbRP2JNcQRD5SP9A	2019/9/11 10:02	文件夾
📁	m7F_Gtw8Ts-26zY5AHykGA	2019/9/11 10:02	文件夾
📁	MpLSEfXHT6O6HjqLciakRQ	2019/9/11 10:02	文件夾
📁	njxz6gLvSv2rS6NIckmBQg	2019/9/11 10:02	文件夾
📁	z26iYq5oSKaT0SfvZxC0GA	2019/9/11 10:02	文件夾

圖 5-7

indices 資料夾中的內容為索引資訊，每個索引有一個隨機字串的名稱，顯然該資料夾下的索引數量與本機中的索引數量相同。隨機開啟一個索引為 3MJyBU9yQ-m0xyjN1NQDCQ 的資料夾，如圖 5-8 所示。

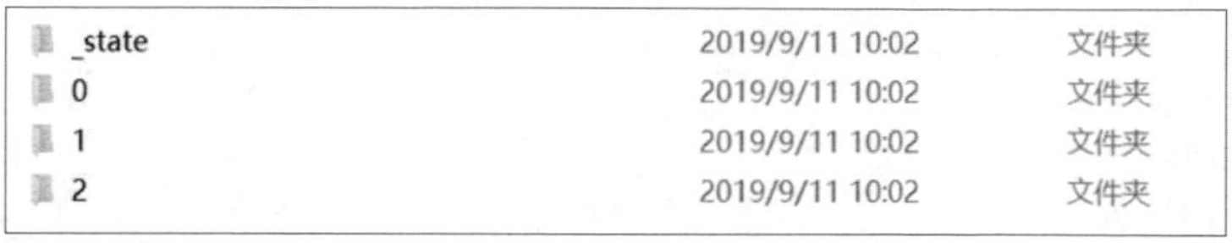

_state	2019/9/11 10:02	文件夹
0	2019/9/11 10:02	文件夹
1	2019/9/11 10:02	文件夹
2	2019/9/11 10:02	文件夹

圖 5-8

在圖 5-8 中，有兩種子資料夾：_state 類別資料夾和分片類別資料夾。其中，0、1、2 為分片資料夾。

_state 資料夾包含了 indices / {index-name} / state / state- {version} .st 路徑中的索引狀態檔案；0、1、2。分片資料夾包含與索引的第一、二、三個分片相關的資料（即分片 0、分片 1 和分片 2）。

開啟分片 0 的目錄，如圖 5-9 所示。

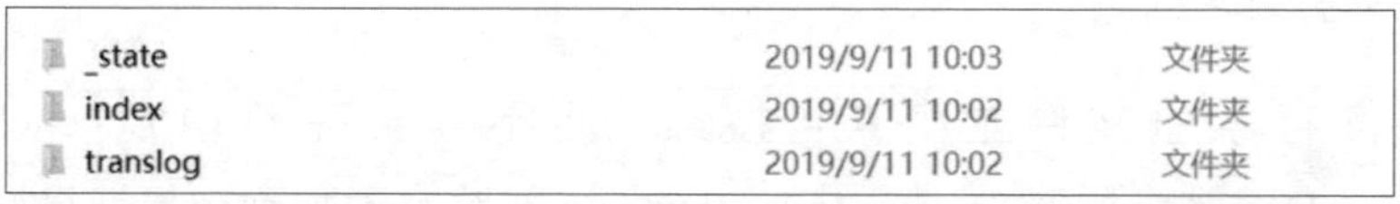

_state	2019/9/11 10:03	文件夹
index	2019/9/11 10:02	文件夹
translog	2019/9/11 10:02	文件夹

圖 5-9

在圖 5-9 中，分片 0 目錄包含分片相關的狀態檔案，其中包含版本控制及有關分片是主分片還是備份的資訊，如圖 5-10 所示。

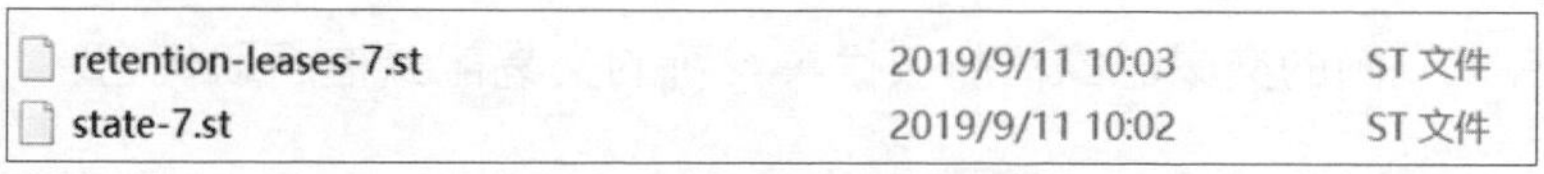

retention-leases-7.st	2019/9/11 10:03	ST 文件
state-7.st	2019/9/11 10:02	ST 文件

圖 5-10

此外，還有本分片下的索引資訊和 translog 記錄檔資訊。其中，translog 記錄檔資訊是 Elasticsearch 的交易記錄檔，在每個分片 translog 目錄中的字首 translog 中存在，如圖 5-11 所示。

translog.ckp	2019/9/11 10:02	CKP 文件	1 KB
translog-2.ckp	2019/9/6 10:29	CKP 文件	1 KB
translog-2.tlog	2019/9/6 10:29	TLOG 文件	1 KB
translog-3.ckp	2019/9/7 13:40	CKP 文件	1 KB
translog-3.tlog	2019/9/7 13:40	TLOG 文件	1 KB
translog-4.ckp	2019/9/10 10:37	CKP 文件	1 KB
translog-4.tlog	2019/9/10 10:37	TLOG 文件	1 KB
translog-5.ckp	2019/9/10 10:52	CKP 文件	1 KB
translog-5.tlog	2019/9/10 10:52	TLOG 文件	1 KB
translog-6.ckp	2019/9/10 14:12	CKP 文件	1 KB
translog-6.tlog	2019/9/10 14:12	TLOG 文件	1 KB
translog-7.ckp	2019/9/10 17:55	CKP 文件	1 KB
translog-7.tlog	2019/9/10 17:55	TLOG 文件	1 KB
translog-8.ckp	2019/9/10 18:05	CKP 文件	1 KB
translog-8.tlog	2019/9/10 18:05	TLOG 文件	1 KB
translog-9.tlog	2019/9/11 10:02	TLOG 文件	1 KB

圖 5-11

在 Elasticsearch 中，交易記錄檔用於確保安全地將資料索引到 Elasticsearch，而無須為每個文件執行低階 Lucene 提交。當提交 Lucene 索引時，會在 Lucene 等級建立一個新的 segment，即執行 fsync()，會產生大量磁碟 I／O，進一步影響效能。

為了能儲存索引文件並使其可搜索，而不需要完整地 Lucene 提交，Elasticsearch 將其增加到 Lucene IndexWriter，並將其附加到交易記錄檔中。

這樣，在每個 refresh_interval 之後，它將在 Lucene 索引上呼叫 reopen()，這使資料可以在不需要提交的情況下就能進行搜索。這是 Lucene 近即時搜索 API 的一部分。當 IndexWriter 最後自動更新交易記錄檔或由於顯性更新操作而提交時，先前的交易記錄檔將被捨棄，新的交易記錄檔將取代它。

如果需要恢復，則首先恢復在 Lucene 中寫入磁碟的 segments，然後重放交易記錄檔，以防止遺失尚未完全提交到磁碟的操作。

5.9 基礎知識連結

在開發過程中，與 Elasticsearch 相似，且同為文件型儲存的中介軟體是 MongoDB。

MongoDB 是一個以分散式檔案儲存為基礎的資料庫，以 C++ 為基礎撰寫，主要在為 Web 應用提供可擴充的高性能資料儲存解決方案。

MongoDB 與關聯式資料庫類似，也是非關聯式資料庫當中功能最豐富的。

那麼，Elasticsearch 和 MongoDB 有什麼異同之處嗎？它們的使用場景有什麼區別？

Elasticsearch 和 MongoDB 的相同之處在於兩者都是以 JSON 格式進行資料儲存，都支援對文件資料的增刪改查，即 CRUD 操作。

此外，兩者都使用了分片和複製技術，都支援處理超大規模資料。

Elasticsearch 和 MongoDB 的不同之處也有很多。

（1）開發語言不同。Elasticsearch 是以 Java 撰寫為基礎；而 MongoDB 是以 C++ 撰寫為基礎。

（2）分片方式不同。Elasticsearch 以 Hash 模式進行分片；而 MongoDB 為基礎的分片模式除了 Hash 模式，還有 Range 模式。

（3）叢集的設定方式不同。Elasticsearch 天然是分散式的，主副分片自動分配和複製；而 MongoDB 需要手動設定。

（4）全文檢索的便捷程度不同。Elasticsearch 全文檢索功能強大，欄位自動索引；而 MongoDB 僅支援有限的欄位檢索，且需人工索引。

在使用場景上，Elasticsearch 適用於全文檢索場景，而 MongoDB 適用於資料大量儲存的場景。

5.10 小結

本章主要介紹了在 Java 進階用戶端中對文件操作 API 的使用，如文件在索引中的增刪改查操作，還介紹了文件在索引中存在性檢查 API 的使用和文件詞向量 API 的使用。

進階用戶端文件實戰二

襟三江而帶五湖
控蠻荊而引甌越

在第 5 章中，主要介紹了編號（1）至（7）的進階用戶端文件 API 的使用。在本章中，將介紹編號（8）至（13）的進階用戶端文件 API 的使用。

（1）Single document APIs──單文件操作 API。

（2）Index API──文件索引 API。

（3）Get API──文件取得 API。

（4）Exists API──文件存在性判斷 API。

（5）Delete API──文件刪除 API。

（6）Update API──文件更新 API。

（7）Term Vectors API──詞向量 API。

（8）Bulk API──批次處理 API。

（9）Multi-Get API──多文件取得 API。

（10）ReIndex API──重新索引 API。

（11）Update By Query API──查詢更新 API。

（12）Delete By Query API──查詢刪除 API。

（13）Multi Term Vectors API──多詞條向量 API。

6.1 批次請求

為了讓使用者能夠對文件進行高效的增刪改查，Elasticsearch 提供了批次請求介面，即 Bulk API。

■1 建置批次請求

在使用批次請求前，需要建置批次請求，即 BulkRequest。BulkRequest 可用於透過單次請求執行多個操作請求，如文件索引、文件更新、文件刪除等操作。在 BulkRequest 中需要增加至少一個操作，當然，BulkRequest 還支援在其中增加不同類型的操作。

BulkRequest 的建置程式如下所示：

```
// 建置 BulkRequest
  public void buildBulkRequest(String indexName, String field) {
    /*
     * 方式 1：增加同型請求
     */
    BulkRequest request = new BulkRequest();
    // 增加第一個 IndexRequest
    request.add(new IndexRequest(indexName).id("1").source(XContentType.JSON,
        field,
      " 事實上，自今年年初開始，美聯儲就已傳遞出貨幣政策或將轉向的跡象 "));
    // 增加第二個 IndexRequest
    request.add(new IndexRequest(indexName).id("2").source(XContentType.JSON,
        field,
      " 自 6 月起，市場對於美聯儲降息的預期愈發強烈 "));
    // 增加第三個 IndexRequest
    request.add(new IndexRequest(indexName).id("3").source(XContentType.JSON,
        field,
      " 從此前美聯儲降息歷程來看，美聯儲降息將開啟全球各國央行的降息視窗 "));
    /*
     * 方式 2：增加異型請求
     */
    // 增加一個 DeleteRequest
```

```
    request.add(new DeleteRequest(indexName, "3"));
    // 增加一個 UpdateRequest
    request.add(new UpdateRequest(indexName, "2").doc(XContentType.JSON, field,
        "自今年初美聯儲暫停加息以來，全世界的降息大幕就已拉開，不僅包含新興經濟體，
        發達經濟體也加入降息陣營，僅 7 月份一個月內，就有 6 國央行降息"));
    // 增加一個 IndexRequest
    request.add(new IndexRequest(indexName).id("4").source(XContentType.JSON,
        field,
        "在此次美聯儲降息後，央行或不會立即跟進降息"));
}
```

BulkRequest 也有一些可選參數供使用者進行必要的設定。可選參數有逾時設定、資料更新策略、分片備份數量、全域管線標識、全域路由等。

在 buildBulkRequest 方法中增加以下程式：

```
/*
 * 以下是可選參數的設定
 */
    // 設定逾時
    request.timeout(TimeValue.timeValueMinutes(2));
    request.timeout("2m");

    // 設定資料更新策略
    request.setRefreshPolicy(WriteRequest.RefreshPolicy.WAIT_UNTIL);
    request.setRefreshPolicy("wait_for");

    // 設定在繼續執行索引 / 更新 / 刪除操作之前必須處於活動狀態的分片備份數
    request.waitForActiveShards(2);
    request.waitForActiveShards(ActiveShardCount.ALL);

    // 用於所有子請求的全域 pipelineid，即全域管線標識
    request.pipeline("pipelineId");

    // 用於所有子請求的全域路由 ID
request.routing("routingId");
```

❷ 執行批次請求

在 BulkRequest 建置後，即可執行批次請求。與文件索引請求類似，批次請求
也有同步和非同步兩種執行方式。

☑ 同步方式

當以同步方式執行批次請求時，用戶端會等待 Elasticsearch 伺服器傳回的查詢
結果 BulkResponse。在收到 BulkResponse 後，用戶端會繼續執行相關的邏輯
程式。以同步方式執行的程式如下所示：

```java
// 以同步方式執行 BulkRequest
public void executeBulkRequest(String indexName, String field) {
  BulkRequest request = new BulkRequest();
  // 增加第一個 IndexRequest
  request.add(new IndexRequest(indexName).id("1").source(XContentType.JSON,
      field,
    "事實上，自今年年初開始，美聯儲就已傳遞出貨幣政策或將轉向的跡象"));
  // 增加第二個 IndexRequest
  request.add(new IndexRequest(indexName).id("2").source(XContentType.JSON,
      field,
    "自 6 月起，市場對於美聯儲降息的預期愈發強烈"));
  // 增加第三個 IndexRequest
  request.add(new IndexRequest(indexName).id("3").source(XContentType.JSON,
      field,
    "從此前美聯儲降息歷程來看，美聯儲降息將開啟全球各國央行的降息視窗"));
  try {
    BulkResponse bulkResponse = restClient.bulk(request,
RequestOptions.DEFAULT);
  } catch (Exception e) {
    e.printStackTrace();
  }
  // 關閉 Elasticsearch 連接
  closeEs();
}
```

非同步方式

當以非同步方式執行批次請求時，進階用戶端不必同步等待請求結果的傳回，可以直接向介面呼叫方傳回非同步介面執行成功的結果。

為了處理非同步傳回的回應資訊或處理在請求執行過程中引發的例外資訊，使用者需要指定監聽器。以非同步方式執行的程式如下所示：

```
client.bulkAsync(request, RequestOptions.DEFAULT, listener);
```

其中，listener 為監聽器。

當然，在非同步請求執行過程中可能會出現例外，例外的處理與同步方式執行情況相同。

在非同步請求處理後，如果請求執行成功，則呼叫 ActionListener 類別中的 onResponse 方法進行相關邏輯的處理；如果請求執行失敗，則呼叫 ActionListener 類別中的 onFailure 方法進行相關邏輯的處理。

以非同步方式執行的程式如下所示：

```
// 以非同步方式執行 BulkRequest
public void executeBulkRequestAsync(String indexName, String field) {
  BulkRequest request = new BulkRequest();
  // 增加第一個 IndexRequest
  request.add(new IndexRequest(indexName).id("1").source(XContentType.JSON,
      field,
      "事實上，自今年年初開始，美聯儲就已傳遞出貨幣政策或將轉向的跡象"));
  // 增加第二個 IndexRequest
  request.add(new IndexRequest(indexName).id("2").source(XContentType.JSON,
      field,
      "自 6 月起，市場對於美聯儲降息的預期愈發強烈"));
  // 增加第三個 IndexRequest
  request.add(new IndexRequest(indexName).id("3").source(XContentType.JSON,
      field,
      "從此前美聯儲降息歷程來看，美聯儲降息將開啟全球各國央行的降息視窗"));
  // 建置監聽器
```

```
ActionListener<BulkResponse> listener = new ActionListener<BulkResponse>() {
  @Override
  public void onResponse(BulkResponse bulkResponse) {
  }
  @Override
  public void onFailure(Exception e) {
  }
};
try {
  restClient.bulkAsync(request, RequestOptions.DEFAULT, listener);
} catch (Exception e) {
  e.printStackTrace();
}
// 關閉 Elasticsearch 連接
closeEs();
}
```

❸ 解析批次請求的回應結果

不論同步方式，還是非同步方式，用戶端均會收到批次請求的傳回結果
BulkResponse。BulkResponse 中包含有關已執行操作的資訊，並允許使用者對
每個請求的結果進行反覆運算解析。以同步執行方式為例，BulkResponse 的
解析程式共分為三層，分別是 Controller 層、Service 層和 ServiceImpl 實現層。

其中，在 Controller 層的 MeetHighElasticSearchController 類別中增加以下程
式：

```
// 以同步方式執行 BulkRequest
@RequestMapping("/index/bulk")
public String bulkGetInHighElasticSearch(String indexName, String field) {
  if (Strings.isEmpty(indexName) || Strings.isEmpty(field)) {
    return "Parameters are error!";
  }
  meetHighElasticSearchService.executeBulkRequest(indexName, field);
  return "Bulk Get In High ElasticSearch Client Successed!";
}
```

在 Service 層的 MeetHighElasticSearchService 類別中增加以下程式：

```
// 以同步方式執行 BulkRequest
  public void executeBulkRequest(String indexName, String field);
```

在 ServiceImpl 實現層的 MeetHighElasticSearchServiceImpl 類別中增加以下程式：

```
// 以同步方式執行 BulkRequest
  public void executeBulkRequest(String indexName, String field) {
    BulkRequest request = new BulkRequest();
    // 增加第一個 IndexRequest
    request.add(new IndexRequest(indexName).id("1").source(XContentType.JSON,
        field,
        "事實上，自今年年初開始，美聯儲就已傳遞出貨幣政策或將轉向的跡象"));
    // 增加第二個 IndexRequest
    request.add(new IndexRequest(indexName).id("2").source(XContentType.JSON,
        field,
        "自 6 月起，市場對於美聯儲降息的預期愈發強烈"));
    // 增加第三個 IndexRequest
    request.add(new IndexRequest(indexName).id("3").source(XContentType.JSON,
        field,
        "從此前美聯儲降息歷程來看，美聯儲降息將開啟全球各國央行的降息視窗"));
    try {
      BulkResponse bulkResponse = restClient.bulk(request, RequestOptions.
            DEFAULT);
      // 解析 BulkResponse
      processBulkResponse(bulkResponse);
    } catch (Exception e) {
      e.printStackTrace();
    }
    // 關閉 Elasticsearch 連接
    closeEs();
  }
  // 解析 BulkResponse
  private void processBulkResponse(BulkResponse bulkResponse) {
    if (bulkResponse == null) {
      return;
```

```
        }
    for (BulkItemResponse bulkItemResponse : bulkResponse) {
      DocWriteResponse itemResponse = bulkItemResponse.getResponse();
      switch (bulkItemResponse.getOpType()) {
        // 索引狀態
        case INDEX:
          // 索引產生
        case CREATE:
          IndexResponse indexResponse = (IndexResponse) itemResponse;
          String index = indexResponse.getIndex();
          String id = indexResponse.getId();
          long version = indexResponse.getVersion();
          log.info("create id is " + id + ", index is " + index + ",version is
              " + version);
          break;
        // 索引更新
        case UPDATE:
          UpdateResponse updateResponse = (UpdateResponse) itemResponse;
          break;
        // 索引刪除
        case DELETE:
          DeleteResponse deleteResponse = (DeleteResponse) itemResponse;
      }
    }
  }
```

隨後編輯專案，在專案根目錄下輸入以下指令：

```
mvn clean package
```

透過以下指令啟動專案服務：

```
java -jar ./target/esdemo-0.0.1-SNAPSHOT.jar
```

在專案服務啟動後，在瀏覽器中透過呼叫以下介面檢視批次請求結果的解析情況：

```
http://localhost:8080/springboot/es/high/index/bulk?indexName=ultraman&field
=content
```

在請求執行後，瀏覽器中顯示內容如下所示，證明介面執行成功：

```
Bulk Get In High ElasticSearch Client Successed!
```

在伺服器中輸出的內容如下所示。

```
2019-08-01 19:49:18.811  INFO 72252 --- [nio-8080-exec-2] n.e.s.i.MeetHighElasticSearchServiceImpl : create id is 1, ind
ex is ultraman,version is 5
2019-08-01 19:49:18.814  INFO 72252 --- [nio-8080-exec-2] n.e.s.i.MeetHighElasticSearchServiceImpl : create id is 2, ind
ex is ultraman,version is 1
2019-08-01 19:49:18.819  INFO 72252 --- [nio-8080-exec-2] n.e.s.i.MeetHighElasticSearchServiceImpl : create id is 3, ind
ex is ultraman,version is 1
```

三個請求在同一次批次請求中已經成功執行。因為文件 1 已經進行了 5 次處理，此時 version 為 5；而文件 2、3 剛剛進行索引，因此 version 為 1。

6.2 批次處理器

除批次請求介面 Bulk API 外，Elasticsearch 還提供了 BulkProcessor 進行批次操作處理。BulkProcessor 提供了一個應用程式類別，來簡化批次 Bulk API 的使用。BulkProcessor 允許將索引、更新、刪除文件的操作增加到處理器中透明地執行。

為了執行批次處理請求，BulkProcessor 需要依賴以下元件。

（1）RestHighLevelClient：此用戶端用於執行 BulkRequest 和檢索 BulkResponse。
（2）BulkProcessor.Listener： 在 每 次 執 行 BulkRequest 之 前 和 之 後， 或 當 BulkRequest 失敗時，都會呼叫此監聽器。

BulkProcessor 的建置程式如下所示：

```
// 建置 BulkProcessor
  public void buildBulkRequestWithBulkProcessor(String indexName, String
      field) {
    BulkProcessor.Listener listener = new BulkProcessor.Listener() {
      @Override
      public void beforeBulk(long executionId, BulkRequest request) {
        // 批次處理前的動作
      }
```

```
    @Override
    public void afterBulk(long executionId, BulkRequest request,
        BulkResponse response) {
      // 批次處理後的動作
    }
    @Override
    public void afterBulk(long executionId, BulkRequest request, Throwable
        failure) {
      // 批次處理後的動作
    }
};
BulkProcessor bulkProcessor = BulkProcessor.builder((request, bulkListener)
    -> restClient
      .bulkAsync(request, RequestOptions.DEFAULT, bulkListener),
          listener).build();
  }
```

在 BulkProcessor 中，使用者還可以根據目前增加的運算元設定更新批次請求的時間、根據目前增加的操作大小設定更新批次請求的時間、設定允許執行的平行處理請求數、設定更新間隔，以及設定後退策略等。如上述程式所述，BulkProcessor.Builder 提供了設定 BulkProcessor 批次處理器處理請求執行的方法，我們在上述方法 buildBulkRequestWithBulkProcessor 中增加以下程式：

```
/*
 * BulkProcessor 的設定
 */
    BulkProcessor.Builder builder = BulkProcessor.builder((request,
bulkListener)
        -> restClient
        .bulkAsync(request, RequestOptions.DEFAULT, bulkListener), listener);

    // 根據目前增加的運算元，設定更新批次請求的時間（預設值為 1000，使用 -1 表示禁用）
    builder.setBulkActions(500);

    // 根據目前增加的操作大小，設定更新批次請求的時間（預設為 5MB，使用 -1 表示禁用）
    builder.setBulkSize(new ByteSizeValue(1L, ByteSizeUnit.MB));
```

```
// 設定允許執行的平行處理請求數（預設為 1，當使用 0 時表示僅允許執行單一請求）
builder.setConcurrentRequests(0);

// 設定更新間隔（預設為未設定）
builder.setFlushInterval(TimeValue.timeValueSeconds(10L));

// 設定一個固定的後退策略，該策略最初等待 1s，最多重試 3 次
builder.setBackoffPolicy(BackoffPolicy.constantBackoff(TimeValue.
timeValueSeconds(1L), 3));
```

BulkProcessor 類別提供了一個簡單的介面，它可以根據請求數量或在指定的時間段後自動地更新批次操作。

在建立完 BulkProcessor 後，使用者就可以向其增加請求了。我們在上述方法 buildBulkRequestWithBulkProcessor 中增加以下程式：

```
/**
 * 增加索引請求
 */
    IndexRequest one = new IndexRequest(indexName).id("6").source
(XContentType.JSON, "title",
        "8 月 1 日，中國空軍發佈強軍宣傳片《初心伴我去戰鬥》，透過殲 -20、轟 -6K 等新型
        戰機練兵備戰的震撼場景，展現新時代空軍發展的新氣象，彰顯中國空軍維護國家主
        權、保衛國家安全、保障和平發展的意志和能力。");
    IndexRequest two = new IndexRequest(indexName).id("7").source
(XContentType.JSON, "title",
        " 在 2 分鐘的宣傳片中，中國空軍現役先進戰機悉數亮相，包含殲 -20、殲 -16、
        殲 -11、殲 -10B/C、蘇 -35、蘇 -27、轟 -6K 等機型 ");
    IndexRequest three = new IndexRequest(indexName).id("8").source
(XContentType.JSON, "title",
        " 宣傳片發佈正逢八一建軍節，而今年是新中國成立 70 周年，也是人民空軍成立 70 周
        年。70 年來，中國空軍在各領域取得全面發展，戰略打擊、戰略預警、空天防禦和戰略
        投送等能力得到顯著進步。");
    bulkProcessor.add(one);
    bulkProcessor.add(two);
bulkProcessor.add(three);
```

上述程式中的三個文件索引請求將由 BulkProcessor 執行，它負責為每個批次請求呼叫 BulkProcessor.Listener，設定的監聽器提供存取 BulkRequest 和 BulkResponse 的方法。程式為分三層，分別是 Controller 層、Service 層和 ServiceImpl 實現層。

在 Controller 層的 MeetHighElasticSearchController 類別中增加以下程式：

```
// 以同步方式執行 BulkRequest
@RequestMapping("/index/bulkProcessor")
public String bulkProcessorGetInHighElasticSearch(String indexName, String
    field) {
  if (Strings.isEmpty(indexName) || Strings.isEmpty(field)) {
    return "Parameters are error!";
  }
  meetHighElasticSearchService.executeBulkRequestWithBulkProcessor
(indexName, field);
  return "BulkProcessor Get In High ElasticSearch Client Successed!";
}
```

在 Service 層的 MeetHighElasticSearchService 類別中增加以下程式：

```
// 以同步方式執行 BulkRequest
public void executeBulkRequestWithBulkProcessor(String indexName, String
field);
```

在 ServiceImpl 實現層的 MeetHighElasticSearchServiceImpl 類別中增加以下程式：

```
// 以同步方式執行 BulkRequest
public void executeBulkRequestWithBulkProcessor(String indexName, String
    field) {
  BulkProcessor.Listener listener = new BulkProcessor.Listener() {
    @Override
    public void beforeBulk(long executionId, BulkRequest request) {
      // 批次處理前的動作
      int numberOfActions = request.numberOfActions();
      log.info("Executing bulk " + executionId + " with " + numberOfActions
        + " requests");
```

```java
      }
      @Override
      public void afterBulk(long executionId, BulkRequest request,
        BulkResponse response) {
        // 批次處理後的動作
        if (response.hasFailures()) {
          log.info("Bulk " + executionId + " executed with failures");
        } else {
          log.info("Bulk " + executionId + " completed in " +
              response.getTook().getMillis()
            + " milliseconds");
        }
      }
      @Override
      public void afterBulk(long executionId, BulkRequest request, Throwable
          failure) {
        // 批次處理後的動作
        log.error("Failed to execute bulk", failure);
      }
    };
    BulkProcessor bulkProcessor = BulkProcessor.builder((request, bulkListener) ->
        restClient
      .bulkAsync(request, RequestOptions.DEFAULT, bulkListener), listener).
build();
    /**
     * 增加索引請求
     */
    IndexRequest one = new IndexRequest(indexName).id ("6").source
(XContentType.JSON, "title",
        "8月1日，中國空軍發佈強軍宣傳片《初心伴我去戰鬥》，透過殲-20、轟-6K等新型
        戰機練兵備戰的震撼場景，展現新時代空軍發展的新氣象，彰顯中國空軍維護國家主
        權、保衛國家安全、保障和平發展的意志和能力。");
    IndexRequest two = new IndexRequest(indexName).id("7").source
(XContentType.JSON, "title",
        "在2分鐘的宣傳片中，中國空軍現役先進戰機悉數亮相，包含殲-20、殲-16、
        殲-11、殲-10B/C、蘇-35、蘇-27、轟-6K等機型");
    IndexRequest three = new IndexRequest(indexName).id("8").source
```

```
(XContentType.JSON, "title",
        " 宣傳片發佈正逢八一建軍節,而今年是新中國成立 70 周年,也是人民空軍成立 70 周
        年。70 年來,中國空軍在各領域取得全面發展,戰略打擊、戰略預警、空天防禦和戰略
        投送等能力得到顯著進步。");
    bulkProcessor.add(one);
    bulkProcessor.add(two);
    bulkProcessor.add(three);
}
```

隨後編輯專案,在專案根目錄下輸入以下指令:

```
mvn clean package
```

透過以下指令啟動專案服務:

```
java -jar ./target/esdemo-0.0.1-SNAPSHOT.jar
```

在專案服務啟動後,在瀏覽器中呼叫以下介面檢視批次處理器對多請求的處
理情況:

```
http://localhost:8080/springboot/es/high/index/bulkProcessor?indexName=
ultraman&field=content
```

在介面呼叫成功後,即可完成上述三個文件索引的批次寫入。隨後,呼叫文
件查詢介面檢視文件的索引情況,如選擇索引名稱為 ultraman、文件編號為 6
的文件的索引,介面呼叫情況如下所示:

```
localhost:8080/springboot/es/high/index/get?indexName=ultraman&document=6
```

伺服器的輸出內容如下所示,可見批次操作的請求已經執行成功:

```
2019-08-02 11:06:49.909  INFO 80140 --- [nio-8080-exec-6] n.e.s.i.MeetHighElasticSearchServiceImpl : id is 6, index is u
ltraman
```

6.3 MultiGet 批次處理實戰

Elasticsearch 不僅提供了批次請求介面 Bulk API,還提供了批次取得 API。批次
取得 API 可以合併多個請求,以達到減少每個請求單獨處理所需的網路負擔。

如果使用者需要從 Elasticsearch 中檢索多個文件，則與一個一個的檢索相比，更快的方法是在一個請求中使用 MultiGet 或 MGet API。

顧名思義，MultiGet API 是在單一 HTTP 請求中同時執行多個 Get 請求。

❶ 建置 MultiGet 批次處理請求

在執行 MultiGet 批次處理請求前，需要建置 MultiGet 批次處理請求，即 MultiGetRequest。MultiGetRequest 在初始化時為空，需要增加 MultiGetRequest.Item 以設定要分析的內容。MultiGetRequest.Item 有兩個必選參數，即索引名稱和文件 ID。建置 MultiGetRequest 的程式如下所示：

在 MeetHighElasticSearchServiceImpl 類別中增加以下程式：

```
// 建置 MultiGetRequest
public void buildMultiGetRequest(String indexName, String[] documentIds) {
  if (documentIds == null || documentIds.length <= 0) {
    return;
  }
  MultiGetRequest request = new MultiGetRequest();
  for (String documentId : documentIds) {
    // 增加請求
    request.add(new MultiGetRequest.Item(indexName, documentId));
  }
}
```

同理，MultiGetRequest 也有一些可選參數供使用者設定。可選參數主要有禁用來源檢索、為特定欄位設定來源包含關係、為特定欄位設定來源排除關係、為特定儲存欄位設定檢索、設定路由、設定版本和版本類型、設定偏好值、設定即時標示、設定在檢索文件之前執行更新策略等。

由此可見，MultiGet API 的可選參數列表與 Get API 的可選參數列表相同。

MultiGetRequest 的可選參數設定程式如下所示，在 buildMultiGetRequest 方法中增加以下程式：

```
/*
 * 可選參數使用介紹
 */
    // 禁用來源檢索，在預設情況下啟用
    request.add(new MultiGetRequest.Item(indexName, documentIds[0])
        .fetchSourceContext(FetchSourceContext.DO_NOT_FETCH_SOURCE));
    // 為特定欄位設定來源包含關係
    String[] excludes = Strings.EMPTY_ARRAY;
    String[] includes = {"title", "content"};
    FetchSourceContext fetchSourceContext = new FetchSourceContext(true,
        includes, excludes);
    request.add(
        new MultiGetRequest.Item(indexName, documentIds[0]).
fetchSourceContext(fetchSourceContext));
    // 為特定欄位設定來源排除關係
    fetchSourceContext = new FetchSourceContext(true, includes, excludes);
    request.add(
        new MultiGetRequest.Item(indexName, documentIds[0]).
            fetchSourceContext(fetchSourceContext));
    // 為特定儲存欄位設定檢索（要求欄位在索引中單獨儲存欄位）
    try {
      request.add(new MultiGetRequest.Item(indexName, documentIds
          [0]).storedFields("title"));
      MultiGetResponse response = restClient.mget(request, RequestOptions.
          DEFAULT);
      MultiGetItemResponse item = response.getResponses()[0];
      // 檢索 title 儲存欄位（要求該欄位單獨儲存在索引中）
      String value = item.getResponse().getField("title").getValue();
      log.info("value is " + value);
    } catch (Exception e) {
      e.printStackTrace();
    } finally {
      // 關閉 Elasticsearch 連接
      closeEs();
    }
    // 設定路由
    request.add(new MultiGetRequest.Item(indexName, documentIds [0]).routing
```

```
                ("routing"));
    // 設定版本和版本類型
    request.add(new MultiGetRequest.Item(indexName, documentIds[0])
        .versionType(VersionType.EXTERNAL).version(10123L));
    // 設定偏好值
    request.preference("title");
    // 將即時標示設定為假（預設為真）
    request.realtime(false);
    // 在檢索文件之前執行更新（預設為 false）
request.refresh(true);
```

☑ 執行 MultiGet 批次處理請求

在 MultiGetRequest 建置後，即可執行 MultiGet 批次處理請求。與文件索引請求類似，MultiGet 批次處理請求也有同步和非同步兩種執行方式。

✍ 同步方式

當以同步方式執行 MultiGet 批次處理請求時，用戶端會等待 Elasticsearch 伺服器傳回的查詢結果 MultiGetResponse。在收到 MultiGetResponse 後，用戶端會繼續執行相關的邏輯程式。以同步方式執行的核心程式如下所示：

```
// 以同步方式執行 MultiGetRequest
  public void executeMultiGetRequest(String indexName, String[] documentIds) {
    if (documentIds == null || documentIds.length <= 0) {
      return;
    }
    MultiGetRequest request = new MultiGetRequest();
    for (String documentId : documentIds) {
      // 增加請求
      request.add(new MultiGetRequest.Item(indexName, documentId));
    }
    try {
      MultiGetResponse response = restClient.mget(request,RequestOptions.
          DEFAULT);
    } catch (Exception e) {
      e.printStackTrace();
    } finally {
```

```
        // 關閉 Elasticsearch 連接
        closeEs();
    }
}
```

非同步方式

當以非同步方式執行 MultiGet 批次處理請求時，進階用戶端不必同步等待請求結果的傳回，可以直接向介面呼叫方傳回非同步介面執行成功的結果。

為了處理非同步傳回的回應資訊或處理在請求執行過程中引發的例外資訊，使用者需要指定監聽器。

在非同步請求處理後，如果請求執行成功，則呼叫 ActionListener 類別中的 onResponse 方法進行相關邏輯的處理；如果請求執行失敗，則呼叫 ActionListener 類別中的 onFailure 方法進行相關邏輯的處理。

以非同步方式執行的程式如下所示：

```
// 以非同步方式執行 MultiGetRequest
  public void executeMultiGetRequestAsync(String indexName, String[]
      documentIds) {
    if (documentIds == null || documentIds.length <= 0) {
      return;
    }
    MultiGetRequest request = new MultiGetRequest();
    for (String documentId : documentIds) {
      // 增加請求
      request.add(new MultiGetRequest.Item(indexName, documentId));
    }
    // 增加 ActionListener
    ActionListener listener = new ActionListener<MultiGetResponse>() {
      @Override
      public void onResponse(MultiGetResponse response) {
      }
      @Override
      public void onFailure(Exception e) {
      }
```

```
  };
  // 執行批次取得
  try {
    MultiGetResponse response = restClient.mget(request, RequestOptions.
        DEFAULT);
  } catch (Exception e) {
    e.printStackTrace();
  } finally {
    // 關閉 Elasticsearch 連接
    closeEs();
  }
}
```

當然，在非同步請求執行過程中可能會出現例外，對例外的處理與同步方式執行情況相同。

3 解析 MultiGet 批次處理請求的回應結果

不論同步方式，還是非同步方式，在執行 MultiGet 批次處理後，用戶端均會對傳回結果 MultiGetResponse 進行處理和解析。

MultiGetResponse 中包含了 MultiGetItemResponse 的清單，清單中的元素順序與請求的順序相同。如果 Get 請求執行成功，則包含 GetResponse；如果 Get 請求執行失敗，則 MultiGetItemResponse 中包含 MultiGetResponse.failure。

以同步執行方式為例，MultiGetResponse 的解析程式分為三層，分別是 Controller 層、Service 層和 ServiceImpl 實現層。

在 Controller 層的 MeetHighElasticSearchController 類別中增加以下程式：

```
// 以同步方式執行 MultiGetreques
  @RequestMapping("/index/multiget")
  public String multigetInHighElasticSearch(String indexName, String
      documentId) {
    if (Strings.isEmpty(indexName) || Strings.isEmpty(documentId)) {
      return "Parameters are error!";
    }
```

```
// 將 field( 英文逗點分隔的 ) 轉化成 String[]
List<String> documentIds = Splitter.on(",").splitToList(documentId);
meetHighElasticSearchService.executeMultiGetRequest(indexName,
    documentIds.toArray(new String[documentIds.size()]));
return "MultiGet In High ElasticSearch Client Successed!";
}
```

在 Service 層的 **MeetHighElasticSearchService** 類別中增加以下程式：

```
// 以同步方式執行 MultiGetRequest
public void executeMultiGetRequest(String indexName, String[] documentIds);
```

在 ServiceImpl 實現層的 **MeetHighElasticSearchServiceImpl** 類別中增加以下程式：

```
// 以同步方式執行 MultiGetRequest
public void executeMultiGetRequest(String indexName, String[] documentIds)
  {
  if (documentIds == null || documentIds.length <= 0) {
    return;
  }
  MultiGetRequest request = new MultiGetRequest();
  for (String documentId : documentIds) {
    // 增加請求
    request.add(new MultiGetRequest.Item(indexName, documentId));
  }
  try {
    MultiGetResponse response = restClient.mget(request, RequestOptions.
        DEFAULT);
    // 解析 MultiGetResponse
    processMultiGetResponse(response);
  } catch (Exception e) {
    e.printStackTrace();
  } finally {
    // 關閉 Elasticsearch 連接
    closeEs();
  }
}
```

```
// 解析 MultiGetResponse
private void processMultiGetResponse(MultiGetResponse multiResponse) {
  if (multiResponse == null) {
    return;
  }
  MultiGetItemResponse[] responses = multiResponse.getResponses();
  log.info("responses is " + responses.length);
  for (MultiGetItemResponse response : responses) {
    GetResponse getResponse = response.getResponse();
    String index = response.getIndex();
    String id = response.getId();
    log.info("index is " + index + ";id is " + id);
    if (getResponse.isExists()) {
      long version = getResponse.getVersion();
      // 按字串方式取得內容
      String sourceAsString = getResponse.getSourceAsString();
      // 按 Map 方式取得內容
      Map<String, Object> sourceAsMap = getResponse.getSourceAsMap();
      // 逐位元組陣列方式取得內容
      byte[] sourceAsBytes = getResponse.getSourceAsBytes();
      log.info("version is " + version + ";sourceAsString is " +
          sourceAsString);
    }
  }
}
```

隨後編輯專案，在專案根目錄下輸入以下指令：

```
mvn clean package
```

透過以下指令啟動專案服務：

```
java -jar ./target/esdemo-0.0.1-SNAPSHOT.jar
```

在專案服務啟動後，在瀏覽器中呼叫以下介面檢視 MultiGETRequest 請求的執行情況：

```
http://localhost:8080/springboot/es/high/index/multiget?indexName=
ultraman&documentId=6,7,8
```

伺服器中的輸出結果如下所示：

```
2019-08-02 13:51:33.624 INFO 59948 --- [nio-8080-exec-2] n.e.s.i.MeetHighElasticSearchServiceImpl : responses is 3
2019-08-02 13:51:33.626 INFO 59948 --- [nio-8080-exec-2] n.e.s.i.MeetHighElasticSearchServiceImpl : index is ultraman;i
d is 6
2019-08-02 13:51:33.627 INFO 59948 --- [nio-8080-exec-2] n.e.s.i.MeetHighElasticSearchServiceImpl : index is ultraman;i
d is 7
2019-08-02 13:51:33.628 INFO 59948 --- [nio-8080-exec-2] n.e.s.i.MeetHighElasticSearchServiceImpl : index is ultraman;i
d is 8
```

從輸出結果可以看出，三個文件的索引查詢請求均執行成功。

而在發出請求的瀏覽器頁面上會顯示以下內容，表示介面執行成功：

```
MultiGet In High ElasticSearch Client Successed!
```

6.4 文件 ReIndex 實戰

文件 ReIndex，可以譯為文件重新索引，用於從一個或更多的索引中複製相關的文件到一個新的索引中進行索引重建。因此，文件 ReIndex 請求需要一個現有的來源索引和一個可能存在或不存在的目標索引。需要指出的是，文件 ReIndex 不嘗試設定目標索引，它不會複製來源索引的設定。因此，使用者需要在執行文件 ReIndex 操作之前設定目標索引，包含設定對映、分片計數、備份等。

■1 建置文件 ReIndex 請求

在執行文件 ReIndex 請求前，需要建置文件 ReIndex 請求，即 ReindexRequest。ReindexRequest 的建置程式如下所示：

```java
// 建置 ReindexRequest
  public void bulidReindexRequest(String fromIndex, String toIndex) {
    ReindexRequest request = new ReindexRequest();
    // 增加要從中複製的來源的列表
    request.setSourceIndices("source1", "source2",fromIndex);
    // 增加目標索引
    request.setDestIndex(toIndex);
}
```

同理，ReindexRequest 也有一些可選參數供使用者進行設定。可選參數主要有設定目標索引的版本類型、設定目標索引的操作類型、設定版本衝突時的處理策略、增加查詢來限制文件的策略、設定大小來限制已處理文件的數量、設定 ReIndex 的批次、設定管線模式、設定排序策略、分片設定、捲動設定、逾時設定、ReIndex 後更新索引的策略設定等。

在 bulidReindexRequest 方法中增加可選參數設定的程式如下：

```
/*
 * ReindexRequest 的參數設定
 */
    // 設定目標索引的版本類型
    request.setDestVersionType(VersionType.EXTERNAL);
    // 設定目標索引的操作類型為建立類型
    request.setDestOpType("create");
    // 在預設情況下，版本衝突會中止重新索引處理程序，我們可以用以下方法計算它們
    request.setConflicts("proceed");
    // 透過增加查詢限制文件。下面僅複製使用者欄位設定為 kimchy 的文件
    request.setSourceQuery(new TermQueryBuilder("user", "kimchy"));
    // 透過設定大小限制已處理文件的數量
    request.setSize(10);

    // 在預設情況下，ReIndex 使用 1000 個批次。可以使用 sourceBatchSize 更改批大小
    request.setSourceBatchSize(100);
    // 指定管線模式
    request.setDestPipeline("my_pipeline");
    // 如果需要用到來源索引中的一組特定文件，則需要使用 sort。建議最好選擇更具選擇性
    // 的查詢，而非進行大小和排序
    request.addSortField("field1", SortOrder.DESC);
    request.addSortField("field2", SortOrder.ASC);
    // 使用切片捲動對 uid 進行切片。使用 setslices 指定要使用的切片數
    request.setSlices(2);
    // 使用 scroll 參數控制 "search context"，保持活動的時間
    request.setScroll(TimeValue.timeValueMinutes(10));
    // 設定逾時
    request.setTimeout(TimeValue.timeValueMinutes(2));
    // 呼叫 reindex 後更新索引
    request.setRefresh(true);
```

在上述可選參數的設定中，versionType 的設定可以像索引 API 一樣設定 dest 元素來控制樂觀平行處理控制。如果省略 versionType 或將其設定為 Internal，就會導致 Elasticsearch 盲目地將文件轉存到目標中。如果將 versionType 設定為 external，則 Elasticsearch 會保留原始檔案中的版本，並更新目標索引中版本比原始檔案索引中版本舊的所有文件。

當 opType 設定為 cause_reindex 時，會在目標索引中建立缺少的文件。所有現有文件都將導致版本衝突。預設 opType 是 index。

② 執行文件 ReIndex 請求

在 ReindexRequest 建置後，即可執行文件重新索引請求。與文件索引請求類似，文件重新索引請求也有同步和非同步兩種執行方式。

◿ 同步方式

當以同步方式執行文件重新索引請求時，用戶端會等待 Elasticsearch 伺服器傳回的查詢結果 BulkByScrollResponse。在收到 BulkByScrollResponse 後，用戶端會繼續執行相關的邏輯程式。以同步方式執行的程式如下所示：

```
// 以同步方式執行 ReindexRequest
public void executeReindexRequest(String fromIndex, String toIndex) {
  ReindexRequest request = new ReindexRequest();
  // 增加要從中複製的來源的列表
  request.setSourceIndices(fromIndex);
  // 增加目標索引
  request.setDestIndex(toIndex);
  try {
    BulkByScrollResponse bulkResponse = restClient.reindex(request,
        RequestOptions.DEFAULT);
  } catch (Exception e) {
    e.printStackTrace();
  } finally {
    // 關閉 Elasticsearch 連接
    closeEs();
  }
}
```

非同步方式

當以非同步方式執行文件重新索引請求時，進階用戶端不必同步等待請求結果的傳回，可以直接向介面呼叫方傳回非同步介面執行成功的結果。

為了處理非同步傳回的回應資訊或處理在請求執行過程中引發的例外資訊，使用者需要指定監聽器。以非同步方式執行的程式如下所示：

```
client.reindexAsync(request, RequestOptions.DEFAULT, listener);
```

其中，listener 為監聽器。

在非同步請求處理後，如果請求執行成功，則呼叫 ActionListener 類別中的 onResponse 方法進行相關邏輯的處理；如果請求執行失敗，則呼叫 ActionListener 類別中的 onFailure 方法進行相關邏輯的處理。

當然，在非同步請求執行過程中可能會出現例外，例外的處理與同步方式執行情況相同。

以非同步方式執行的完整程式如下所示：

```java
// 以非同步方式執行 ReindexRequest
public void executeReindexRequestAsync(String fromIndex, String toIndex) {
    ReindexRequest request = new ReindexRequest();
    // 增加要從中複製的來源的列表
    request.setSourceIndices(fromIndex);
    // 增加目標索引
    request.setDestIndex(toIndex);
    // 建置監聽器
    ActionListener listener = new ActionListener<BulkByScrollResponse>() {
        @Override
        public void onResponse(BulkByScrollResponse bulkResponse) {
        }
        @Override
        public void onFailure(Exception e) {
        }
    };
    try {
```

```
    restClient.reindexAsync(request, RequestOptions.DEFAULT, listener);
} catch (Exception e) {
    e.printStackTrace();
} finally {
    // 關閉 Elasticsearch 連接
    closeEs();
}
}
```

3 解析 ReIndex 請求的回應結果

不論同步請求，還是非同步請求，用戶端均需對傳回結果 BulkByScrollResponse 進行處理和解析。傳回的 BulkByScrollResponse 中包含有關已執行操作的資訊，並允許按序對每個結果進行反覆運算解析。

BulkByScrollResponse 的解析程式如下所示，以同步請求方式為例，程式分為三層，分別是 Controller 層、Service 層和 ServiceImpl 實現層。

在 Controller 層的 MeetHighElasticSearchController 類別中增加以下程式：

```
// 以同步方式執行 ReindexRequest
@RequestMapping("/index/reindex")
public String reindexInHighElasticSearch(String fromIndexName, String
    toIndexName) {
  if (Strings.isEmpty(fromIndexName) || Strings.isEmpty(toIndexName)) {
    return "Parameters are error!";
  }
  meetHighElasticSearchService.executeReindexRequest(fromIndexName,
      toIndexName);
  return "Reindex In High ElasticSearch Client Successed!";
}
```

在 Service 層的 MeetHighElasticSearchService 類別中增加以下程式：

```
// 以同步方式執行 ReindexRequest
public void executeReindexRequest(String fromIndex, String toIndex);
```

在 ServiceImpl 實現層的 MeetHighElasticSearchServiceImpl 類別中增加以下程式：

```java
// 以同步方式執行 ReindexRequest
public void executeReindexRequest(String fromIndex, String toIndex) {
  ReindexRequest request = new ReindexRequest();
  // 增加要從中複製的來源的列表
  request.setSourceIndices(fromIndex);
  // 增加目標索引
  request.setDestIndex(toIndex);
  try {
    BulkByScrollResponse bulkResponse = restClient.reindex(request,
        RequestOptions.DEFAULT);
    // 解析 BulkByScrollResponse
    processBulkByScrollResponse(bulkResponse);
  } catch (Exception e) {
    e.printStackTrace();
  } finally {
    // 關閉 Elasticsearch 連接
    closeEs();
  }
}
// 解析 BulkByScrollResponse
private void processBulkByScrollResponse(BulkByScrollResponse bulkResponse) {
  if (bulkResponse == null) {
    return;
  }
  // 取得總耗時
  TimeValue timeTaken = bulkResponse.getTook();
  log.info("time is " + timeTaken.getMillis());
  // 檢查請求是否逾時
  boolean timedOut = bulkResponse.isTimedOut();
  log.info("timedOut is " + timedOut);
  // 取得已處理的文件總數
  long totalDocs = bulkResponse.getTotal();
  log.info("totalDocs is " + totalDocs);
  // 已更新的文件數
  long updatedDocs = bulkResponse.getUpdated();
  log.info("updatedDocs is " + updatedDocs);
  // 已建立的文件數
```

```
long createdDocs = bulkResponse.getCreated();
log.info("createdDocs is " + createdDocs);
// 已刪除的文件數
long deletedDocs = bulkResponse.getDeleted();
log.info("deletedDocs is " + deletedDocs);
// 已執行的批次數
long batches = bulkResponse.getBatches();
log.info("batches is " + batches);
// 跳過的文件數
long noops = bulkResponse.getNoops();
log.info("noops is " + noops);
// 版本衝突數
long versionConflicts = bulkResponse.getVersionConflicts();
log.info("versionConflicts is " + versionConflicts);
// 重試批次索引操作的次數
long bulkRetries = bulkResponse.getBulkRetries();
log.info("bulkRetries is " + bulkRetries);
// 重試搜索操作的次數
long searchRetries = bulkResponse.getSearchRetries();
log.info("searchRetries is " + searchRetries);
// 請求阻塞的總時間，不包含目前處於休眠狀態的限制時間
TimeValue throttledMillis = bulkResponse.getStatus().getThrottled();
log.info("throttledMillis is " + throttledMillis.getMillis());
// 查詢失敗數量
List<ScrollableHitSource.SearchFailure> searchFailures = bulkResponse.
    getSearchFailures();
log.info("searchFailures is " + searchFailures.size());
// 批次操作失敗數量
List<BulkItemResponse.Failure> bulkFailures = bulkResponse.
    getBulkFailures();
log.info("bulkFailures is " + bulkFailures.size());
}
```

隨後編輯專案，在專案根目錄下輸入以下指令：

```
mvn clean package
```

透過以下指令啟動專案服務：

```
java -jar ./target/esdemo-0.0.1-SNAPSHOT.jar
```

在專案服務啟動後，在瀏覽器中呼叫以下介面檢視文件查詢索引的情況：

```
http://localhost:8080/springboot/es/high/index/reindex?fromIndexName=
ultraman&toIndexName=ultraman1
```

在介面呼叫成功後，如果瀏覽器中顯示內容如下，則證明介面呼叫成功：

```
Reindex In High ElasticSearch Client Successed!
```

伺服器中展示內容如下所示：

```
2019-08-02 14:41:11.193  INFO 80924 --- [nio-8080-exec-1] n.e.s.i.MeetHighElasticSearchServiceImpl : time is 1778
2019-08-02 14:41:11.195  INFO 80924 --- [nio-8080-exec-1] n.e.s.i.MeetHighElasticSearchServiceImpl : timedOut is false
2019-08-02 14:41:11.197  INFO 80924 --- [nio-8080-exec-1] n.e.s.i.MeetHighElasticSearchServiceImpl : totalDocs is 3
2019-08-02 14:41:11.198  INFO 80924 --- [nio-8080-exec-1] n.e.s.i.MeetHighElasticSearchServiceImpl : updatedDocs is 0
2019-08-02 14:41:11.199  INFO 80924 --- [nio-8080-exec-1] n.e.s.i.MeetHighElasticSearchServiceImpl : createdDocs is 3
2019-08-02 14:41:11.201  INFO 80924 --- [nio-8080-exec-1] n.e.s.i.MeetHighElasticSearchServiceImpl : deletedDocs is 0
2019-08-02 14:41:11.204  INFO 80924 --- [nio-8080-exec-1] n.e.s.i.MeetHighElasticSearchServiceImpl : batches is 1
2019-08-02 14:41:11.206  INFO 80924 --- [nio-8080-exec-1] n.e.s.i.MeetHighElasticSearchServiceImpl : noops is 0
2019-08-02 14:41:11.208  INFO 80924 --- [nio-8080-exec-1] n.e.s.i.MeetHighElasticSearchServiceImpl : versionConflicts is 0
2019-08-02 14:41:11.212  INFO 80924 --- [nio-8080-exec-1] n.e.s.i.MeetHighElasticSearchServiceImpl : bulkRetries is 0
2019-08-02 14:41:11.213  INFO 80924 --- [nio-8080-exec-1] n.e.s.i.MeetHighElasticSearchServiceImpl : searchRetries is 0
2019-08-02 14:41:11.215  INFO 80924 --- [nio-8080-exec-1] n.e.s.i.MeetHighElasticSearchServiceImpl : throttledMillis is 0
2019-08-02 14:41:11.217  INFO 80924 --- [nio-8080-exec-1] n.e.s.i.MeetHighElasticSearchServiceImpl : searchFailures is 0
2019-08-02 14:41:11.219  INFO 80924 --- [nio-8080-exec-1] n.e.s.i.MeetHighElasticSearchServiceImpl : bulkFailures is 0
```

從展示內容可以看出，文件重建索引操作已經成功，名稱為 ultraman1 的索引中已經複製了索引名稱為 ultraman 的資料。

這時，在瀏覽器中透過下面的介面檢視名稱為 ultraman1 的索引中的文件資訊：

```
http://localhost:8080/springboot/es/high/index/get?indexName=
ultraman1&document=1
```

伺服器中展示內容如下所示：

```
2019-08-02 14:42:51.300  INFO 65956 --- [nio-8080-exec-2] n.e.s.i.MeetHighElasticSearchServiceImpl : id is 1, index is ultraman1
2019-08-02 14:42:51.383  INFO 65956 --- [nio-8080-exec-2] n.e.s.i.MeetHighElasticSearchServiceImpl : version is 1, sourceAsString is {"content":"事实上，自今年年初开始，美联储就已传递出货币政策或将转向的迹象"}
```

可見，ultraman 的重建索引已經成功。此時在新索引 ultraman1 中，文件 1 的版本編號碼為 1，為新增狀態。而舊索引 ultraman 中的版本與此不同。我們可以透過下面的介面來檢視：

```
http://localhost:8080/springboot/es/high/index/get?indexName=
ultraman&document=1
```

此時，伺服器中展示內容如下所示。

2019-08-02 14:46:03.580 INFO 80476 --- [nio-8080-exec-9] n.e.s.i.MeetHighElasticSearchServiceImpl : id is 1, index is u
ltraman
2019-08-02 14:46:03.653 INFO 80476 --- [nio-8080-exec-9] n.e.s.i.MeetHighElasticSearchServiceImpl : version is 5, sourc
eAsString is {"content":"事实上，自今年年初开始，美联储就已传递出货币政策或将转向的迹象"}

編號為 1 的文件在舊索引 ultraman 中的版本編號為 5。

6.5 文件查詢時更新實戰

在文件查詢時更新介面，以更新索引中的文件。與 update 的更新方式不同，查詢時更新（update_by_query）是在不更改來源的情況下對索引中的每個文件進行更新，用可以在文件查詢時更新介面來修改欄位或新增欄位。

▌1 建置文件查詢時更新請求

在執行文件查詢時更新操作前，需要建置文件查詢時更新的請求，即 UpdateByQueryRequest。UpdateByQueryRequest 的必選參數只有一個，即需要執行查詢時更新操作的索引名稱。

建置 UpdateByQueryRequest 的 程 式 如 下 所 示， 我 們 在 MeetHighElasticSearchServiceImpl 類別中增加以下程式：

```
// 建置 UpdateByQueryRequest
  public void buildUpdateByQueryRequest(String indexName) {
    UpdateByQueryRequest request = new UpdateByQueryRequest(indexName);
  }
```

同理，UpdateByQueryRequest 也有可選參數供使用者進行設定。可選參數有版本衝突處理策略、限制查詢準則設定、設定大小來限制已處理文件的數量、設定批次處理的大小、設定管線名稱、設定捲動模式、設定全域路由、設定逾時、設定操作後的更新策略和設定索引選項等。

UpdateByQueryRequest 的可選參數設定程式如下所示，我們將這部分程式增加在 MeetHighElasticSearchServiceImpl 類別的 buildUpdateByQueryRequest 方法中：

```
// 建置 UpdateByQueryRequest
  public void buildUpdateByQueryRequest(String indexName) {
    UpdateByQueryRequest request = new UpdateByQueryRequest(indexName);

    /*
     * 設定 UpdateByQueryRequest
     */

    // 在預設情況下，版本衝突將中止 UpdateByQueryRequest 處理程序，但我們可以使用以下
    // 方法來計算它們
    request.setConflicts("proceed");

    // 透過增加查詢準則來限制。下面僅更新欄位設定為 niudong 的文件
    request.setQuery(new TermQueryBuilder("user", "niudong"));

    // 設定大小來限制已處理文件的數量
    request.setSize(10);

    // 在預設情況下，UpdateByQueryRequest 使用的批數為 1000。可以使用 setBatchSize
    // 更改批大小
    request.setBatchSize(100);

    // 指定管線模式
    request.setPipeline("my_pipeline");

    // 設定分片捲動來平行化
    request.setSlices(2);

    // 使用捲動參數控制 " 搜索上下文 "，保持連接的時間
    request.setScroll(TimeValue.timeValueMinutes(10));

    // 如果提供路由，那麼路由將被複製到捲動查詢，進一步限制與該路由值符合的分片處理
    request.setRouting("=cat");
```

```
// 設定等待請求的逾時
request.setTimeout(TimeValue.timeValueMinutes(2));

// 呼叫 update by query 後更新索引
request.setRefresh(true);

// 設定索引選項
request.setIndicesOptions(IndicesOptions.LENIENT_EXPAND_OPEN);

}
```

❷ 執行文件查詢時更新請求

在 UpdateByQueryRequest 建置後，即可執行文件查詢時更新請求。與文件索引請求類似，文件查詢時更新請求也有同步和非同步兩種執行方式。

◪ 同步方式

當以同步方式執行文件查詢時更新請求時，用戶端會等待 Elasticsearch 伺服器傳回的查詢結果 UpdateByQueryResponse。在收到 UpdateByQueryResponse 後，用戶端會繼續執行相關的邏輯程式。以同步方式執行的程式如下所示：

```
// 以同步方式執行 UpdateByQueryRequest
  public void executeUpdateByQueryRequest(String indexName) {
    UpdateByQueryRequest request = new UpdateByQueryRequest(indexName);
    try {
      BulkByScrollResponse bulkResponse = restClient.updateByQuery(request,
RequestOptions.DEFAULT);
    } catch (Exception e) {
      e.printStackTrace();
    } finally {
      // 關閉 Elasticsearch 連接
      closeEs();
    }
  }
```

▨ 非同步方式

當以非同步方式執行文件查詢時更新請求時,進階用戶端不必同步等待請求
結果的傳回,可以直接向介面呼叫方傳回非同步介面執行成功的結果。

為了處理非同步傳回的回應資訊或處理在請求執行過程中引發的例外資訊,
使用者需要指定監聽器。

以非同步方式執行的程式如下所示:

```
// 以非同步方式執行 UpdateByQueryRequest
public void executeUpdateByQueryRequestAsync(String indexName) {
  UpdateByQueryRequest request = new UpdateByQueryRequest(indexName);
  // 增加監聽器
  ActionListener listener = new ActionListener<BulkByScrollResponse>() {
    @Override
    public void onResponse(BulkByScrollResponse bulkResponse) {
    }
    @Override
    public void onFailure(Exception e) {
    }
  };
  try {
    restClient.updateByQueryAsync(request, RequestOptions.DEFAULT,
        listener);
  } catch (Exception e) {
    e.printStackTrace();
  } finally {
    // 關閉 Elasticsearch 連接
    closeEs();
  }
}
```

在非同步處理完成後,如果請求執行成功,則呼叫 ActionListener 類別
中的 onResponse 方法進行相關邏輯的處理;如果請求執行失敗,則呼叫
ActionListener 類別中的 onFailure 方法進行相關邏輯的處理。

❸ 解析文件查詢時更新請求的回應結果

不論同步方式，還是非同步方式，在文件查詢時更新請求執行後，用戶端均需要對傳回結果 UpdateByQueryResponse 進行處理和解析。UpdateByQueryResponse 中包含了有關已執行操作的資訊，並允許使用者按序對每個結果進行反覆運算解析。

下面以同步方式執行為例，展示對文件查詢時更新請求的傳回結果 UpdateByQueryResponse 的解析。程式共分為三層，分別是 Controller 層、Service 層和 ServiceImpl 實現層。

在 Controller 層的 MeetHighElasticSearchController 類別中增加以下程式：

```
// 以同步方式執行 UpdateByQueryResponse
@RequestMapping("/index/updateByQuery")
public String updateByQueryInHighElasticSearch(String indexName) {
  if (Strings.isEmpty(indexName)) {
    return "Parameters are error!";
  }
  meetHighElasticSearchService.executeUpdateByQueryRequest(indexName);
  return "UpdateByQuery In High ElasticSearch Client Successed!";
}
```

在 Service 層的 MeetHighElasticSearchService 類別中增加以下程式：

```
// 以同步方式執行 UpdateByQueryRequest
public void executeUpdateByQueryRequest(String indexName);
```

在 ServiceImpl 實現層的 MeetHighElasticSearchServiceImpl 類別中增加以下程式：

```
// 以同步方式執行 UpdateByQueryRequest
public void executeUpdateByQueryRequest(String indexName) {
  UpdateByQueryRequest request = new UpdateByQueryRequest(indexName);
  try {
    BulkByScrollResponse bulkResponse = restClient.updateByQuery(request,
        RequestOptions.DEFAULT);
    // 處理 BulkByScrollResponse
```

```
    processBulkByScrollResponse(bulkResponse);
  } catch (Exception e) {
    e.printStackTrace();
  } finally {
    // 關閉 Elasticsearch 連接
    closeEs();
  }
}
```

其中，processBulkByScrollResponse 方法與文件 ReIndex 介面中的名稱相同方法相同。

隨後編輯專案，在專案根目錄下輸入以下指令：

```
mvn clean package
```

透過以下指令啟動專案服務：

```
java -jar ./target/esdemo-0.0.1-SNAPSHOT.jar
```

在專案服務啟動後，在瀏覽器中呼叫以下介面檢視文件查詢時更新請求的執行情況：

```
http://localhost:8080/springboot/es/high/index/updateByQuery?indexName=
ultraman1
```

在介面執行後，我們可以在伺服器中看到如下所示內容：

```
2019-08-02 15:43:26.988  INFO 80032 --- [nio-8080-exec-3] n.e.s.i.MeetHighElasticSearchServiceImpl : time is 201
2019-08-02 15:43:26.991  INFO 80032 --- [nio-8080-exec-3] n.e.s.i.MeetHighElasticSearchServiceImpl : timedOut is false
2019-08-02 15:43:26.994  INFO 80032 --- [nio-8080-exec-3] n.e.s.i.MeetHighElasticSearchServiceImpl : totalDocs is 3
2019-08-02 15:43:26.997  INFO 80032 --- [nio-8080-exec-3] n.e.s.i.MeetHighElasticSearchServiceImpl : updatedDocs is 3
2019-08-02 15:43:27.000  INFO 80032 --- [nio-8080-exec-3] n.e.s.i.MeetHighElasticSearchServiceImpl : createdDocs is 0
2019-08-02 15:43:27.003  INFO 80032 --- [nio-8080-exec-3] n.e.s.i.MeetHighElasticSearchServiceImpl : deletedDocs is 0
2019-08-02 15:43:27.005  INFO 80032 --- [nio-8080-exec-3] n.e.s.i.MeetHighElasticSearchServiceImpl : batches is 1
2019-08-02 15:43:27.009  INFO 80032 --- [nio-8080-exec-3] n.e.s.i.MeetHighElasticSearchServiceImpl : noops is 0
2019-08-02 15:43:27.013  INFO 80032 --- [nio-8080-exec-3] n.e.s.i.MeetHighElasticSearchServiceImpl : versionConflicts is 0
2019-08-02 15:43:27.017  INFO 80032 --- [nio-8080-exec-3] n.e.s.i.MeetHighElasticSearchServiceImpl : bulkRetries is 0
2019-08-02 15:43:27.019  INFO 80032 --- [nio-8080-exec-3] n.e.s.i.MeetHighElasticSearchServiceImpl : searchRetries is 0
2019-08-02 15:43:27.022  INFO 80032 --- [nio-8080-exec-3] n.e.s.i.MeetHighElasticSearchServiceImpl : throttledMillis is 0
2019-08-02 15:43:27.026  INFO 80032 --- [nio-8080-exec-3] n.e.s.i.MeetHighElasticSearchServiceImpl : searchFailures is 0
2019-08-02 15:43:27.031  INFO 80032 --- [nio-8080-exec-3] n.e.s.i.MeetHighElasticSearchServiceImpl : bulkFailures is 0
```

從中可以看到，索引名稱為 ultraman1 的 3 個文件進行了 3 次更新。

此時，如果瀏覽器中顯示內容如下所示，則證明介面執行成功：

```
UpdateByQuery In High ElasticSearch Client Successed!
```

6.6 文件查詢時刪除實戰

除文件查詢時更新外，Elasticsearch 還提供了在查詢時刪除的介面。

■ 建置文件查詢時刪除請求

在 執 行 文 件 查 詢 時 刪 除 請 求 前， 需 要 先 建 置 查 詢 時 刪 除 請 求， 即
DeleteByQueryRequest。DeleteByQueryRequest 有一個必選參數，即執行查詢
時刪除操作的現有索引名稱。

建置 DeleteByQueryRequest 的程式如下所示：

```
// 建置 DeleteByQueryRequest
  public void buildDeleteByQueryRequest (String indexName) {
    DeleteByQueryRequest  request = new DeleteByQueryRequest (indexName);
}
```

同理，DeleteByQueryRequest 也有一些可選參數供使用者進行設定。可選參數
有版本衝突處理策略、限制查詢準則設定、設定大小來限制已處理文件的數
量、設定批次處理的大小、設定管線名稱、設定捲動模式、設定全域路由、
設定逾時、設定操作後的更新策略、設定索引選項等。

DeleteByQueryRequest 的可選參數設定程式如下所示，我們將這部分程式增加
到 MeetHighElasticSearchServiceImpl 類別的 buildDeleteByQueryRequest 方法
中：

```
// 建置 DeleteByQueryRequest
  public void buildDeleteByQueryRequest (String indexName) {
    DeleteByQueryRequest  request = new DeleteByQueryRequest (indexName);
    /*
     * 設定 DeleteByQueryRequest
     */
    // 在預設情況下，版本衝突將中止 DeleteByQueryRequest 處理程序，但我們可以使用以下
    // 方法來計算它們
    request.setConflicts("proceed");
    // 透過增加查詢準則進行限制。例如僅刪除使用者欄位設定為 niudong 的文件
```

```
   request.setQuery(new TermQueryBuilder("user", "niudong"));
   // 設定大小，限制已處理文件的數量
   request.setSize(10);
   // 在預設情況下，UpdateByQueryRequest 使用的批數為 1000。可以使用 setBatchSize
   // 更改批大小
   request.setBatchSize(100);
   // 設定分片捲動，以實現平行化
   request.setSlices(2);
   // 使用捲動參數控制 " 搜索上下文 "，保持連接的時間
   request.setScroll(TimeValue.timeValueMinutes(10));
   // 如果提供路由，那麼路由將被複製到捲動查詢，進一步限制與該路由值符合的分片處理
   request.setRouting("=cat");
   // 設定等待請求的逾時
   request.setTimeout(TimeValue.timeValueMinutes(2));
   // 呼叫 update by query 後更新索引
   request.setRefresh(true);
   // 設定索引選項
   request.setIndicesOptions(IndicesOptions.LENIENT_EXPAND_OPEN);
}
```

❷ 執行文件查詢時刪除請求

在 DeleteByQueryRequest 建置後，即可執行文件查詢時刪除請求。與文件索引請求類似，文件查詢時刪除請求也有同步和非同步兩種執行方式。

▨ 同步方式

當以同步方式執行文件查詢時刪除請求時，用戶端會等待 Elasticsearch 伺服器傳回的查詢結果 DeleteByQueryResponse。在收到 DeleteByQueryResponse 後，用戶端會繼續執行相關的邏輯程式。以同步方式執行的程式如下所示：

```
// 以同步方式執行 DeleteByQueryRequest
  public void executeDeleteByQueryRequest(String indexName) {
    DeleteByQueryRequest request = new DeleteByQueryRequest(indexName);
    try {
      BulkByScrollResponse bulkResponse = restClient.deleteByQuery(request,
          RequestOptions.DEFAULT);
    } catch (Exception e) {
```

```
        e.printStackTrace();
    } finally {
    // 關閉 Elasticsearch 連接
    closeEs();
    }
}
```

非同步方式

當以非同步方式執行 DeleteByQueryRequest 時，進階用戶端不必同步等待請求結果的傳回，可以直接向介面呼叫方傳回非同步介面執行成功的結果。

為了處理非同步傳回的回應資訊或處理在請求執行過程中引發的例外資訊，使用者需要指定監聽器。以非同步方式執行的程式如下所示：

```
// 以非同步方式執行 DeleteByQueryRequest
  public void executeDeleteByQueryRequestAsync(String indexName) {
    DeleteByQueryRequest request = new DeleteByQueryRequest(indexName);
    // 建置監聽器
    ActionListener listener = new ActionListener<BulkByScrollResponse>() {
      @Override
      public void onResponse(BulkByScrollResponse bulkResponse) {
      }
      @Override
      public void onFailure(Exception e) {
      }
    };
    // 執行 DeleteByQuery
    try {
      restClient.deleteByQueryAsync(request, RequestOptions.DEFAULT,
listener);
    } catch (Exception e) {
      e.printStackTrace();
    } finally {
      // 關閉 Elasticsearch 連接
      closeEs();
    }
  }
```

在非同步請求處理後，如果請求執行成功，則呼叫 ActionListener 類別中的 onResponse 方法進行相關邏輯的處理；如果請求執行失敗，則呼叫 ActionListener 類別中的 onFailure 方法進行相關邏輯的處理。

當然，在非同步請求執行過程中可能會出現例外，例外的處理與同步方式執行情況相同。

3 解析文件查詢時刪除請求的回應結果

不論同步方式，還是非同步方式，用戶端均需要對文件查詢時刪除請求的回應結果 DeleteByQueryResponse 進行處理和解析。DeleteByQueryResponse 中包含有關已執行操作的資訊，並允許使用者按序對每個結果進行反覆運算解析。

以同步方式執行文件查詢時刪除請求為例，回應結果 DeleteByQueryResponse 的解析範例程式如下所示。程式共分為三層，分別是 Controller 層、Service 層和 ServiceImpl 實現層。

以同步方式執行的程式如下所示：

在 Controller 層的 MeetHighElasticSearchController 類別中增加以下程式：

```
// 以同步方式執行 DeleteByQueryRequest
@RequestMapping("/index/deleteByQuery")
public String deleteByQueryInHighElasticSearch(String indexName) {
  if (Strings.isEmpty(indexName)) {
    return "Parameters are error!";
  }
  meetHighElasticSearchService.executeDeleteByQueryRequest(indexName);
  return "DeleteByQuery In High ElasticSearch Client Successed!";
}
```

在 Service 層的 MeetHighElasticSearchService 類別中增加以下程式：

```
// 以同步方式執行 DeleteByQueryRequest
public void executeDeleteByQueryRequest(String indexName);
```

在 ServiceImpl 實現層的 MeetHighElasticSearchServiceImpl 類別中增加以下程式：

```
// 以同步方式執行 DeleteByQueryRequest
  public void executeDeleteByQueryRequest(String indexName) {
    DeleteByQueryRequest request = new DeleteByQueryRequest(indexName);
    // 透過增加查詢準則進行限制。舉例來說，僅刪除欄位 content 設定為 niudong 的文件
    request.setQuery(new TermQueryBuilder("content", "niudong"));
    try {
      BulkByScrollResponse bulkResponse = restClient.deleteByQuery(request,
          RequestOptions.DEFAULT);
      // 處理 BulkByScrollResponse
      processBulkByScrollResponse(bulkResponse);
    } catch (Exception e) {
      e.printStackTrace();
    } finally {
      // 關閉 Elasticsearch 連接
      closeEs();
    }
  }
```

隨後編輯專案，在專案根目錄下輸入以下指令：

```
mvn clean package
```

透過以下指令啟動專案服務：

```
java -jar ./target/esdemo-0.0.1-SNAPSHOT.jar
```

在專案服務啟動後，在瀏覽器中呼叫以下介面檢視文件查詢時刪除請求的執行情況：

```
http://localhost:8080/springboot/es/high/index/deleteByQuery?indexName=
ultraman1
```

在介面執行後，如果瀏覽器中顯示內容如下所示，則表示介面執行成功：

```
DeleteByQuery In High ElasticSearch Client Successed!
```

此時，在伺服器中輸出內容如下所示：

```
2019-08-02 16:01:07.911  INFO 82916 --- [nio-8080-exec-1] n.e.s.i.MeetHighElasticSearchServiceImpl : time is 51
2019-08-02 16:01:07.913  INFO 82916 --- [nio-8080-exec-1] n.e.s.i.MeetHighElasticSearchServiceImpl : timedOut is false
2019-08-02 16:01:07.918  INFO 82916 --- [nio-8080-exec-1] n.e.s.i.MeetHighElasticSearchServiceImpl : totalDocs is 0
2019-08-02 16:01:07.920  INFO 82916 --- [nio-8080-exec-1] n.e.s.i.MeetHighElasticSearchServiceImpl : updatedDocs is 0
2019-08-02 16:01:07.923  INFO 82916 --- [nio-8080-exec-1] n.e.s.i.MeetHighElasticSearchServiceImpl : createdDocs is 0
2019-08-02 16:01:07.924  INFO 82916 --- [nio-8080-exec-1] n.e.s.i.MeetHighElasticSearchServiceImpl : deletedDocs is 0
2019-08-02 16:01:07.926  INFO 82916 --- [nio-8080-exec-1] n.e.s.i.MeetHighElasticSearchServiceImpl : batches is 0
2019-08-02 16:01:07.932  INFO 82916 --- [nio-8080-exec-1] n.e.s.i.MeetHighElasticSearchServiceImpl : noops is 0
2019-08-02 16:01:07.936  INFO 82916 --- [nio-8080-exec-1] n.e.s.i.MeetHighElasticSearchServiceImpl : versionConflicts is 0
2019-08-02 16:01:07.940  INFO 82916 --- [nio-8080-exec-1] n.e.s.i.MeetHighElasticSearchServiceImpl : bulkRetries is 0
2019-08-02 16:01:07.944  INFO 82916 --- [nio-8080-exec-1] n.e.s.i.MeetHighElasticSearchServiceImpl : searchRetries is 0
2019-08-02 16:01:07.954  INFO 82916 --- [nio-8080-exec-1] n.e.s.i.MeetHighElasticSearchServiceImpl : throttledMillis is 0
2019-08-02 16:01:07.960  INFO 82916 --- [nio-8080-exec-1] n.e.s.i.MeetHighElasticSearchServiceImpl : searchFailures is 0
2019-08-02 16:01:07.962  INFO 82916 --- [nio-8080-exec-1] n.e.s.i.MeetHighElasticSearchServiceImpl : bulkFailures is 0
```

由於沒有 content 為 niudong 的文件，所以查詢時刪除的文件為 0。

此時，我們執行文件編號為 6、7、8 的三個文件查詢，驗證索引名稱為 ultraman1 的索引中僅有的這三個文件是否依然存在，可以透過以下介面驗證：

```
http://localhost:8080/springboot/es/high/index/multiget?indexName=
ultraman1&documentId=6,7,8
```

在請求執行後，我們可以在伺服器中看到如下所示的內容輸出，表示文件編號為 6、7、8 的三個文件依然存在：

```
2019-08-02 16:04:40.120  INFO 82568 --- [nio-8080-exec-1] n.e.s.i.MeetHighElasticSearchServiceImpl : responses is 3
2019-08-02 16:04:40.123  INFO 82568 --- [nio-8080-exec-1] n.e.s.i.MeetHighElasticSearchServiceImpl : index is ultraman1;
id is 6
2019-08-02 16:04:40.137  INFO 82568 --- [nio-8080-exec-1] n.e.s.i.MeetHighElasticSearchServiceImpl : index is ultraman1;
id is 7
2019-08-02 16:04:40.143  INFO 82568 --- [nio-8080-exec-1] n.e.s.i.MeetHighElasticSearchServiceImpl : index is ultraman1;
id is 8
```

6.7 取得文件索引的多詞向量

與其他介面類別似，詞向量介面也有批次實現的方式，即多詞向量介面。多詞向量介面允許使用者一次取得多個詞向量資訊。

1 建置多詞向量請求

在執行多詞向量請求前，我們需要建置多詞向量請求，即 MultitemVectorsRequest。Elasticsearch 提供了兩種建置方法，分別是：

（1）建立一個空的 MultiTermVectorsRequest，然後在其增加單一 Term Vectors
請求。
（2）在所有詞向量請求共用相同參數（如索引和其他設定）時，可以使用所
有必要的設定建立範本 TermVectorsRequest，並且可以將此範本請求連同
執行這些請求的所有文件 ID，傳遞給 MultitemVectorsRequest 物件。

兩種建置多終端向量請求的程式如下所示：

```
// 建置 MultiTermVectorsRequest
public void buildMultiTermVectorsRequest(String indexName, String[]
    documentIds, String field) {
// 方法 1：建立一個空的 MultiTermVectorsRequest，在其增加單一 Term Vectors 請求
MultiTermVectorsRequest request = new MultiTermVectorsRequest();
for (String documentId : documentIds) {
  TermVectorsRequest tvrequest = new TermVectorsRequest(indexName,
      documentId);
  tvrequest.setFields(field);
  request.add(tvrequest);
}

// 方法 2：所有詞向量請求共用相同參數（如索引和其他設定）
TermVectorsRequest tvrequestTemplate = new TermVectorsRequest(indexName,
    "1");
tvrequestTemplate.setFields(field);
String[] ids = {"1", "2"};
request = new MultiTermVectorsRequest(ids, tvrequestTemplate);
}
```

❷ 執行多詞向量請求

在 MultitemVectorsRequest 建置後，即可執行多詞向量請求。與文件索引請求
類似，多詞向量請求也有同步和非同步兩種執行方式。

▨ 同步方式

當以同步方式執行多詞向量請求時，用戶端會等待 Elasticsearch 伺服器傳回的
查詢結果 MultiTermVectorsResponse。在收到 MultiTermVectorsResponse 後，

用戶端會繼續執行相關的邏輯程式。以同步方式執行的程式如下所示：

```
// 以同步方式執行 MultiTermVectorsRequest
  public void executeMultiTermVectorsRequest(String indexName, String[]
 documentIds, String field) {
    // 方法 1：建立一個空的 MultiTermVectorsRequest，在其增加單一 Term Vectors 請求
    MultiTermVectorsRequest request = new MultiTermVectorsRequest();
    for (String documentId : documentIds) {
      TermVectorsRequest tvrequest = new TermVectorsRequest(indexName,
          documentId);
      tvrequest.setFields(field);
      request.add(tvrequest);
    }
    try {
      MultiTermVectorsResponse response = restClient.mtermvectors(request,
RequestOptions.DEFAULT);
    } catch (Exception e) {
      e.printStackTrace();
    } finally {
      // 關閉 Elasticsearch 連接
      closeEs();
    }
  }
```

✎ 非同步方式

當以非同步方式執行多詞向量請求時，進階用戶端不必同步等待請求結果的傳回，可以直接向介面呼叫方傳回非同步介面執行成功的結果。

為了處理非同步傳回的回應資訊或處理在請求執行過程中引發的例外資訊，使用者需要指定監聽器。以非同步方式執行的程式如下所示：

```
// 以非同步方式執行 MultiTermVectorsRequest
  public void executeMultiTermVectorsRequestAsync(String indexName, String[]
      documentIds,
    String field) {
    // 方法 1：建立一個空的 MultiTermVectorsRequest，在其增加單一 Term Vectors 請求
    MultiTermVectorsRequest request = new MultiTermVectorsRequest();
```

```
for (String documentId : documentIds) {
  TermVectorsRequest tvrequest = new TermVectorsRequest(indexName,
      documentId);
  tvrequest.setFields(field);
  request.add(tvrequest);
}
// 建置監聽器
ActionListener listener = new ActionListener<MultiTermVectorsResponse>() {
  @Override
  public void onResponse(MultiTermVectorsResponse mtvResponse) {
  }
  @Override
  public void onFailure(Exception e) {
  }
};
try {
  restClient.mtermvectorsAsync(request, RequestOptions.DEFAULT, listener);
} catch (Exception e) {
  e.printStackTrace();
} finally {
  // 關閉 Elasticsearch 連接
  closeEs();
}
}
```

在非同步請求處理後，如果請求執行成功，則呼叫 ActionListener 類別
中的 onResponse 方法進行相關邏輯的處理；如果請求執行失敗，則呼叫
ActionListener 類別中的 onFailure 方法進行相關邏輯的處理。

當然，在非同步請求執行過程中可能會出現例外，例外的處理與同步方式執
行情況相同。

❸ 解析多詞向量請求的回應結果

不論同步方式，還是非同步方式，用戶端均需要對多詞向量請求的查詢結果
MultiTermVectorsResponse 進行處理和解析。MultitermVectorsResponse 允許使

用者取得術語向量回應列表，每個回應都可以按照詞向量 API 中的描述進行檢查。

下面以同步方式執行多詞向量請求為例，展示傳回結果 MultiTermVectorsResponse 的解析方法。解析 MultiTermVectorsResponse 的程式。程式共分為三層，分別是 Controller 層、Service 層和 ServiceImpl 實現層。

其中，在 Controller 層的 MeetHighElasticSearchController 類別中增加以下程式：

```
// 以同步方式執行 MultiTermVectorsRequest
@RequestMapping("/index/multiterm")
public String multitermInHighElasticSearch(String indexName, String
    documentId, String field) {
  if (Strings.isEmpty(indexName) || Strings.isEmpty(documentId) ||
    Strings.isEmpty(field)) {
    return "Parameters are error!";
  }
  // 將 field( 英文逗點分隔的 ) 轉化成 String[]
  List<String> documentIds = Splitter.on(",").splitToList(documentId);
  meetHighElasticSearchService.executeMultiTermVectorsRequest(indexName,
      documentIds.toArray(new String[documentIds.size()]), field);
  return "MultiTermVectorsRequest In High ElasticSearch Client Successed!";
}
```

在 Service 層的 MeetHighElasticSearchService 類別中增加以下程式：

```
// 以同步方式執行 MultiTermVectorsRequest
  public void executeMultiTermVectorsRequest(String indexName, String[]
documentIds, String field);
```

在 ServiceImpl 實現層的 MeetHighElasticSearchServiceImpl 類別中增加以下程式：

```
// 以同步方式執行 MultiTermVectorsRequest
  public void executeMultiTermVectorsRequest(String indexName, String[]
      documentIds, String field) {
    // 方法 1：建立一個空的 MultiTermVectorsRequest，在其增加單一 term vectors 請求
```

```
MultiTermVectorsRequest request = new MultiTermVectorsRequest();
for (String documentId : documentIds) {
  TermVectorsRequest tvrequest = new TermVectorsRequest(indexName,
      documentId);
  tvrequest.setFields(field);
  request.add(tvrequest);
}
try {
  MultiTermVectorsResponse response = restClient.mtermvectors(request,
      RequestOptions.DEFAULT);
  // 解析 MultiTermVectorsResponse
  processMultiTermVectorsResponse(response);
} catch (Exception e) {
  e.printStackTrace();
} finally {
  // 關閉 Elasticsearch 連接
  closeEs();
  }
}
// 解析 MultiTermVectorsResponse
private void processMultiTermVectorsResponse(MultiTermVectorsResponse
    response) {
  if (response == null) {
    return;
  }
  List<TermVectorsResponse> tvresponseList = response.
      getTermVectorsResponses();
  if (tvresponseList == null) {
    return;
  }
  log.info("tvresponseList size is " + tvresponseList.size());
  for (TermVectorsResponse tvresponse : tvresponseList) {
    String id = tvresponse.getId();
    String index = tvresponse.getIndex();
    log.info("id size is " + id + "; index is " + index);
  }
}
```

隨後編輯專案，在專案根目錄下輸入以下指令：

```
mvn clean package
```

透過以下指令啟動專案服務：

```
java -jar ./target/esdemo-0.0.1-SNAPSHOT.jar
```

在專案服務啟動後，在瀏覽器中呼叫以下介面取得多詞向量請求的執行情況：

```
http://localhost:8080/springboot/es/high/index/multiterm?indexName=
ultraman1&documentId=1&field=content
```

此時，在伺服器中輸出的內容如下所示：

```
2019-08-02 16:28:00.535  INFO 81804 --- [io-8080-exec-10] n.e.s.i.MeetHighElasticSearchServiceImpl : tvresponseList size
is 1
2019-08-02 16:28:00.540  INFO 81804 --- [io-8080-exec-10] n.e.s.i.MeetHighElasticSearchServiceImpl : id size is 1; index
is ultraman1
```

Elasticsearch 伺服器傳回了一個命中的文件。

如果在瀏覽器中輸出以下內容，則表示介面執行成功：

```
MultiTermVectorsRequest In High ElasticSearch Client Successed!
```

6.8 文件處理過程解析

下面介紹在 Elasticsearch 內部是如何將文件分片儲存的。

6.8.1 Elasticsearch 文件分片儲存

一個索引一般由多個分片群組成，當使用者執行增加、刪除、修改文件操作時，Elasticsearch 需要決定把這個文件儲存在哪個分片，這個過程就稱為資料路由。

當把文件路由到分片時需要使用路由演算法，Elasticsearch 中的路由演算法如下所示：

```
shard = hash(routing) % number_of_primary_shards
```

下面透過範例展示文件路由到分片的過程。

假設某個索引由 3 個主分片組成，使用者每次對文件進行增刪改查時，都有一個 routing 值，預設是該文件 ID 的值。隨後對這個 routing 值使用 Hash 函數進行計算，計算出的值再和主分片個數取餘數，餘數的設定值範圍永遠是（0 ～ number_of_primary_shards - 1）之間，文件知道應該儲存在哪個對應的分片。

需要指出的是，雖然 routing 值預設是文件 ID 的值，但 Elasticsearch 也支援使用者手動指定一個值。手動指定對於負載平衡及提升批次讀取的效能有一定的幫助。

正是 Elasticsearch 的這種路由機制，主分片的個數在索引建立之後不能修改。因為修改索引主分片數目會直接導致路由規則出現嚴重問題，部分資料將無法被檢索。

那麼讀者每次對文件進行增刪改查時，主分片和備份分片是如何運行的呢？

假設讀者本機有三個節點的叢集，該叢集包含一個名叫作 niudong 的索引，並擁有兩個主分片，每個主分片有兩個備份分片。

在 Elasticsearch 中，出於資料安全和災難恢復等因素考慮，相同的分片不會放在同一個節點上，所以此時的叢集如圖 6-1 所示。

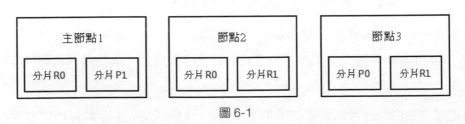

圖 6-1

在圖 6-1 中，有一個標記為主節點的節點，也稱之為請求節點。一般來說，使用者能夠發送請求給叢集中的任意一個節點。每個節點都有能力處理使用者提交的任意請求，每個節點都知道任意文件所在的節點，所以也可以將請求轉發到需要的節點。

當使用者執行新增索引、更新和刪除請求等寫入操作時，文件必須在主分片成功完成請求，才能複製到相關的備份分片，分片資料同步過程如圖 6-2 所示。

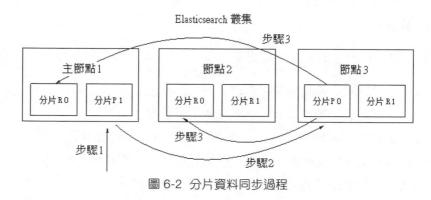

圖 6-2 分片資料同步過程

用戶端發送了一個索引或刪除的請求給主節點 1。此時，主節點會進行以下處理：

（1）主節點 1 透過請求中文件的 ID 值判斷出該文件應該被儲存在哪個分片，如現在已經判斷出需要分片編號為 0 的分片在節點 3 中，則主節點 1 會把這個請求轉發到節點 3 處。

（2）節點 3 在分片編號為 0 的主分片執行請求。如果請求執行成功，則節點 3 將平行地將該請求發給分片編號為 0 的所有備份上，即圖 6-2 中位於主節點 1 和節點 2 中的備份分片。

　　如果所有的備份分片都成功地執行了請求，那麼將向節點 3 回覆一個請求執行成功的確認訊息。當節點 3 收到所有備份節點的確認資訊後，會向用戶端傳回一個成功的回應訊息。

（3）當用戶端收到成功的回應資訊時，文件的操作就已經被應用於主分片和所有的備份分片，此時操作就會生效。

6.8.2 Elasticsearch 的資料分區

一般來説，搜尋引擎有兩種資料分區方式，即以文件為基礎的分區方式和以詞條為基礎的分區方式。Elasticsearch 使用的是以文件為基礎的分區方式。

以文件為基礎的分區（Document Based Partitioning）指的是每個文件只存一個分區，每個分區持有整個文件集的子集。這裡説的分區是指一個功能完整的倒排索引。

以文件為基礎的分區的優點整理如下：

（1）每個分區都可以獨立地處理查詢。

（2）可以非常方便地增加以文件為單位的索引資訊。

（3）在搜索過程中網路負擔很小，每個節點可以分別獨立地執行搜索，執行完之後只需傳迴文件的 ID 和評分資訊即可。而呈現給使用者的結果集是在執行分散式搜索的節點上執行合併操作實現的。

以文件為基礎的分區的缺點也很明顯，如果查詢需要在所有的分區上執行，則它將執行 $O(K \times N)$ 次磁碟操作（K 是詞條 term 的數量，N 是分區的數量）。

從實用性角度來看，以文件為基礎的分區方式已經被證明是一個建置大型的分散式資訊檢索系統的行之有效的方法，因此 Elasticsearch 使用的是以文件為基礎的分區方式。

以詞條為基礎的分區（Term Based Partitioning）指的是每個分區擁有一部斷詞條，詞條裡面包含了與該詞條相關的整個 index 的文件資料。

目前也有一些以詞條分區為基礎的搜尋引擎系統，如 Riak Search、Lucandra 和 Solandra。

以詞條為基礎的分區的優點整理如下：

（1）部分分區執行查詢。一般來説，使用者只需要在很小的部分分區上執行查詢就可以了。舉個實例，假如使用者有 3 個 term 詞條的查詢，則 Elasticsearch 在搜索時將至多命中 3 個分區。如果夠幸運，這 3 個 term

詞條都儲存在同一個分區中，那麼使用者只需存取一個分區即可。在搜索的背後，使用者無須知道 Elasticsearch 中實際的分區數量。

（2）時間複雜度低。一般而言，當對應 K 個 term 詞條的查詢時，使用者只需執行 $O(K)$ 次磁碟尋找即可。

以詞條為基礎的分區的缺點整理如下：

（1）最主要的缺點是將失去 Lucene Segment 概念裡面很多固有的結構。
而對於比較複雜的查詢，搜索過程中的網路負擔將變得非常高，並且可能使得系統可用性大幅降低，特別是植入前置搜索或模糊搜索的場景。

（2）取得每個文件的資訊將變得非常困難。由於是按詞條分區儲存的，如果使用者想取得文件的一部分資料做進一步的控制，或取得每個文件的這些資料，都將變得非常困難，因為這種分區方式使得文件的資料被分散到了不同的地方，所以實現植入評分、自訂評分等都將變得難以實現。

6.9 基礎知識連結

如前文所述，Elasticsearch 使用了樂觀鎖來解決資料一致性問題。當使用者對文件操作時，並不需要對文件作加鎖、解鎖的操作，只需指定要操作的版本即可。當版本編號一致時，Elasticsearch 會允許該操作順利進行；當版本編號衝突時，Elasticsearch 會提示衝突並拋出例外 VersionConflictEngineException。在 Elasticsearch 中，文件的版本編號的設定值範圍為 1 到 2^{63}-1。

其實，樂觀鎖不僅是一種鎖的類型，更是一種設計思想，這種設計思想在軟體開發過程中十分常見。

這裡先簡介一下樂觀鎖。在樂觀鎖的思想中，會認為資料一般不會引發衝突，因此在資料更新時，才會檢測是否存在資料衝突。在檢測時，如果發現資料衝突，則傳回衝突結果，以便讀者自主決定如何去做。

和樂觀鎖對應的是悲觀鎖，在悲觀鎖的思想中，會認為資料一般會引發衝突。也就是說，在讀取資料時寫資料操作常常也正在進行，因此在讀取資料前需要上鎖，沒有拿到鎖的讀者或處理程序只能等待鎖的釋放。悲觀鎖在關聯式資料庫中有大量應用，我們常見的行鎖、表鎖、讀取鎖、寫入鎖等都是悲觀鎖。

樂觀鎖除在 Elasticsearch 中有應用外，在關聯式資料庫和 Java 中都有應用。

在關聯式資料庫中，如果某張表對應的應用場景是讀多寫少的場景，則可以使用樂觀鎖。採用樂觀鎖控制資料庫表後，表中會新增一列欄位，一般稱之為 version，version 用於記錄行資料的版本。當資料初始寫入時，版本預設為 1；每當對資料有修改操作時，version 都會加 1。在修改操作過程中，資料庫會比較資料地區的 version 和目前資料庫的 version，如果相同，則新資料才會寫入。

在 Java 中，樂觀鎖思想的實現就是 CAS 技術，即 Compare and Swap。在 JDK 1.5 中新增的 java.util.concurrent 就是建立在 CAS 之上的，開發人員常用的 AtomicInteger 也是其中之一。

在 Java 中，當多個執行緒嘗試使用 CAS 同時更新同一個變數時，只有一個執行緒能成功更新變數的值，而其他執行緒都會失敗。失敗的執行緒並不會被暫停，而是被告知失敗，並可再次嘗試。

在實作方式上，CAS 操作包含三個運算元，即記憶體位置（V）、預期原值（A）和新值 (B)。如果記憶體位置的值與預期原值相同，則處理器會自動將該位置值更新為新值；不然處理器不做任何操作。

6.10 小結

本章主要介紹了文件進階 API 的使用，均為批次操作介面，如文件批次請求介面、批次處理器的使用、MultiGet 批次處理方法的使用、文件 ReIndex 介面的使用、文件查詢時更新介面的使用、文件查詢時刪除介面的使用和取得文件索引多詞向量的介面的使用。

搜索實戰

前面介紹了在進階用戶端中對文件的各種操作，整體來說，這些介面都是圍繞著文件索引展開的，下面我們將圍繞文件搜索這個核心過程展開。

在 Elasticsearch 中，進階用戶端支援以下搜索 API：

（1）Search API

（2）Search Scroll API

（3）Clear Scroll API

（4）Search Template API

（5）Multi-Search-Template API

（6）Multi-Search API

（7）Field Capabilities API

（8）Ranking Evaluation API

（9）Explain API

（10）Count API

這些搜索 API 允許讀者執行搜索查詢並傳回比對查詢的搜索命中結果，它們可以跨一個或多個索引，以及跨一個或多個類型來執行。

7.1 搜索 API

搜索 API，即 Search API，允許使用者執行一個搜索查詢並傳回與查詢符合的搜索點擊，使用者可以使用簡單的查詢字串作為參數或使用請求主體提供查詢。

最基本的搜索 API 是空搜索，空搜索不指定任何的查詢準則，只傳回叢集索引中的所有文件。在實際開發中，基本不會用到空搜索。

在常見的搜索請求發出前，使用者需要建置搜索請求，即 SearchRequest。

❶ 建置搜索請求

SearchRequest 可用於與搜索文件、聚合、Suggest 有關的任何操作，另外還提供了請求反白顯示結果文件的方法。

SearchRequest 的建置需要依賴 SearchSourceBuilder。在基本查詢中，使用者需要建置查詢 Query，並把它增加到查詢 Request 中。SearchRequest 的建置程式如下所示：

```
// 建置 SearchRequest
  public void buildSearchRequest() {
    SearchRequest searchRequest = new SearchRequest();

    // 大多數搜索參數都增加到 SearchSourceBuilder 中，它為進入搜索請求主體的所有內容
    // 提供 setter
    SearchSourceBuilder searchSourceBuilder = new SearchSourceBuilder();

    // 在 searchSourceBuilder 中增加 " 全部比對 " 查詢
    searchSourceBuilder.query(QueryBuilders.matchAllQuery());

    // 將 searchSourceBuilder 增加到 searchRequest 中
    searchRequest.source(searchSourceBuilder);
  }
```

同理，SearchRequest 也有可選參數供使用者進行設定。SearchRequest 的可選
參數有限制請求類型、設定路由參數、設定 IndicesOptions 和使用首選參數等。

在 SearchRequest 中設定可選參數的程式如下所示：

```
// 建置 SearchRequest
  public void buildSearchRequest() {
    SearchRequest searchRequest = new SearchRequest();
    // 大多數搜索參數都增加到 SearchSourceBuilder 中，它為進入搜索請求主體的所有內容
    // 提供 setter
    SearchSourceBuilder searchSourceBuilder = new SearchSourceBuilder();
    // 在 searchSourceBuilder 中增加 " 全部比對 " 查詢
    searchSourceBuilder.query(QueryBuilders.matchAllQuery());
    // 將 searchSourceBuilder 增加到 serchRequest 中
    searchRequest.source(searchSourceBuilder);
    /*
     * 可選參數設定
     */
    // 在索引上限制請求
    searchRequest = new SearchRequest("posts");

    // 設定路由參數
    searchRequest.routing("routing");

    // 設定 IndicesOptions 控制方法
    searchRequest.indicesOptions(IndicesOptions.lenientExpandOpen());

    // 使用首選參數，舉例來說，執行搜索以首選本地分片。預設值是在分片之間隨機的
    searchRequest.preference("_local");
  }
```

✍ 使用 SearchSourceBuilder

除 SearchRequest 的 建 置 需 要 依 賴 SearchSourceBuilder 外，事 實 上，大
多 數 控 制 搜 索 行 為 的 選 項 都 可 以 在 SearchSourceBuilder 上 進 行 設 定。
SearchSourceBuilder 包含了與 REST API 搜索請求正文中相同的選項。

SearchSourceBuilder 常用的搜索選項有設定查詢準則、設定搜索結果索引的起始位址（預設為 0）、設定要傳回的搜索命中數的大小（預設為 10）、設定一個可選的逾時（以便控制允許搜索的時間）等。

在 SearchSourceBuilder 中設定常見搜索選項的程式如下所示：

```
/*
 * 使用 SearchSourceBuilder
 */
    SearchSourceBuilder sourceBuilder = new SearchSourceBuilder();
    // 設定查詢準則
    sourceBuilder.query(QueryBuilders.termQuery("content", " 貨幣 "));
    // 設定搜索結果索引的起始位址，預設為 0
    sourceBuilder.from(0);
    // 設定要傳回的搜索命中數的大小，預設為 10
    sourceBuilder.size(5);
    // 設定一個可選的逾時，控制允許搜索的時間
    sourceBuilder.timeout(new TimeValue(60, TimeUnit.SECONDS));

    // 將 searchSourceBuilder 增加到 searchRequest 中
    searchRequest.source(sourceBuilder);
```

▨ 產生查詢

搜索查詢是以 QueryBuilder 建立為基礎。在 Elasticsearch 中，每個搜索查詢都需要用到 QueryBuilder。QueryBuilder 是以 Elasticsearch DSL 實現為基礎。

QueryBuilder 既可以使用其建置函數來建立，也可以使用 QueryBuilders 工具類別來建立。QueryBuilders 工具類別提供了流式程式設計格式來建立 QueryBuilder。

QueryBuilder 的建置程式如下所示，我們在 buildSearchRequest 方法中增加以下程式：

```
/*
 * 搜索查詢 MatchQueryBuilder 的使用
 */
```

```
// 方法 1
    MatchQueryBuilder matchQueryBuilder = new MatchQueryBuilder("content",
        "貨幣");
    // 建立 QueryBuilder，提供設定搜索查詢選項的方法
    // 比較對查詢啟用模糊比對
    matchQueryBuilder.fuzziness(Fuzziness.AUTO);

    // 在比對查詢上設定字首長度
    matchQueryBuilder.prefixLength(3);

    // 設定最大擴充選項以控制查詢的模糊過程
    matchQueryBuilder.maxExpansions(10);
// 方法 2
    matchQueryBuilder = QueryBuilders.matchQuery("content", "貨幣").fuzziness
(Fuzziness.AUTO).prefixLength(3).maxExpansions(10);
```

上述程式展示了建置函數的方式，以及如何使用 QueryBuilders 工具類別建立
QueryBuilder。在 QueryBuilder 中以 content 欄位中的貨幣為查詢物件。

無論用於建立 QueryBuilder 的方法是什麼，使用者都必須將 matchQueryBuilder
增加到 searchSourceBuilder 中，程式如下所示：

```
// 增加 matchQueryBuilder 到 searchSourceBuilder 中
searchSourceBuilder.query(matchQueryBuilder);
```

▨ 設定排序策略

在建置搜索請求時，Elasticsearch 還支援使用者設定搜索結果的排序策略。一
般我們在使用搜尋引擎時，對搜索結果常見的排序策略有按時間排序和按相
關性排序兩種。按時間排序的搜索結果如圖 7-1 所示。

SearchSourceBuilder 允許使用者增加一個或多個 SortBuilder 實例。有 4 種特殊
的實現類別，分別是 FieldSortBuilder、ScoreSortBuilder、GeoDistanceSortBuilder
和 ScriptSortBuilder。

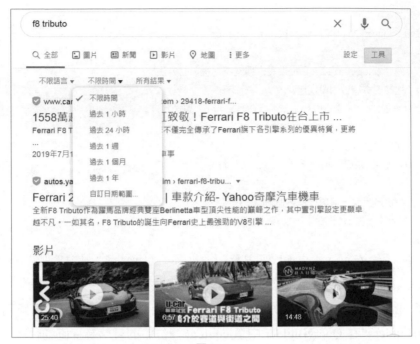

圖 7-1

在 SearchSourceBuilder 中設定 SortBuilder 實例的程式如下所示,程式增加在前文中使用的 buildSearchRequest 方法中:

```
/*
 * 指定排序方法
 */
    // 按分數降冪排序(預設)
    sourceBuilder.sort(new ScoreSortBuilder().order(SortOrder.DESC));
    // 按 ID 欄位昇冪排序
    sourceBuilder.sort(new FieldSortBuilder("_id").order(SortOrder.ASC));
```

☑ 來源篩選

在預設情況下,搜索請求一般會傳回文件來源的內容。不過,與 REST API 一樣,使用者可以覆蓋此行為,即使用者可以完全關閉來源檢索。

在 SearchSourceBuilder 實例中設定來源搜索開關的程式如下所示,程式增加在前文中的 buildSearchRequest 方法中:

```
/*
 * 來源篩選的方法使用
 */
    sourceBuilder.fetchSource(false);
```

該方法還接受一個或多個萬用字元模式的陣列，以更細粒度的方式控制哪些欄位被包含或被排除。相關程式如下所示，程式增加在前文中使用的 buildSearchRequest 方法中：

```
/*
 * 來源篩選的方法使用
 */
    sourceBuilder.fetchSource(false);

    // 該方法還接受一個或多個萬用字元模式的陣列，以更細的粒度控制哪些欄位被包含或排除
    String[] includeFields = new String[] {"title", "innerObject.*"};
    String[] excludeFields = new String[] {"user"};
sourceBuilder.fetchSource(includeFields, excludeFields);
```

▨ 請求反白顯示

在搜索查詢請求中，使用者還可以設定請求的反白顯示。

使用者可以在 SearchSourceBuilder 上設定 HighlightBuilder 來達到反白顯示搜索結果的目的。

透過在 HighlightBuilder 中增加一個或多個 HighlightBuilder.Field 實例，就可以為每個欄位定義不同的反白顯示行為。

相關程式如下所示，程式增加在前文中使用的 buildSearchRequest 方法中：

```
/*
 * 設定請求反白顯示
 */
HighlightBuilder highlightBuilder = new HighlightBuilder();
// 為 title 欄位建立欄位反白
HighlightBuilder.Field highlightTitle = new HighlightBuilder.Field("title");
// 設定欄位反白類型
```

```
highlightTitle.highlighterType("unified");
// 將 highlightTitle 增加到 highlightBuilder 中
highlightBuilder.field(highlightTitle);
// 增加第二個反白顯示欄位
HighlightBuilder.Field highlightUser = new HighlightBuilder.Field("user");
highlightBuilder.field(highlightUser);
searchSourceBuilder.highlighter(highlightBuilder);
```

▨ 請求聚合

在搜索查詢請求中，使用者還可以設定請求聚合結果。

在取得請求聚合前，需要先建立聚合建置元 AggregationBuilder，然後將 AggregationBuilder 增加到 SearchSourceBuilder 中。

在下面的程式中，我們建立了公司名稱的術語聚合，並對公司中員工的平均年齡進行了子聚合，相關程式如下所示。程式增加在前文中使用的 buildSearchRequest 方法中：

```
/*
 * 聚合請求的使用
 */
TermsAggregationBuilder aggregation =
        AggregationBuilders.terms("by_company").field("company.keyword");
    aggregation.subAggregation(AggregationBuilders.avg("average_age").field
("age"));
searchSourceBuilder.aggregation(aggregation);
```

▨ 建議請求

在搜索查詢請求中，使用者還可以設定請求結果。請求在搜尋引擎中非常常見，舉例來説，我們在搜索框中輸入 2019，則會列出 10 條請求提示結果，如圖 7-2 所示。

在 Elasticsearch 中，想要在搜索請求中增加請求，則需要使用 SuggestBuilder 工廠類別。SuggestBuilder 工廠類別是 SuggestionBuilder 類別的實現類別之一，它的特性是簡單好用。

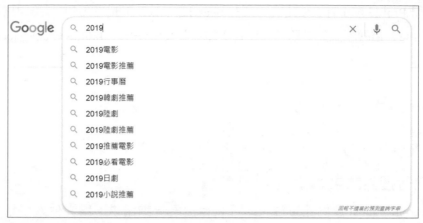

圖 7-2

SuggestBuilder 工 廠 類 別 需 要 增 加 到 頂 級 SuggestBuilder 中，並 將 頂 級 SuggestBuilder 增加到 SearchSourceBuilder 中。

相關程式如下所示，程式增加在前文中使用的 buildSearchRequest 方法中：

```
/*
 * Suggestions 建議請求的使用
 */
    // 在 TermSuggestionBuilder 中為 content 欄位增加貨幣的 Suggestions
    SuggestionBuilder termSuggestionBuilder =
        SuggestBuilders.termSuggestion("content").text(" 貨幣 ");
    SuggestBuilder suggestBuilder = new SuggestBuilder();

    // 增加建議產生器並命名
    suggestBuilder.addSuggestion("suggest_user", termSuggestionBuilder);

    // 將 suggestBuilder 增加到 searchSourceBuilder 中
    searchSourceBuilder.suggest(suggestBuilder);
```

☒ 分析查詢和聚合

Elasticsearch 從 2.2 版本開始提供 Profile API，以供使用者檢索、聚合、過濾執行時間和其他細節資訊，幫助讀者分析每次檢索各個環節所用的時間。

在使用 Profile API 時，使用者必須在 SearchSourceBuilder 實例中將設定標示設定為 true。相關程式如下所示，程式增加在前文中使用的 buildSearchRequest 方法中：

```
/*
 * 分析查詢和聚合 API 的使用
 */
    searchSourceBuilder.profile(true);
```

2 執行搜索請求

在 SearchRequest 實例建置後，即可執行搜索查詢請求。與文件索引請求類似，搜索的查詢請求也有同步和非同步兩種執行方式。

同步方式

當以同步方式執行搜索查詢請求時，用戶端會等待 Elasticsearch 伺服器傳回的查詢結果 SearchResponse。在收到 SearchResponse 後，用戶端會繼續執行相關的邏輯程式。以同步方式執行的程式如下所示：

```
// 參數化建置 SearchRequest
  public SearchRequest buildSearchRequest(String filed, String text) {
    SearchRequest searchRequest = new SearchRequest();
    // 大多數搜索參數都增加到 SearchSourceBuilder 中，它為進入搜索請求主體的所有內容
    // 提供 setter
    SearchSourceBuilder searchSourceBuilder = new SearchSourceBuilder();
    // 在 searchSourceBuilder 中增加 " 全部比對 " 查詢
    searchSourceBuilder.query(QueryBuilders.matchAllQuery());
    // 將 searchSourceBuilder 增加到 searchRequest 中
    searchRequest.source(searchSourceBuilder);
    /*
     * 使用 SearchSourceBuilder
     */
    // 設定查詢準則
    searchSourceBuilder.query(QueryBuilders.termQuery(filed, text));
    // 設定搜索結果索引的起始位址，預設為 0
    searchSourceBuilder.from(0);
    // 設定要傳回的搜索命中數的大小，預設為 10
```

```
  searchSourceBuilder.size(5);
  // 設定一個可選的逾時，控制允許搜索的時間
  searchSourceBuilder.timeout(new TimeValue(60, TimeUnit.SECONDS));
  // 將 SearchSourceBuilder 增加到 SearchRequest 中
  searchRequest.source(searchSourceBuilder);
  /*
   * 設定請求反白顯示
   */
  HighlightBuilder highlightBuilder = new HighlightBuilder();
  // 為 title 欄位建立欄位反白
  HighlightBuilder.Field highlightTitle = new HighlightBuilder.Field(filed);
  // 設定欄位反白類型
  highlightTitle.highlighterType("unified");
  // 將 highlightTitle 增加到 highlightBuilder 中
  highlightBuilder.field(highlightTitle);
  searchSourceBuilder.highlighter(highlightBuilder);
  /*
   * 建議請求的使用
   */
  // 在 TermSuggestionBuilder 中為 content 欄位增加貨幣的 Suggestions
  SuggestionBuilder termSuggestionBuilder = SuggestBuilders.termSuggestion
      (filed).text(text);
  SuggestBuilder suggestBuilder = new SuggestBuilder();
  // 增加建議產生器並命名
  suggestBuilder.addSuggestion("suggest_user", termSuggestionBuilder);
  // 將 suggestBuilder 增加到 searchSourceBuilder 中
  searchSourceBuilder.suggest(suggestBuilder);
  return searchRequest;
}
// 以同步方式執行 SearchRequest
public void executeSearchRequest() {
  SearchRequest searchRequest = buildSearchRequest("content", "貨幣");
  // 執行
  try {
    SearchResponse searchResponse = restClient.search(searchRequest,
        RequestOptions. DEFAULT);
    log.info(searchResponse.toString());
```

```
  } catch (Exception e) {
    e.printStackTrace();
  } finally {
    // 關閉 Elasticsearch 連接
    closeEs();
  }
}
```

在 Service 層的 SearchService 類別中增加以下程式：

```
// 以同步方式執行 SearchRequest
  public void executeSearchRequest();
```

☑ 非同步方式

當以非同步方式執行 SearchRequest 時，進階用戶端不必同步等待請求結果的傳回，可以直接向介面呼叫方傳回非同步介面執行成功的結果。

為了處理非同步傳回的回應資訊或處理在請求執行過程中引發的例外資訊，讀者需要指定監聽器。以非同步方式執行的程式如下所示，在 ServiceImpl 實現層的 SearchServiceImpl 類別中增加以下程式：

```
// 以非同步方式執行 SearchRequest
  public void executeSearchRequestAsync() {
    SearchRequest searchRequest = buildSearchRequest("content", " 貨幣 ");
    // 建置監聽器
    ActionListener<SearchResponse> listener = new ActionListener
        <SearchResponse>() {
      @Override
      public void onResponse(SearchResponse searchResponse) {
        log.info("response is " + searchResponse.toString());
      }
      @Override
      public void onFailure(Exception e) {
      }
    };
    // 執行
    try {
```

```
      restClient.searchAsync(searchRequest, RequestOptions.DEFAULT, listener);
   } catch (Exception e) {
      e.printStackTrace();
   } finally {
      // 關閉 Elasticsearch 連接
      closeEs();
   }
 }
```

在執行搜索請求後，如果執行成功，則呼叫 ActionListener 類別中的
onResponse 方法進行相關邏輯的處理；如果執行失敗，則呼叫 ActionListener
類別中的 onFailure 方法進行相關邏輯的處理。

當然，在非同步請求執行過程中可能會出現例外，例外的處理與同步方式執
行情況相同。

3 解析搜索請求的回應結果

不論同步方式，還是非同步方式，用戶端均需對搜索請求執行的傳回結果
SearchResponse 進行處理和解析。SearchResponse 中提供了有關搜索執行本身
及對傳迴文件存取的詳細資訊，如 HTTP 狀態碼、執行時間，或請求是提前
終止還是逾時等。

相關程式如下所示，程式增加在 SearchServiceImpl 類別中：

```
// 解析 SearchResponse
 private void processSearchResponse(SearchResponse searchResponse) {
   if (searchResponse == null) {
     return;
   }
   // 取得 HTTP 狀態碼
   RestStatus status = searchResponse.status();
   // 取得請求執行時間
   TimeValue took = searchResponse.getTook();
   // 取得請求是否提前終止
   Boolean terminatedEarly = searchResponse.isTerminatedEarly();
   // 取得請求是否逾時
```

```
   boolean timedOut = searchResponse.isTimedOut();
   log.info("status is " + status + ";took is " + took + ";terminatedEarly
        is " + terminatedEarly
      + ";timedOut is " + timedOut);
 }
```

☒ 搜索請求相關的分片

回應結果 SearchResponse 提供了有關搜索影響的分片總數，以及成功與失敗分片的統計資訊，並提供了有關分片等級執行的資訊。對於請求執行失敗的資訊，可以透過解析 ShardSearchFailure 實例陣列元素來處理。

該部分的程式如下所示，程式增加在 ServiceImpl 實現層 SearchServiceImpl 類別的 processSearchResponse 方法內：

```
// 檢視搜索影響的分片總數
   int totalShards = searchResponse.getTotalShards();
   // 搜索成功的分片的統計資訊
   int successfulShards = searchResponse.getSuccessfulShards();
   // 搜索失敗的分片的統計資訊
   int failedShards = searchResponse.getFailedShards();
   log.info("totalShards is " + totalShards + ";successfulShards is " +
        successfulShards
      + ";failedShards is " + failedShards);
   for (ShardSearchFailure failure : searchResponse.getShardFailures()) {
     log.info("fail is " + failure.toString());
   }
```

☒ 取得搜索結果

取得結果回應中包含的搜索結果 SearchHits，取得程式如下所示：

```
SearchHits hits = searchResponse.getHits();
```

SearchHits 中提供了有關所有搜索結果的全部資訊，如點擊總數或最高分數等，這些內容的取得程式如下所示：

```
TotalHits totalHits = hits.getTotalHits();
 // 搜索結果的總量數
```

```
long numHits = totalHits.value;
// 搜索結果的相關性資料
TotalHits.Relation relation = totalHits.relation;
float maxScore = hits.getMaxScore();
```

SearchHits 的 解 析 程 式 如 下 所 示 ， 程 式 增 加 在 ServiceImpl 實 現 層 SearchServiceImpl 類別中的 processSearchResponse 方法內：

```
// 取得回應中包含的搜索結果
    SearchHits hits = searchResponse.getHits();
    // SearchHits 提供了有關所有結果的全部資訊，如點擊總數或最高分數
    TotalHits totalHits = hits.getTotalHits();
    // 點擊總數
    long numHits = totalHits.value;
    // 最高分數
    float maxScore = hits.getMaxScore();
    log.info("numHits is " + numHits + ";maxScore is " + maxScore);
```

對 SearchHits 的解析還可以透過檢查 SearchHit 陣列實現，程式如下所示：

```
SearchHit[] searchHits = hits.getHits();
for (SearchHit hit : searchHits) {
        // 解析 SearchHit
}
```

SearchHits 提供了對文件基本資訊的存取，如索引名稱、文件 ID 和每次搜索的得分，文件基本資訊可以透過以下程式取得：

```
// 取得索引名稱
String index = hit.getIndex();
// 取得文件 ID
String id = hit.getId();
// 取得搜索的得分
float score = hit.getScore();
```

透過檢查 SearchHit 陣列來解析 SearchHits 的全部程式如下所示，程式增加在 ServiceImpl 實 現 層 SearchServiceImpl 類 別 中 的 processSearchResponse 方 法 內：

```
// 巢狀結構在 SearchHits 中的是可以反覆運算的單一搜索結果
    SearchHit[] searchHits = hits.getHits();
    for (SearchHit hit : searchHits) {
    // SearchHits 提供對基本資訊的存取，如索引、文件 ID 和每次搜索的得分
    String index = hit.getIndex();
    String id = hit.getId();
    float score = hit.getScore();
    log.info("docId is " + id + ";docIndex is " + index + ";docScore is " +
        score);
}
```

此外，Elasticsearch 允許以簡單的 JSON 字串或鍵值對形式傳回文件來源。在這個對映中，正常欄位由欄位名稱作為鍵值，包含欄位值；而多值欄位作為物件列表傳回；巢狀結構物件作為另一個鍵值傳回。這部分內容的程式如下所示，程式增加在 ServiceImpl 實現層 SearchServiceImpl 類別中的 processSearchResponse 方法內：

```
// 以 JSON 字串形式傳回文件來源
    String sourceAsString = hit.getSourceAsString();
    // 以鍵值對的形式傳回文件來源
    Map<String, Object> sourceAsMap = hit.getSourceAsMap();
    String documentTitle = (String) sourceAsMap.get("title");
    List<Object> users = (List<Object>) sourceAsMap.get("user");
    Map<String, Object> innerObject = (Map<String, Object>) sourceAsMap.get
        ("innerObject");
    log.info(
        "sourceAsString is " + sourceAsString + ";sourceAsMap size is " +
            sourceAsMap.size());
```

▨ 搜索結果反白顯示

如果對搜索結果展示有需求，則使用者可以從傳回結果中的每筆資料結果，即 SearchHits，自己取得搜索反白顯示的文字片段。

SearchHits 提供了對 HighlightField 實例中欄位名稱的對映存取，每個實例包含一個或多個反白顯示的文字片段。

程式如下所示，程式增加在 ServiceImpl 實現層 SearchServiceImpl 類別中的 processSearchResponse 方法內：

```
// 反白顯示
    Map<String, HighlightField> highlightFields = hit.getHighlightFields();
    HighlightField highlight = highlightFields.get("content");
    // 取得包含反白顯示欄位內容的或多個片段
    Text[] fragments = highlight.fragments();
    String fragmentString = fragments[0].string();
    log.info("fragmentString is " + fragmentString);
```

☑ 搜索聚合結果

使用者可以透過 SearchResponse 實例取得搜索的聚合結果，首先取得聚合樹的根，即聚合物件，然後按照名稱取得搜索聚合結果。

取得搜索聚合結果的程式如下所示，程式增加在 ServiceImpl 實現層 SearchServiceImpl 類別中的 processSearchResponse 方法內：

```
// 聚合搜索
    Aggregations aggregations = searchResponse.getAggregations();
    // 按 content 聚合
    Terms byCompanyAggregation = aggregations.get("by_content");
    // 取得 Elastic 為關鍵字的 buckets
    Bucket elasticBucket = byCompanyAggregation.getBucketByKey("Elastic");
    // 取得平均年齡的子聚合
    Avg averageAge = elasticBucket.getAggregations().get("average_age");
    double avg = averageAge.getValue();
    log.info("avg is " + avg);
```

☑ 解析 Suggestions 結果

首先在 SearchResponse 實例中使用 Suggest 物件作為進入點，然後檢索巢狀結構的 Suggest 物件。

對 Suggestions 結果的解析程式如下所示，程式增加在 ServiceImpl 實現層 SearchServiceImpl 類別中的 processSearchResponse 方法內：

```
// Suggest 搜索
  Suggest suggest = searchResponse.getSuggest();
  // 按 content 搜索 Suggest
  TermSuggestion termSuggestion = suggest.getSuggestion("content");
  for (TermSuggestion.Entry entry : termSuggestion.getEntries()) {
    for (TermSuggestion.Entry.Option option : entry) {
      String suggestText = option.getText().string();
      log.info("suggestText is " + suggestText);
    }
  }
}
```

☑ 搜索回應結果的解析綜合範例

前面我們分別介紹了搜索回應結果 SearchResponse 中包含的各種資訊的解析，下面透過一段程式綜合展示 SearchResponse 的解析。程式以同步方式執行搜索請求，共分為三層，分別是 Controller 層、Service 層和 ServiceImpl 實現層。

在 Controller 層中新增 SearchController 類別，程式如下所示：

```
package com.niudong.esdemo.controller;
import org.springframework.beans.factory.annotation.Autowired;
import org.springframework.web.bind.annotation.RequestMapping;
import org.springframework.web.bind.annotation.RestController;
import com.niudong.esdemo.service.SearchService;
@RestController
@RequestMapping("/springboot/es/search")
public class SearchController {
  @Autowired
  private SearchService searchService;

  // 以同步方式執行 SearchRequest
  @RequestMapping("/sr")
  public String executeSearchRequest() {
    searchService.executeSearchRequest();

    return "Execute SearchRequest success!";
```

```
  }

}
```

在 Service 層中，增加新類別 SearchService，程式如下所示：

```
package com.niudong.esdemo.service;
public interface SearchService {
  // 以同步方式執行 SearchRequest
  public void executeSearchRequest();
}
```

在 ServiceImpl 實現層中，增加新類別 SearchServiceImpl，核心程式如下所示：

```
// 以同步方式執行 SearchRequest
  public void executeSearchRequest() {
    SearchRequest searchRequest = buildSearchRequest("content", " 貨幣 ");
    // 執行
    try {
      SearchResponse searchResponse = restClient.search(searchRequest,
          RequestOptions.DEFAULT);
      log.info(searchResponse.toString());
      // 解析 SearchResponse
      processSearchResponse(searchResponse);
    } catch (Exception e) {
      e.printStackTrace();
    } finally {
      // 關閉 Elasticsearch 的連接
      closeEs();
    }
  }
  // 解析 SearchResponse
  private void processSearchResponse(SearchResponse searchResponse) {
    if (searchResponse == null) {
      return;
    }
    // 取得 HTTP 狀態碼
```

```
RestStatus status = searchResponse.status();
// 取得請求執行時間
TimeValue took = searchResponse.getTook();
// 取得請求是否提前終止
Boolean terminatedEarly = searchResponse.isTerminatedEarly();
// 取得請求是否逾時
boolean timedOut = searchResponse.isTimedOut();
log.info("status is " + status + ";took is " + took + ";terminatedEarly
    is " + terminatedEarly
    + ";timedOut is " + timedOut);
// 檢視搜索影響的分片總數
int totalShards = searchResponse.getTotalShards();
// 執行搜索成功的分片的統計資訊
int successfulShards = searchResponse.getSuccessfulShards();
// 執行搜索失敗的分片的統計資訊
int failedShards = searchResponse.getFailedShards();
log.info("totalShards is " + totalShards + ";successfulShards is " +
    successfulShards
    + ";failedShards is " + failedShards);
for (ShardSearchFailure failure : searchResponse.getShardFailures()) {
  log.info("fail is " + failure.toString());
}
// 取得回應中包含的搜索結果
SearchHits hits = searchResponse.getHits();
// SearchHits 提供了相關結果的全部資訊，如點擊總數或最高分數
TotalHits totalHits = hits.getTotalHits();
// 點擊總數
long numHits = totalHits.value;
// 最高分數
float maxScore = hits.getMaxScore();
log.info("numHits is " + numHits + ";maxScore is " + maxScore);
// 巢狀結構在 SearchHits 中的是可以反覆運算的單一搜索結果
SearchHit[] searchHits = hits.getHits();
for (SearchHit hit : searchHits) {
  // SearchHit 提供了對基本資訊的存取，如索引、文件 ID 和每次搜索的得分
  String index = hit.getIndex();
```

```
    String id = hit.getId();
    float score = hit.getScore();
    log.info("docId is " + id + ";docIndex is " + index + ";docScore is " +
        score);
    // 以 JSON 字串形式傳回文件來源
    String sourceAsString = hit.getSourceAsString();
    // 以鍵值對形式傳回文件來源
    Map<String, Object> sourceAsMap = hit.getSourceAsMap();
    String documentTitle = (String) sourceAsMap.get("title");
    List<Object> users = (List<Object>) sourceAsMap.get("user");
    Map<String, Object> innerObject = (Map<String, Object>) sourceAsMap.
        get("innerObject");
    log.info(
        "sourceAsString is " + sourceAsString + ";sourceAsMap size is " +
            sourceAsMap.size());
    // 反白顯示
    Map<String, HighlightField> highlightFields = hit.getHighlightFields();
    HighlightField highlight = highlightFields.get("content");
    // 取得包含反白顯示欄位內容的或多個片段
    Text[] fragments = highlight.fragments();
    String fragmentString = fragments[0].string();
    log.info("fragmentString is " + fragmentString);
}
// 聚合搜索
Aggregations aggregations = searchResponse.getAggregations();
if (aggregations == null) {
  return;
}
// 按 content 聚合
Terms byCompanyAggregation = aggregations.get("by_content");
// 取得以 Elastic 為關鍵字的 Bucket
Bucket elasticBucket = byCompanyAggregation.getBucketByKey("Elastic");
// 取得平均年齡的子聚合
Avg averageAge = elasticBucket.getAggregations().get("average_age");
double avg = averageAge.getValue();
log.info("avg is " + avg);
```

```
    // 搜索
    Suggest suggest = searchResponse.getSuggest();
    if (suggest == null) {
      return;
    }
    // 按 content 搜索 Suggest
    TermSuggestion termSuggestion = suggest.getSuggestion("content");
    for (TermSuggestion.Entry entry : termSuggestion.getEntries()) {
      for (TermSuggestion.Entry.Option option : entry) {
        String suggestText = option.getText().string();
        log.info("suggestText is " + suggestText);
      }
    }
    // 在搜索分時析結果
    Map<String, ProfileShardResult> profilingResults = searchResponse.
        getProfileResults();
    if (profilingResults == null) {
      return;
    }
    for (Map.Entry<String, ProfileShardResult> profilingResult :
        profilingResults.entrySet()) {
      String key = profilingResult.getKey();
      ProfileShardResult profileShardResult = profilingResult.getValue();
      log.info("key is " + key + ";profileShardResult is " +
          profileShardResult.toString());
    }
  }
```

隨後編輯專案，在專案根目錄下輸入以下指令：

```
mvn clean package
```

透過以下指令啟動專案服務：

```
java -jar ./target/esdemo-0.0.1-SNAPSHOT.jar
```

在專案服務啟動後，在瀏覽器中呼叫以下介面檢視搜索回應結果的解析情況：

```
http://localhost:8080/springboot/es/search/sr
```

此時，在伺服器中輸出如下所示內容：

```
2019-08-03 14:19:51.511  INFO 74120 --- [nio-8080-exec-1] c.n.e.service.impl.SearchServiceImpl      : {"took":4,"timed_out
t":false,"_shards":{"total":2,"successful":2,"skipped":0,"failed":0},"hits":{"total":{"value":0,"relation":"eq"},"max_sc
ore":null,"hits":[]},"suggest":{"suggest_user":[{"text":"货","offset":0,"length":1,"options":[]},{"text":"币","offset":1
,"length":1,"options":[]}]}}
2019-08-03 14:19:51.513  INFO 74120 --- [nio-8080-exec-1] c.n.e.service.impl.SearchServiceImpl      : status is OK;took i
s 4ms;terminatedEarly is null;timedOut is false
2019-08-03 14:19:51.526  INFO 74120 --- [nio-8080-exec-1] c.n.e.service.impl.SearchServiceImpl      : totalShards is 2;su
ccessfulShards is 2;failedShards is 0
2019-08-03 14:19:51.527  INFO 74120 --- [nio-8080-exec-1] c.n.e.service.impl.SearchServiceImpl      : numHits is 0;maxSco
re is NaN
```

如果在瀏覽器中顯示以下內容，則證明介面呼叫成功：

```
Execute SearchRequest success!
```

7.2 捲動搜索

捲動搜索 API，即 Search Scroll API，可透過搜索請求，取得大量搜索結果。捲動搜索有點類似資料庫中的分頁查詢。

1 建置捲動搜索請求

為了使用捲動搜索，需要按循序執行以下步驟。

步驟 1：初始化捲動搜索的上下文資訊。

在執行捲動搜索 API 時，捲動搜索階段的初始化必須帶有捲動搜索參數的搜索請求，即 SearchRequest。

在執行該搜索請求時，Elasticsearch 會檢測到捲動搜索參數的存在，並在對應的時間間隔內保持搜索上下文活動。

帶有捲動搜索參數的搜索請求的建置程式如下所示，程式在 ServiceImpl 實現層的 ScrollSearchServiceImpl 類別中：

```
// 建置 SearchRequest
  public void buildAndExecuteScrollSearchRequest (String indexName, int size)
{
    // 設定索引名稱
    SearchRequest searchRequest = new SearchRequest(indexName);
```

```
SearchSourceBuilder searchSourceBuilder = new SearchSourceBuilder();
searchSourceBuilder.query(QueryBuilders.matchQuery("title",
    "Elasticsearch"));

// 建立 SearchRequest 及對應的 SearchSourceBuilder。還可以選擇設定大小以控制
// 一次檢索多少結果
searchSourceBuilder.size(size);
searchRequest.source(searchSourceBuilder);

// 設定捲動間隔
searchRequest.scroll(TimeValue.timeValueMinutes(1L));
try {
   SearchResponse searchResponse = restClient.search(searchRequest,
RequestOptions.DEFAULT);

   // 讀取傳回的捲動 ID，該 ID 指向保持活動狀態的搜索上下文，並在後續搜索捲動呼叫
   // 中被需要
   String scrollId = searchResponse.getScrollId();
   // 檢索第一批搜索結果
   SearchHits hits = searchResponse.getHits();

} catch (Exception e) {
   e.printStackTrace();
} finally {
   // 關閉 Elasticsearch 連接
   closeEs();
   }
}
```

步驟 2：檢索所有相關文件。

首先在 SearchScrollRequest 中設定上文提及的捲動識別符號和新的捲動間隔；其次在設定好 SearchScrollRequest 後，將其傳送給 searchScroll 方法。

在請求發出後，Elasticsearch 伺服器會傳回另一批帶有新的捲動識別符號的結果。依次類推，使用者需要在新的 SearchScrollRequest 中設定前文提及的捲動識別符號和新的捲動間隔，以便取得下一批次的結果。

這個過程會在一個循環中重複執行，直到不再有任何結果傳回。這表示捲動搜索已經完成，所有符合的文件都已被檢索。

該部分的程式如下所示，程式增加在前文提及的 buildAndExecuteScrollSearchRequest 方法中：

```
// 設定捲動識別符號
    SearchScrollRequest scrollRequest = new SearchScrollRequest(scrollId);
    scrollRequest.scroll(TimeValue.timeValueSeconds(30));
    SearchResponse searchScrollResponse =
        restClient.scroll(scrollRequest, RequestOptions.DEFAULT);

    // 讀取新的捲動 ID，該 ID 指向保持活動狀態的搜索上下文，並在後續搜索捲動呼叫中
    // 被需要
    scrollId = searchScrollResponse.getScrollId();
    // 檢索另一批搜索結果
    hits = searchScrollResponse.getHits();
```

此時，**buildAndExecuteScrollSearchRequest** 方法中的全部程式如下所示：

```
// 建置 SearchRequest
  public void buildAndExecuteScrollSearchRequest(String indexName, int size) {
    // 索引名稱
    SearchRequest searchRequest = new SearchRequest(indexName);
    SearchSourceBuilder searchSourceBuilder = new SearchSourceBuilder();
    searchSourceBuilder.query(QueryBuilders.matchQuery("title",
        "Elasticsearch"));
    // 建立 SearchRequest 及對應的 SearchSourceBuilder。還可以選擇設定大小，以控制
    // 一次檢索多少結果
    searchSourceBuilder.size(size);
    searchRequest.source(searchSourceBuilder);
    // 設定捲動間隔
    searchRequest.scroll(TimeValue.timeValueMinutes(1L));
    try {
      SearchResponse searchResponse = restClient.search(searchRequest,
          RequestOptions.DEFAULT);
      // 讀取傳回的捲動 ID，該 ID 指向保持活動狀態的搜索上下文，並在後續搜索捲動呼叫中
      // 被需要
```

```
    String scrollId = searchResponse.getScrollId();
    // 檢索第一批搜索結果
    SearchHits hits = searchResponse.getHits();
    while (hits != null && hits.getHits().length != 0) {
        // 設定捲動識別符號
        SearchScrollRequest scrollRequest = new SearchScrollRequest(scrollId);
        scrollRequest.scroll(TimeValue.timeValueSeconds(30));
        SearchResponse searchScrollResponse =
            restClient.scroll(scrollRequest, RequestOptions.DEFAULT);

        // 讀取新的捲動 ID，該 ID 指向保持活動狀態的搜索上下文，並在後續搜索捲動呼叫中
        // 被需要
        scrollId = searchScrollResponse.getScrollId();
        // 檢索另一批搜索結果
        hits = searchScrollResponse.getHits();

        log.info("scrollId is " + scrollId);
        log.info(
            "total hits is " + hits.getTotalHits().value + ";now hits is " +
                hits.getHits().length);
    }
} catch (Exception e) {
    e.printStackTrace();
} finally {
    // 關閉 Elasticsearch 連接
    closeEs();
}
}
```

▨ SearchScrollRequest 的可選參數

在 SearchScrollRequest 中，除捲動識別符號外，還提供了可選參數供使用者進行設定。SearchScrollRequest 提供的主要可選參數是捲動搜索的過期時間，程式如下所示：

```
// 設定捲動搜索的過期時間
scrollRequest.scroll(TimeValue.timeValueSeconds(60L));
scrollRequest.scroll("60s");
```

在實際開發中，如果讀者沒有為 SearchScrollRequest 設定捲動識別符號，則一旦初始捲動時間過期（即初始搜索請求中設定的捲動時間過期），則捲動搜索的上下文也會過期。

2 清除捲動搜索的上下文

在捲動搜索請求執行後，使用者可以使用 Clear Scroll API 刪除最後一個捲動識別符號，以釋放捲動搜索的上下文。當捲動搜索逾時到期時，這個過程也會自動發生。一般在捲動搜索階段後，需立即清除捲動搜索的上下文。

在執行清除捲動搜索上下文的請求之前，需要建置清除捲動搜索請求，即 ClearScrollRequest。ClearScrollRequest 需要把捲動識別符號作為參數輸入，建置 ClearScrollRequest 的程式如下所示：

```java
// 建置 ClearScrollRequest
public void buildClearScrollRequest(String scrollId){
  ClearScrollRequest request = new ClearScrollRequest();
  // 增加單一捲動識別符號
  request.addScrollId(scrollId);
}
```

在建置 ClearScrollRequest 時，不僅可以設定單一捲動識別符號，還可以設定多個捲動識別符號。我們繼續在上述方法中增加以下程式：

```java
// 建置 ClearScrollRequest
public void buildClearScrollRequest(String scrollId){
  ClearScrollRequest request = new ClearScrollRequest();
  // 增加單一捲動識別符號
  request.addScrollId(scrollId);

  // 增加多個捲動識別符號
  List<String> scrollIds  = new ArrayList<>();
  scrollIds.add(scrollId);
  request.setScrollIds(scrollIds);
}
```

3 執行清除捲動搜索請求

在 ClearScrollRequest 建置後，即可執行清除捲動搜索請求了。與文件索引請求類似，清除捲動搜索請求也有同步和非同步兩種執行方式。

☑ 同步方式

當以同步方式執行清除捲動搜索請求時，用戶端會等待 Elasticsearch 伺服器傳回的查詢結果 ClearScrollResponse。在收到 ClearScrollResponse 後，用戶端會繼續執行相關的邏輯程式。以同步方式執行的程式如下所示：

```
// 以同步方式執行清除捲動搜索請求
public void executeClearScrollRequest(String scrollId) {
  ClearScrollRequest request = new ClearScrollRequest();
  // 增加單一捲動識別符號
  request.addScrollId(scrollId);
  try {
    ClearScrollResponse response = restClient.clearScroll(request,
        RequestOptions.DEFAULT);
  } catch (Exception e) {
    e.printStackTrace();
  } finally {
    // 關閉 Elasticsearch 連接
    closeEs();
  }
}
```

☑ 非同步方式

當以非同步方式執行清除捲動搜索請求時，進階用戶端不必同步等待請求結果的傳回，可以直接向介面呼叫方傳回非同步介面執行成功的結果。

為了處理非同步傳回的回應資訊或處理在請求執行過程中引發的例外資訊，使用者需要指定監聽器。

在非同步請求處理後，如果請求執行成功，則呼叫 ActionListener 類別中的 onResponse 方法進行相關邏輯的處理；如果請求執行失敗，則呼叫 ActionListener 類別中的 onFailure 方法進行相關邏輯的處理。

以非同步方式執行的程式如下所示：

```
// 以非同步方式執行清除捲動搜索請求
 public void executeClearScrollRequestAsync(String scrollId) {
  ClearScrollRequest request = new ClearScrollRequest();
  // 增加單一捲動識別符號
  request.addScrollId(scrollId);
  // 增加監聽器
  ActionListener<ClearScrollResponse> listener = new ActionListener
      <ClearScrollResponse>() {
    @Override
    public void onResponse(ClearScrollResponse clearScrollResponse) {
    }
    @Override
    public void onFailure(Exception e) {
    }
  };

  try {
    restClient.clearScrollAsync(request, RequestOptions.DEFAULT, listener);
  } catch (Exception e) {
    e.printStackTrace();
  } finally {
    // 關閉 Elasticsearch 連接
    closeEs();
  }
 }
```

當然，在非同步請求執行過程中可能會出現例外，例外的處理與同步方式執行情況相同。

▉ 解析清除捲動搜索請求的回應結果

不論同步方式，還是非同步方式，用戶端均需要對請求的傳回結果進行處理和解析。ClearScrollResponse 中含有已發佈的捲動搜索上下文的資訊。

解析清除捲動搜索請求的核心程式如下所示：

```
// 如果請求成功，則會傳回 true 的結果
boolean success = response.isSucceeded();
// 傳回已釋放的搜索上下文數
int released = response.getNumFreed();
```

完整程式如下所示：

```
// 以同步方式執行清除捲動搜索請求
  public void executeClearScrollRequest(String scrollId) {
    ClearScrollRequest request = new ClearScrollRequest();
    // 增加單一捲動識別符號
    request.addScrollId(scrollId);
    try {
      ClearScrollResponse response = restClient.clearScroll(request,
          RequestOptions. DEFAULT);
      // 如果請求成功，則會傳回 true 的結果
      boolean success = response.isSucceeded();
      // 傳回已釋放的搜索上下文數
      int released = response.getNumFreed();
      log.info("success is " + success + ";released is  " + released);
    } catch (Exception e) {
      e.printStackTrace();
    } finally {
      // 關閉 Elasticsearch 連接
      closeEs();
    }
  }
```

5 執行捲動搜索請求

在 SearchScrollRequest 建置後，即可執行捲動搜索請求。與文件索引請求類似，捲動搜索請求也有同步和非同步兩種執行方式。

☑ 同步方式

當以同步方式執行捲動搜索請求時，用戶端會等待 Elasticsearch 伺服器傳回的查詢結果 SearchResponse。在收到 SearchResponse 後，用戶端會繼續執行相關的邏輯程式。以同步方式執行的程式如下所示：

```
SearchResponse searchResponse = client.scroll(scrollRequest,
RequestOptions. DEFAULT);
```

以同步方式執行的全部程式詳見上文中的 **buildAndExecuteScrollSearchRequest** 方法。

▨ 非同步方式

當以非同步方式執行捲動搜索請求時，進階用戶端不必同步等待請求結果的傳回，可以直接向介面呼叫方傳回非同步介面執行成功的結果。

為了處理非同步傳回的回應資訊或處理在請求執行過程中引發的例外資訊，使用者需要指定監聽器。以非同步方式執行的程式如下所示：

```
client.scrollAsync(scrollRequest, RequestOptions.DEFAULT, scrollListener);
```

其中，**scrollListener** 是監聽器。

在非同步請求處理後，如果請求執行成功，則呼叫 ActionListener 類別中的 onResponse 方法進行相關邏輯的處理；如果請求執行失敗，則呼叫 ActionListener 類別中的 onFailure 方法進行相關邏輯的處理。

以非同步方式執行的程式如下所示：

```
// 建置 SearchRequest
  public void buildAndExecuteScrollSearchRequestAsync(String indexName, int
      size) {
    // 索引名稱
    SearchRequest searchRequest = new SearchRequest(indexName);
    SearchSourceBuilder searchSourceBuilder = new SearchSourceBuilder();
    searchSourceBuilder.query(QueryBuilders.matchQuery("title",
        "Elasticsearch"));
    // 建立 SearchRequest 及對應的 SearchSourceBuilder。還可以選擇設定大小，以控制
    // 一次檢索多少結果
    searchSourceBuilder.size(size);
    searchRequest.source(searchSourceBuilder);
    // 設定捲動間隔
    searchRequest.scroll(TimeValue.timeValueMinutes(1L));
```

```
try {
  SearchResponse searchResponse = restClient.search(searchRequest,
      RequestOptions.DEFAULT);
  // 讀取傳回的捲動 ID，該 ID 指向保持活動狀態的搜索上下文，並在後續搜索捲動呼叫中
  // 被需要
  String scrollId = searchResponse.getScrollId();
  // 檢索第一批搜索結果
  SearchHits hits = searchResponse.getHits();
  // 設定監聽器
  ActionListener<SearchResponse> scrollListener = new ActionListener
      <SearchResponse>() {
    @Override
    public void onResponse(SearchResponse searchResponse) {
    // 讀取新的捲動 ID，該 ID 指向保持活動狀態的搜索上下文，並在後續搜索捲動呼叫中
    // 被需要
      String scrollId = searchResponse.getScrollId();
      // 檢索另一批搜索結果
      SearchHits hits = searchResponse.getHits();
      log.info("scrollId is " + scrollId);
      log.info("total hits is " + hits.getTotalHits().value + ";now hits
          is "
        + hits.getHits().length);
    }
    @Override
    public void onFailure(Exception e) {
    }
  };
  while (hits != null && hits.getHits().length != 0) {
    // 設定捲動識別符號
    SearchScrollRequest scrollRequest = new SearchScrollRequest(scrollId);
    scrollRequest.scroll(TimeValue.timeValueSeconds(30));
    // 非同步執行
    restClient.scrollAsync(scrollRequest, RequestOptions.DEFAULT,
        scrollListener);
  }
} catch (Exception e) {
  e.printStackTrace();
```

```
    } finally {
      // 關閉 Elasticsearch 連接
      closeEs();
    }
  }
```

當然，在非同步請求執行過程中可能會出現例外，例外的處理與同步方式執行情況相同。

6 解析捲動搜索請求的回應結果

捲動搜索請求與搜索 API 有相同的回應結果，即 SearchResponse 物件。

解析捲動搜索請求結果的程式共分為三層，分別是 Controller 層、Service 層和 ServiceImpl 實現層。

在 Controller 層中新增 ScrollSearchController 類別。ScrollSearchController 類別用於儲存有關捲動搜索的請求呼叫。ScrollSearchController 類別中的程式如下所示：

```java
package com.niudong.esdemo.controller;
import org.elasticsearch.common.Strings;
import org.springframework.beans.factory.annotation.Autowired;
import org.springframework.web.bind.annotation.RequestMapping;
import org.springframework.web.bind.annotation.RestController;
import com.niudong.esdemo.service.ScrollSearchService;
@RestController
@RequestMapping("/springboot/es/scrollsearch")
public class ScrollSearchController {
  @Autowired
  private ScrollSearchService scrollSearchService;
  // 以同步方式執行 SearchScrollRequest
  @RequestMapping("/sr")
  public String executeSearchRequest(String indexName, int size) {
    // 參數驗證
    if (Strings.isNullOrEmpty(indexName) || size <= 0) {
      return "Parameters are wrong!";
    }
```

```
        scrollSearchService.buildAndExecuteScrollSearchRequest(indexName, size);
        return "Execute SearchScrollRequest success!";
    }
}
```

在 Service 層中新增 ScrollSearchService 介面類別，該類別用於描述捲動搜索的 Service 層介面。ScrollSearchService 類別中的程式如下所示：

```
package com.niudong.esdemo.service;
public interface ScrollSearchService {
    // 建置 SearchRequest
    public void buildAndExecuteScrollSearchRequest(String indexName, int size);
}
```

在 ServiceImpl 實現層中新增 ScrollSearchServiceImpl 介面實現類別，該類別用於描述捲動搜索的 ServiceImpl 層介面的實作方式。ScrollSearchServiceImpl 類別中的部分程式如下所示：

```
package com.niudong.esdemo.service.impl;
import java.util.ArrayList;
import java.util.List;
import javax.annotation.PostConstruct;
import org.apache.commons.logging.Log;
import org.apache.commons.logging.LogFactory;
import org.apache.http.HttpHost;
import org.elasticsearch.action.ActionListener;
import org.elasticsearch.action.search.ClearScrollRequest;
import org.elasticsearch.action.search.ClearScrollResponse;
import org.elasticsearch.action.search.SearchRequest;
import org.elasticsearch.action.search.SearchResponse;
import org.elasticsearch.action.search.SearchScrollRequest;
import org.elasticsearch.client.RequestOptions;
import org.elasticsearch.client.RestClient;
import org.elasticsearch.client.RestHighLevelClient;
import org.elasticsearch.common.unit.TimeValue;
import org.elasticsearch.index.query.QueryBuilders;
import org.elasticsearch.search.SearchHits;
import org.elasticsearch.search.builder.SearchSourceBuilder;
```

```
import org.springframework.stereotype.Service;
import com.niudong.esdemo.service.ScrollSearchService;
import ch.qos.logback.classic.Logger;
@Service
public class ScrollSearchServiceImpl implements ScrollSearchService {
  private static Log log = LogFactory.getLog(ScrollSearchServiceImpl.class);
  private RestHighLevelClient restClient;
  // 初始化連接
  @PostConstruct
  public void initEs() {
    restClient = new RestHighLevelClient(RestClient.builder(new HttpHost
        ("localhost", 9200, "http"),
      new HttpHost("localhost", 9201, "http")));
    log.info("ElasticSearch init in service.");
  }
  // 關閉連接
  public void closeEs() {
    try {
      restClient.close();
    } catch (Exception e) {
      e.printStackTrace();
    }
  }
  // 建置 SearchRequest
  public void buildAndExecuteScrollSearchRequest(String indexName, int size)
{
    // 索引名稱
    SearchRequest searchRequest = new SearchRequest(indexName);
    SearchSourceBuilder searchSourceBuilder = new SearchSourceBuilder();
    searchSourceBuilder.query(QueryBuilders.matchQuery("content", " 美聯儲 "));
    // 建立 SearchRequest 及對應的 SearchSourceBuilder。還可以選擇設定大小，以控制
    // 一次檢索多少個結果
    searchSourceBuilder.size(size);
    searchRequest.source(searchSourceBuilder);
    // 設定捲動間隔
    searchRequest.scroll(TimeValue.timeValueMinutes(1L));
    try {
```

```
     SearchResponse searchResponse = restClient.search(searchRequest,
RequestOptions.DEFAULT);
     // 讀取傳回的捲動 ID，該 ID 指向保持活動狀態的搜索上下文，並在後續搜索捲動呼叫中
     // 被需要
     String scrollId = searchResponse.getScrollId();
     // 檢索第一批搜索結果
     SearchHits hits = searchResponse.getHits();
     while (hits != null && hits.getHits().length != 0) {
        // 設定捲動識別符號
        SearchScrollRequest scrollRequest = new SearchScrollRequest(scrollId);
        scrollRequest.scroll(TimeValue.timeValueSeconds(30));
        SearchResponse searchScrollResponse =
           restClient.scroll(scrollRequest, RequestOptions.DEFAULT);
        // 讀取新的捲動 ID，該 ID 指向保持活動狀態的搜索上下文，並在後續搜索捲動呼叫中
        // 被需要
        scrollId = searchScrollResponse.getScrollId();
        // 檢索另一批搜索結果
        hits = searchScrollResponse.getHits();
        log.info("scrollId is " + scrollId);
        log.info(
           "total hits is " + hits.getTotalHits().value + ";now hits is " +
                 hits.getHits().length);
     }

     // 清除捲動搜索上下文的資訊
     executeClearScrollRequest(scrollId);
   } catch (Exception e) {
     e.printStackTrace();
   } finally {
     // 關閉 Elasticsearch 的連接
     closeEs();
   }
 }
// 以同步方式執行清除捲動搜索上下文的請求
 public void executeClearScrollRequest(String scrollId) {
   ClearScrollRequest request = new ClearScrollRequest();
   // 增加單一捲動識別符號
```

```
     request.addScrollId(scrollId);
     try {
       ClearScrollResponse response = restClient.clearScroll(request,
           RequestOptions. DEFAULT);
       // 如果請求成功，則會傳回 true 的結果
       boolean success = response.isSucceeded();
       // 傳回已釋放的搜索上下文數
       int released = response.getNumFreed();
       log.info("success is " + success + ";released is  " + released);
     } catch (Exception e) {
       e.printStackTrace();
     } finally {
       // 關閉 Elasticsearch 連接
       closeEs();
     }
   }
}
```

隨後編輯專案，在專案根目錄中輸入以下指令：

```
mvn clean package
```

透過以下指令啟動專案服務：

```
java -jar ./target/esdemo-0.0.1-SNAPSHOT.jar
```

在專案服務啟動後，在瀏覽器中呼叫以下介面檢視捲動搜索請求的執行情況：

```
http://localhost:8080//springboot/es/scrollsearch/sr?indexName=ultraman&size=1
```

介面執行後，在伺服器中輸出如下所示內容：

```
2019-09-02 15:02:45.295  INFO 31140 --- [nio-8080-exec-9] o.s.web.servlet.DispatcherServlet        : Completed initializ
ation in 13 ms
2019-09-02 15:02:48.017  INFO 31140 --- [nio-8080-exec-9] c.n.e.s.impl.ScrollSearchServiceImpl     : scrollId is DXF1ZXJ
5QW5kRmVOY2gBAAAAAAAAAEWM2dOUmJFTF9US1NvY3BnUm5jjMndqQQ==
2019-09-02 15:02:48.017  INFO 31140 --- [nio-8080-exec-9] c.n.e.s.impl.ScrollSearchServiceImpl     : total hits is 3;now
 hits is 1
2019-09-02 15:02:48.026  INFO 31140 --- [nio-8080-exec-9] c.n.e.s.impl.ScrollSearchServiceImpl     : scrollId is DXF1ZXJ
5QW5kRmVOY2gBAAAAAAAAAEWM2dOUmJFTF9US1NvY3BnUm5jjMndqQQ==
2019-09-02 15:02:48.027  INFO 31140 --- [nio-8080-exec-9] c.n.e.s.impl.ScrollSearchServiceImpl     : total hits is 3;now
 hits is 1
2019-09-02 15:02:48.040  INFO 31140 --- [nio-8080-exec-9] c.n.e.s.impl.ScrollSearchServiceImpl     : scrollId is DXF1ZXJ
5QW5kRmVOY2gBAAAAAAAAAEWM2dOUmJFTF9US1NvY3BnUm5jjMndqQQ==
2019-09-02 15:02:48.041  INFO 31140 --- [nio-8080-exec-9] c.n.e.s.impl.ScrollSearchServiceImpl     : total hits is 3;now
 hits is 0
2019-09-02 15:02:48.064  INFO 31140 --- [nio-8080-exec-9] c.n.e.s.impl.ScrollSearchServiceImpl     : success is true;rel
eased is  1
```

與搜索內容相關的資料共 3 筆，每次捲動搜索數量為 1，因此共輸出三次捲動搜索的內容，與預期相同。

此時，如果在呼叫介面的瀏覽器中顯示以下內容，則表示介面呼叫成功：

```
Execute SearchScrollRequest success!
```

7.3 批次搜索

在 Elasticsearch 中，不僅提供了單次搜索查詢介面，還提供了批次搜索查詢介面，即批次搜索 API（MultiSearch API），支援在單次 HTTP 請求中並存執行多個搜索請求。

1 建置批次搜索請求

在執行批次搜索請求前，需要建置批次搜索請求，即 MultiSearchRequest。在初始化 MultiSearchRequest 時，搜索請求為空，因此使用者需要把要執行的所有搜索添到 MultiSearchRequest 中。

建置 MultiSearchRequest 的程式如下所示：

```
// 建置 MultiSearchRequest
  public void buildMultiSearchRequest() {
    MultiSearchRequest request = new MultiSearchRequest();
    // 建置搜索請求物件 1
    SearchRequest firstSearchRequest = new SearchRequest();
    SearchSourceBuilder searchSourceBuilder = new SearchSourceBuilder();
    searchSourceBuilder.query(QueryBuilders.matchQuery("user", "niudong1"));
    firstSearchRequest.source(searchSourceBuilder);
    // 將搜索請求物件 1 增加到 MultiSearchRequest 中
    request.add(firstSearchRequest);
    // 建置搜索請求物件 2
    SearchRequest secondSearchRequest = new SearchRequest();
    searchSourceBuilder = new SearchSourceBuilder();
    searchSourceBuilder.query(QueryBuilders.matchQuery("user", "niudong2"));
    secondSearchRequest.source(searchSourceBuilder);
```

```
    // 將搜索請求物件 2 增加到 MultiSearchRequest 中
    request.add(secondSearchRequest);
}
```

✍ 可選參數

在 MultiSearchRequest 的建置過程中，有一些可選參數供使用者進行設定。
MultiSearchRequest 中的 SearchRequest 支援搜索的所有可選參數，如設定索
引名稱等，程式如下所示：

```
SearchRequest searchRequest = new SearchRequest("ultraman");
```

❷ 執行批次搜索請求

在 MultiSearchRequest 建置後，即可執行批次搜索請求。與文件索引請求類
似，批次搜索請求也有同步和非同步兩種執行方式。

✍ 同步方式

當以同步方式執行批次搜索請求時，用戶端會等待 Elasticsearch 伺服器傳回的
查詢結果 MultiSearchResponse。在收到 MultiSearchResponse 後，用戶端會繼
續執行相關的邏輯程式。以同步方式執行的程式如下所示：

```
MultiSearchResponse response = client.msearch(request, RequestOptions.DEFAULT);
```

以同步方式執行的完整程式如下所示：

```
// 建置 MultiSearchRequest
public MultiSearchRequest buildMultiSearchRequest(String field, String[]
        keywords) {
  MultiSearchRequest request = new MultiSearchRequest();
  // 建置搜索請求物件 1
  SearchRequest firstSearchRequest = new SearchRequest();
  SearchSourceBuilder searchSourceBuilder = new SearchSourceBuilder();
  searchSourceBuilder.query(QueryBuilders.matchQuery(field, keywords[0]));
  firstSearchRequest.source(searchSourceBuilder);
  // 將搜索請求物件 1 增加到 MultiSearchRequest 中
  request.add(firstSearchRequest);
  // 建置搜索請求物件 2
```

```
    SearchRequest secondSearchRequest = new SearchRequest();
    searchSourceBuilder = new SearchSourceBuilder();
    searchSourceBuilder.query(QueryBuilders.matchQuery(field,  keywords[1]));
    secondSearchRequest.source(searchSourceBuilder);
    // 將搜索請求物件 2 增加到 MultiSearchRequest 中
    request.add(secondSearchRequest);
    return request;
  }
  // 以同步方式執行 MultiSearchRequest
  public void executeMultiSearchRequest(String field, String[] keywords) {
    // 建置 MultiSearchRequest
    MultiSearchRequest request = buildMultiSearchRequest(field,keywords);
    try {
      MultiSearchResponse response = restClient.msearch(request,
RequestOptions.DEFAULT);
    } catch (Exception e) {
      e.printStackTrace();
    } finally {
      // 關閉 Elasticsearch 連接
      closeEs();
    }
  }
```

☑ 非同步方式

當以非同步方式執行批次搜索請求時，進階用戶端不必同步等待請求結果的傳回，可以直接向介面呼叫方傳回非同步介面執行成功的結果。

為了處理非同步傳回的回應資訊或處理在請求執行過程中引發的例外資訊，使用者需要指定監聽器。以非同步方式執行的核心程式如下所示：

```
 client.msearchAsync(searchRequest, RequestOptions.DEFAULT, listener);
```

其中，listener 為監聽器。

在非同步請求處理後，如果請求執行成功，則呼叫 ActionListener 類別中的 onResponse 方法進行相關邏輯的處理；如果請求執行失敗，則呼叫 ActionListener 類別中的 onFailure 方法進行相關邏輯的處理。

以非同步方式執行的全部程式如下所示：

```java
// 建置 MultiSearchRequest
  public MultiSearchRequest buildMultiSearchRequest(String field, String[]
      keywords) {
    MultiSearchRequest request = new MultiSearchRequest();
    // 建置搜索請求物件 1
    SearchRequest firstSearchRequest = new SearchRequest();
    SearchSourceBuilder searchSourceBuilder = new SearchSourceBuilder();
    searchSourceBuilder.query(QueryBuilders.matchQuery(field, keywords[0]));
    firstSearchRequest.source(searchSourceBuilder);
    // 將搜索請求物件 1 增加到 MultiSearchRequest 中
    request.add(firstSearchRequest);
    // 建置搜索請求物件 2
    SearchRequest secondSearchRequest = new SearchRequest();
    searchSourceBuilder = new SearchSourceBuilder();
    searchSourceBuilder.query(QueryBuilders.matchQuery(field, keywords[1]));
    secondSearchRequest.source(searchSourceBuilder);
    // 將搜索請求物件 2 增加到 MultiSearchRequest 中
    request.add(secondSearchRequest);
    return request;
  }
// 以非同步方式執行 MultiSearchRequest
  public void executeMultiSearchRequestAsync(String field, String[] keywords)
  {
    // 建置 MultiSearchRequest
    MultiSearchRequest request = buildMultiSearchRequest(field, keywords);
    // 建置監聽器
    ActionListener<MultiSearchResponse> listener = new ActionListener
        <MultiSearchResponse>() {
      @Override
      public void onResponse(MultiSearchResponse response) {
      }
      @Override
      public void onFailure(Exception e) {
      }
    };
```

```
   // 以非同步方式執行
   try {
     restClient.msearchAsync(request, RequestOptions.DEFAULT, listener);
   } catch (Exception e) {
     e.printStackTrace();
   } finally {
     // 關閉 Elasticsearch 連接
     closeEs();
   }
 }
```

當然，在非同步請求執行過程中可能會出現例外，例外的處理與同步方式執行情況相同。

3 解析批次搜索請求的回應結果

不論同步方式，還是非同步方式，在批次搜索請求執行後，用戶端均需要對傳回結果 MultiSearchResponse 進行處理和解析。

MultiSearchResponse 中 包 含 MultiSearchResponse.item 物 件 列 表，每 個 MultiSearchResponse.item 物件對應於 MultiSearchRequest 中的每個搜索請求。

如 果 批 次 搜 索 請 求 執 行 失 敗，則 每 個 MultiSearchResponse.item 物 件 的 getFailure 方法中都會包含一個例外資訊；反之，如果請求執行成功，則使用 者 可 以 從 每 個 MultiSearchResponse.item 物 件 的 getResponse 方 法 中 獲 得 SearchResponse，並解析對應的結果資訊。

解 析 MultiSearchResponse 的 程 式 如 下 所 示，程 式 共 分 為 三 層，分 別 是 Controller 層、Service 層和 ServiceImpl 實現層。

在 Controller 層中新增 MultiSearchController 類別，程式如下所示：

```
package com.niudong.esdemo.controller;
import java.util.List;
import org.elasticsearch.common.Strings;
import org.springframework.beans.factory.annotation.Autowired;
import org.springframework.web.bind.annotation.RequestMapping;
```

```
import org.springframework.web.bind.annotation.RestController;
import com.google.common.base.Splitter;
import com.niudong.esdemo.service.MultiSearchService;
@RestController
@RequestMapping("/springboot/es/multisearch")
public class MultiSearchController {
  @Autowired
  private MultiSearchService multiSearchService;
  // 以同步方式執行 MultiSearchRequest
  @RequestMapping("/sr")
  public String executeMultiSearchRequest(String field, String keywords) {
    // 參數驗證
    if (Strings.isNullOrEmpty(field) || Strings.isNullOrEmpty(keywords)) {
      return "Parameters are wrong!";
    }
    // 將英文逗點分隔的字串切分成陣列
    List<String> keywordsList = Splitter.on(",").splitToList(keywords);
    multiSearchService.executeMultiSearchRequest(field,
        keywordsList.toArray(new String[keywordsList.size()]));
    return "Execute MultiSearchRequest success!";
  }
}
```

在 Service 層中新增 MultiSearchService 介面類別，該類別主要用於描述 Service 層的介面方法，程式如下所示：

```
package com.niudong.esdemo.service;
public interface MultiSearchService {
  // 以同步方式執行 MultiSearchRequest
  public void executeMultiSearchRequest(String field, String[] keywords);
}
```

在 ServiceImpl 實現層中新增 MultiSearchServiceImpl 介面實現類別，該類別主要用於描述 Service 實現層的介面實現方法，程式如下所示：

```
package com.niudong.esdemo.service.impl;
import javax.annotation.PostConstruct;
import org.apache.commons.logging.Log;
```

```java
import org.apache.commons.logging.LogFactory;
import org.apache.http.HttpHost;
import org.elasticsearch.action.ActionListener;
import org.elasticsearch.action.search.MultiSearchRequest;
import org.elasticsearch.action.search.MultiSearchResponse;
import org.elasticsearch.action.search.MultiSearchResponse.Item;
import org.elasticsearch.action.search.SearchRequest;
import org.elasticsearch.action.search.SearchResponse;
import org.elasticsearch.client.RequestOptions;
import org.elasticsearch.client.RestClient;
import org.elasticsearch.client.RestHighLevelClient;
import org.elasticsearch.index.query.QueryBuilders;
import org.elasticsearch.search.SearchHit;
import org.elasticsearch.search.SearchHits;
import org.elasticsearch.search.builder.SearchSourceBuilder;
import org.springframework.stereotype.Service;
import com.niudong.esdemo.service.MultiSearchService;
@Service
public class MultiSearchServiceImpl implements MultiSearchService {
    private static Log log = LogFactory.getLog(MultiSearchServiceImpl.class);
    private RestHighLevelClient restClient;
    // 初始化連接
    @PostConstruct
    public void initEs() {
        restClient = new RestHighLevelClient(RestClient.builder(new HttpHost
            ("localhost", 9200, "http"),
            new HttpHost("localhost", 9201, "http")));
        log.info("ElasticSearch init in service.");
    }
    // 關閉連接
    public void closeEs() {
        try {
            restClient.close();
        } catch (Exception e) {
            e.printStackTrace();
        }
    }
```

```
// 建置 MultiSearchRequest
public MultiSearchRequest buildMultiSearchRequest(String field, String[]
     keywords) {
  MultiSearchRequest request = new MultiSearchRequest();
  // 建置搜索請求物件 1
  SearchRequest firstSearchRequest = new SearchRequest();
  SearchSourceBuilder searchSourceBuilder = new SearchSourceBuilder();
  searchSourceBuilder.query(QueryBuilders.matchQuery(field, keywords[0]));
  firstSearchRequest.source(searchSourceBuilder);
  // 將搜索請求物件 1 增加到 MultiSearchRequest 中
  request.add(firstSearchRequest);
  // 建置搜索請求物件 2
  SearchRequest secondSearchRequest = new SearchRequest();
  searchSourceBuilder = new SearchSourceBuilder();
  searchSourceBuilder.query(QueryBuilders.matchQuery(field, keywords[1]));
  secondSearchRequest.source(searchSourceBuilder);
  // 將搜索請求物件 2 增加到 MultiSearchRequest 中
  request.add(secondSearchRequest);
  return request;
}
// 以同步方式執行 MultiSearchRequest
public void executeMultiSearchRequest(String field, String[] keywords) {
  // 建置 MultiSearchRequest
  MultiSearchRequest request = buildMultiSearchRequest(field, keywords);
  try {
    MultiSearchResponse response = restClient.msearch(request,
RequestOptions.DEFAULT);
    // 解析傳回結果 MultiSearchResponse
    processMultiSearchResponse(response);
  } catch (Exception e) {
    e.printStackTrace();
  } finally {
    // 關閉 Elasticsearch 連接
    closeEs();
  }
}
// 解析傳回結果 MultiSearchResponse
```

```java
private void processMultiSearchResponse(MultiSearchResponse response) {
  // 取得傳回結果集合
  Item[] items = response.getResponses();
  // 判斷傳回結果集合是否為空
  if (items == null || items.length <= 0) {
    log.info("items is null.");
    return;
  }
  for (Item item : items) {
    Exception exception = item.getFailure();
    if (exception != null) {
      log.info("eception is " + exception.toString());
    }
    SearchResponse searchResponse = item.getResponse();
    SearchHits hits = searchResponse.getHits();
    if (hits.getTotalHits().value <= 0) {
      log.info("hits.getTotalHits().value is 0.");
      return;
    }
    SearchHit[] hitArray = hits.getHits();
    for (int i = 0; i < hitArray.length; i++) {
      SearchHit hit = hitArray[i];
      log.info("id is " + hit.getId() + ";index is " + hit.getIndex() +
";source is "
          + hit.getSourceAsString());
    }
  }
}
```

隨後編輯專案，在專案根目錄下輸入以下指令：

```
mvn clean package
```

透過以下指令啟動專案服務：

```
java -jar ./target/esdemo-0.0.1-SNAPSHOT.jar
```

在專案服務啟動後，在瀏覽器中呼叫以下介面檢視批次搜索的執行情況：

```
http://localhost:8080/springboot/es/multisearch/sr?field=content&keywords=
美聯儲,空軍
```

在伺服器中輸出如下所示內容：

```
2019-09-02 20:26:24.855  INFO 32668 --- [nio-8080-exec-1] c.n.e.s.impl.MultiSearchServiceImpl    : id is 3;index is ul
traman;source is {"content":"从此前美联储降息历程来看，美联储降息将打开全球各国央行的降息窗口"}
2019-09-02 20:26:24.856  INFO 32668 --- [nio-8080-exec-1] c.n.e.s.impl.MultiSearchServiceImpl    : id is 3;index is ul
traman1;source is {"content":"从此前美联储降息历程来看，美联储降息将打开全球各国央行的降息窗口"}
2019-09-02 20:26:24.857  INFO 32668 --- [nio-8080-exec-1] c.n.e.s.impl.MultiSearchServiceImpl    : id is 2;index is ul
traman;source is {"content":"自6月起，市场对于美联储降息的预期愈发强烈"}
2019-09-02 20:26:24.857  INFO 32668 --- [nio-8080-exec-1] c.n.e.s.impl.MultiSearchServiceImpl    : id is 2;index is ul
traman1;source is {"content":"自6月起，市场对于美联储降息的预期愈发强烈"}
2019-09-02 20:26:24.858  INFO 32668 --- [nio-8080-exec-1] c.n.e.s.impl.MultiSearchServiceImpl    : id is 1;index is ul
traman;source is {"content":"事实上，自今年年初开始，美联储就已传递出货币政策或将转向的迹象"}
2019-09-02 20:26:24.858  INFO 32668 --- [nio-8080-exec-1] c.n.e.s.impl.MultiSearchServiceImpl    : id is 1;index is ul
traman1;source is {"content":"事实上，自今年年初开始，美联储就已传递出货币政策或将转向的迹象"}
```

（編按：本圖為簡體中文介面）

索引名稱 ultraman 和 ultraman1 中僅包含了「美聯儲」關鍵字的資訊，因此共搜索出 6 筆資料，與資料的實際情況相同。

此時，如果瀏覽器頁面中的輸出內容如下，則表示介面執行成功：

```
Execute MultiSearchRequest success!
```

7.4 跨索引欄位搜索

Elasticsearch 不僅提供了在特定索引下的欄位搜索介面，還提供了跨索引的欄位搜索介面，即 Field Capabilities API。

1 建置跨索引欄位搜索請求

在執行跨索引欄位搜索請求前，需要建置跨索引欄位搜索請求，即 FieldCapabilitiesRequest。

FieldCapabilitiesRequest 中包含了要搜索的欄位清單及一個可選的目標索引名稱清單。如果沒有提供目標索引名稱清單，則預設對所有索引執行相關請求。

需要指出的是，欄位清單（即 fields 參數）支援萬用字元的表示方法。舉例來說，text_* 將傳回與運算式符合的所有欄位。

建置 FieldCapabilitiesRequest 的程式如下所示：

```
// 建置 FieldCapabilitiesRequest
  public FieldCapabilitiesRequest buildFieldCapabilitiesRequest() {
    FieldCapabilitiesRequest request =
        new FieldCapabilitiesRequest().fields("content").indices("ultraman",
            "ultraman1");
    return request;
  }
```

同理，FieldCapabilitiesRequest 也有可選參數列表供使用者進行設定。FieldCapabilitiesRequest 中的主要可選參數是 IndicesOptions，用於解析不可用的索引及展開萬用字元運算式。

可選參數設定程式如下所示：

```
// 建置 FieldCapabilitiesRequest
  public FieldCapabilitiesRequest buildFieldCapabilitiesRequest() {
    FieldCapabilitiesRequest request =
        new FieldCapabilitiesRequest().fields("content").indices("ultraman",
"ultraman1");

    // 設定可選參數 IndicesOptions：解析不可用的索引及展開萬用字元運算式
    request.indicesOptions(IndicesOptions.lenientExpandOpen());

    return request;
  }
```

2 執行跨索引欄位搜索請求

在 FieldCapabilitiesRequest 建置後，即可執行跨索引欄位搜索請求。與文件索引請求類似，跨索引欄位搜索請求也有同步和非同步兩種執行方式。

同步方式

當以同步方式執行跨索引欄位搜索請求時，用戶端會等待 Elasticsearch 伺服器傳回的查詢結果 FieldCapabilitiesResponse。在收到 FieldCapabilitiesResponse後，用戶端會繼續執行相關的邏輯程式。以同步方式執行的程式如下所示：

```
// 建置 FieldCapabilitiesRequest
  public FieldCapabilitiesRequest buildFieldCapabilitiesRequest(String field,
      String[] indices) {
    FieldCapabilitiesRequest request =
      new FieldCapabilitiesRequest().fields(field).indices(indices[0],
          indices[1]);
    // 設定可選參數 IndicesOptions：解析不可用的索引及展開萬用字元運算式
    request.indicesOptions(IndicesOptions.lenientExpandOpen());
    return request;
  }
  // 以同步方式執行跨索引欄位搜索請求
  public void executeFieldCapabilitiesRequest(String field, String[] indices)
      {
    // 建置 FieldCapabilitiesRequest
    FieldCapabilitiesRequest request= buildFieldCapabilitiesRequest(field,
indices);
    try {
      FieldCapabilitiesResponse response = restClient.fieldCaps(request,
          RequestOptions.DEFAULT);
    } catch (Exception e) {
      e.printStackTrace();
    } finally {
      // 關閉 Elasticsearch 連接
      closeEs();
    }
  }
```

✍ 非同步方式

當以非同步方式執行跨索引欄位搜索請求時，進階用戶端不必同步等待請求
結果的傳回，可以直接向介面呼叫方傳回非同步介面執行成功的結果。

為了處理非同步傳回的回應資訊或處理在請求執行過程中引發的例外資訊，
使用者需要指定監聽器。以非同步方式執行的核心程式如下所示：

```
client.fieldCapsAsync(request, RequestOptions.DEFAULT, listener);
```

其中，listener 為監聽器。

在非同步請求處理後，如果請求執行成功，則呼叫 ActionListener 類別中的 onResponse 方法進行相關邏輯的處理；如果請求執行失敗，則呼叫 ActionListener 類別中的 onFailure 方法進行相關邏輯的處理。

以非同步方式執行的全部程式如下所示：

```java
// 建置 FieldCapabilitiesRequest
  public FieldCapabilitiesRequest buildFieldCapabilitiesRequest(String field,
    String[] indices) {
    FieldCapabilitiesRequest request =
        new FieldCapabilitiesRequest().fields(field).indices(indices[0],
            indices[1]);
    // 設定可選參數 indicesOptions：解析不可用的索引及展開萬用字元運算式
    request.indicesOptions(IndicesOptions.lenientExpandOpen());
    return request;
  }
// 以非同步方式執行跨索引欄位搜索請求
  public void executeFieldCapabilitiesRequestAsync(String field, String[]
        indices) {
    // 建置 FieldCapabilitiesRequest
    FieldCapabilitiesRequest request = buildFieldCapabilitiesRequest(field,
        indices);
    // 設定監聽器
    ActionListener<FieldCapabilitiesResponse> listener =
        new ActionListener<FieldCapabilitiesResponse>() {
          @Override
          public void onResponse(FieldCapabilitiesResponse response) {
          }
          @Override
          public void onFailure(Exception e) {
          }
        };
    // 執行非同步請求
    try {
      restClient.fieldCapsAsync(request, RequestOptions.DEFAULT, listener);
    } catch (Exception e) {
      e.printStackTrace();
```

```
    } finally {
      // 關閉 Elasticsearch 連接
      closeEs();
    }
  }
```

當然，在非同步請求執行過程中可能會出現例外，例外的處理與同步方式執行情況相同。

3 解析跨索引欄位搜索請求的回應結果

不論同步方式，還是非同步方式，在跨索引欄位搜索請求執行後，用戶端均需要對回應結果 FieldCapabilitiesResponse 進行處理和解析。

FieldCapabilitiesResponse 中包含了每個索引中資料是否可被搜索和聚合的資訊，還包含了被搜索欄位在對應索引中的貢獻值。

以同步請求方式為例，FieldCapabilitiesResponse 中的解析程式共分為三層，分別是 Controller 層、Service 層和 ServiceImpl 實現層。

在 Controller 層中新增 FieldCapabilitiesController，用來撰寫跨索引欄位搜索的介面，程式如下所示：

```
package com.niudong.esdemo.controller;
import java.util.List;
import org.elasticsearch.common.Strings;
import org.springframework.beans.factory.annotation.Autowired;
import org.springframework.web.bind.annotation.RequestMapping;
import org.springframework.web.bind.annotation.RestController;
import com.google.common.base.Splitter;
import com.niudong.esdemo.service.FieldCapabilitiesService;
@RestController
@RequestMapping("/springboot/es/fieldsearch")
public class FieldCapabilitiesController {
  @Autowired
  private FieldCapabilitiesService fieldCapabilitiesService;
```

```
// 以同步方式執行 MultiSearchRequest
@RequestMapping("/sr")
public String executeFieldSearchRequest(String field, String indices) {
  // 參數驗證
  if (Strings.isNullOrEmpty(field) || Strings.isNullOrEmpty(indices)) {
    return "Parameters are wrong!";
  }
  // 將英文逗點分隔的字串切分成陣列
  List<String> indicesList = Splitter.on(",").splitToList(indices);
  fieldCapabilitiesService.executeFieldCapabilitiesRequest(field,
      indicesList.toArray(new String[indicesList.size()]));
  return "Execute FieldSearchRequest success!";
  }
}
```

在 Service 層中新增 FieldCapabilitiesService，用來撰寫跨索引欄位搜索的
Service 介面定義，程式如下所示：

```
package com.niudong.esdemo.service;
public interface FieldCapabilitiesService {
  // 以同步方式執行跨索引欄位搜索請求
  public void executeFieldCapabilitiesRequest(String field, String[]
      indices);
}
```

在 ServiceImpl 實現層新增 FieldCapabilitiesServiceImpl 類別，用來撰寫跨索引
欄位搜索的 Service 介面的實作方式邏輯，程式如下所示：

```
package com.niudong.esdemo.service.impl;
import java.util.Map;
import java.util.Set;
import javax.annotation.PostConstruct;
import org.apache.commons.logging.Log;
import org.apache.commons.logging.LogFactory;
import org.apache.http.HttpHost;
import org.elasticsearch.action.ActionListener;
import org.elasticsearch.action.fieldcaps.FieldCapabilities;
import org.elasticsearch.action.fieldcaps.FieldCapabilitiesRequest;
```

```
import org.elasticsearch.action.fieldcaps.FieldCapabilitiesResponse;
import org.elasticsearch.action.support.IndicesOptions;
import org.elasticsearch.client.RequestOptions;
import org.elasticsearch.client.RestClient;
import org.elasticsearch.client.RestHighLevelClient;
import org.springframework.stereotype.Service;
import com.niudong.esdemo.service.FieldCapabilitiesService;
@Service
public class FieldCapabilitiesServiceImpl implements FieldCapabilitiesService {
  private static Log log = LogFactory.getLog(FieldCapabilitiesServiceImpl.
      class);
  private RestHighLevelClient restClient;
  // 初始化連接
  @PostConstruct
  public void initEs() {
    restClient = new RestHighLevelClient(RestClient.builder(new HttpHost
        ("localhost", 9200, "http"),
      new HttpHost("localhost", 9201, "http")));
    log.info("ElasticSearch init in service.");
  }
  // 關閉連接
  public void closeEs() {
    try {
      restClient.close();
    } catch (Exception e) {
      e.printStackTrace();
    }
  }
}
// 建置 FieldCapabilitiesRequest
  public FieldCapabilitiesRequest buildFieldCapabilitiesRequest(String field,
      String[] indices) {
    FieldCapabilitiesRequest request =
        new FieldCapabilitiesRequest().fields(field).indices(indices[0],
            indices[1]);
    // 設定可選參數 indicesOptions：解析不可用的索引及展開萬用字元運算式
    request.indicesOptions(IndicesOptions.lenientExpandOpen());
    return request;
```

```
}
// 以同步方式執行跨索引欄位搜索請求
public void executeFieldCapabilitiesRequest(String field, String[] indices)
{
  // 建置 FieldCapabilitiesRequest
  FieldCapabilitiesRequest request = buildFieldCapabilitiesRequest(field,
      indices);
  try {
    FieldCapabilitiesResponse response = restClient.fieldCaps(request,
        RequestOptions.DEFAULT);
    // 處理傳回結果 FieldCapabilitiesResponse
    processFieldCapabilitiesResponse(response, field, indices);
  } catch (Exception e) {
    e.printStackTrace();
  } finally {
    // 關閉 Elasticsearch 連接
    closeEs();
  }
}
// 處理傳回結果 FieldCapabilitiesResponse
private void processFieldCapabilitiesResponse(FieldCapabilitiesResponse
    response, String field,
    String[] indices) {
  // 取得欄位中可能含有的類型的對映
  Map<String, FieldCapabilities> fieldResponse = response.getField(field);
  Set<String> set = fieldResponse.keySet();
  // 取得文字欄位類型下的資料
  FieldCapabilities textCapabilities = fieldResponse.get("text");
  // 資料是否可被搜索到
  boolean isSearchable = textCapabilities.isSearchable();
  log.info("isSearchable is " + isSearchable);
  // 資料是否可聚合
  boolean isAggregatable = textCapabilities.isAggregatable();
  log.info("isAggregatable is " + isAggregatable);
  // 取得特定欄位類型下的索引
  String[] indicesArray = textCapabilities.indices();
  if (indicesArray != null) {
```

```
      log.info("indicesArray is " + indicesArray.length);
   }
   // field 欄位不能被搜索到的索引集合
   String[] nonSearchableIndices = textCapabilities.nonSearchableIndices();
   if (nonSearchableIndices != null) {
     log.info("nonSearchableIndices is " + nonSearchableIndices.length);
   }
   // field 欄位不能被聚合到的索引集合
   String[] nonAggregatableIndices = textCapabilities.nonAggregatableIndices();
   if (nonAggregatableIndices != null) {
     log.info("nonAggregatableIndices is " + nonAggregatableIndices.length);
   }
}
```

隨後編輯專案，在專案根目錄下輸入以下指令：

```
mvn clean package
```

透過以下指令啟動專案服務：

```
java -jar ./target/esdemo-0.0.1-SNAPSHOT.jar
```

在專案服務啟動後，在瀏覽器中呼叫以下介面檢視跨索引欄位搜索的執行情況：

```
http://localhost:8080/springboot/es/fieldsearch/sr?field=content&indices=
ultraman,ultraman1
```

介面呼叫後，在伺服器中輸出如下所示內容：

```
2019-09-03 11:52:35.378  INFO 29064 --- [nio-8080-exec-1] c.n.e.s.i.FieldCapabilitiesServiceImpl   : isSearchable is true
2019-09-03 11:52:35.378  INFO 29064 --- [nio-8080-exec-1] c.n.e.s.i.FieldCapabilitiesServiceImpl   : isAggregatable is false
```

在 ultraman 和 ultraman1 兩個索引中，搜索欄位 content 均能搜索到資料，但不能進行聚合。

此時，如果呼叫上述介面的瀏覽器中顯示以下內容，則表示介面呼叫成功：

```
Execute FieldSearchRequest success!
```

7.5 搜索結果的排序評估

Elasticsearch 提供了對搜索結果進行排序評估的介面,即 Ranking Evaluation API。Elasticsearch 提供了 rankeval 方法,對一組搜索請求的結果進行排序評估,以便衡量搜索結果的品質。

首先為搜索請求提供一組手動評級的文件,隨後評估批次搜索請求的品質,並計算搜索相關指標,如傳回結果的平均倒數排名、精度或折扣累積收益等。

■ 建置排序評估請求

在對搜索結果進行排序評估之前,需要建置排序評估請求,即 RankEvalRequest。在建立 RankEvalRequest 之前,需要建立 RankEvalRequest 的依賴物件 RankEvalSpec。RankEvalSpec 用於描述評估規則,使用者需要定義 RankEvalRequest 的計算指標及每個搜索請求的分級文件列表。

此外,在建立排序評估請求時,需要將目標索引名稱和 RankEvalSpec 作為參數。

建立排序評估物件 RankEvalRequest 的程式如下所示:

```
// 建置 RankEvalRequest
  public RankEvalRequest buildRankEvalRequest(String index, String
      documentId, String field, String content) {
    EvaluationMetric metric = new PrecisionAtK();
    List<RatedDocument> ratedDocs = new ArrayList<>();
    // 增加按索引名稱、ID 和分級指定的分級文件
    ratedDocs.add(new RatedDocument(index, documentId, 1));
    SearchSourceBuilder searchQuery = new SearchSourceBuilder();
    // 建立要評估的搜索查詢
    searchQuery.query(QueryBuilders.matchQuery(field, content));
    // 將前三部分合併為 RatedRequest
    RatedRequest ratedRequest = new RatedRequest("content_query", ratedDocs,
        searchQuery);
    List<RatedRequest> ratedRequests = Arrays.asList(ratedRequest);
    // 建立排序評估標準
```

```
RankEvalSpec specification = new RankEvalSpec(ratedRequests, metric);
// 建立排序評估請求
RankEvalRequest request = new RankEvalRequest(specification, new String[]
    {index});
return request;
}
```

2 執行排序評估請求

在 RankEvalRequest 建置後，即可執行排序評估請求。與文件索引請求類似，排序評估請求也有同步和非同步兩種執行方式。

✍ 同步方式

當以同步方式執行排序評估請求時，用戶端會等待 Elasticsearch 伺服器傳回的查詢結果 RankEvalResponse。在收到 RankEvalResponse 後，用戶端會繼續執行相關的邏輯程式。以同步方式執行的核心程式如下所示：

```
RankEvalResponse response = client.rankEval(request, RequestOptions.DEFAULT);
```

以同步方式執行的全部程式如下所示：

```
// 建置 RankEvalRequest
  public RankEvalRequest buildRankEvalRequest(String index, String
    documentId, String field,
    String content) {
  EvaluationMetric metric = new PrecisionAtK();
  List<RatedDocument> ratedDocs = new ArrayList<>();
  // 增加按索引名稱、ID 和分級指定的分級文件
  ratedDocs.add(new RatedDocument(index, documentId, 1));
  SearchSourceBuilder searchQuery = new SearchSourceBuilder();
  // 建立要評估的搜索查詢
  searchQuery.query(QueryBuilders.matchQuery(field, content));
  // 將前三部分合併為 RatedRequest
  RatedRequest ratedRequest = new RatedRequest("content_query", ratedDocs,
      searchQuery);
  List<RatedRequest> ratedRequests = Arrays.asList(ratedRequest);
  // 建立排序評估標準
  RankEvalSpec specification = new RankEvalSpec(ratedRequests, metric);
```

```
  // 建立排序評估請求
  RankEvalRequest request = new RankEvalRequest(specification, new String[]
      {index});
  return request;
}
// 以同步方式執行 RankEvalRequest
public void executeRankEvalRequest(String index, String documentId, String
    field,
    String content) {
  // 建置 RankEvalRequest
  RankEvalRequest request = buildRankEvalRequest(index, documentId, field,
      content);
  try {
    RankEvalResponse response = restClient.rankEval(request, RequestOptions.
        DEFAULT);
  } catch (Exception e) {
    e.printStackTrace();
  } finally {
    // 關閉 Elasticsearch 連接
    closeEs();
  }
}
```

▨ 非同步方式

當以非同步方式執行排序評估請求時,進階用戶端不必同步等待請求結果的傳回,可以直接向介面呼叫方傳回非同步介面執行成功的結果。

為了處理非同步傳回的回應資訊或處理在請求執行過程中引發的例外資訊,使用者需要指定監聽器。以非同步方式執行的核心程式如下所示:

```
client.rankEvalAsync(request, RequestOptions.DEFAULT, listener);
```

其中,listener 為監聽器。

在非同步請求處理後,如果請求執行成功,則呼叫 ActionListener 類別中的 onResponse 方法進行相關邏輯的處理;如果請求執行失敗,則呼叫 ActionListener 類別中的 onFailure 方法進行相關邏輯的處理。

以非同步方式執行的程式如下所示：

```
// 建置 RankEvalRequest
  public RankEvalRequest buildRankEvalRequest(String index, String documentId,
      String field,
    String content) {
  EvaluationMetric metric = new PrecisionAtK();
  List<RatedDocument> ratedDocs = new ArrayList<>();
  // 增加按索引名稱、ID 和分級指定的分級文件
  ratedDocs.add(new RatedDocument(index, documentId, 1));
  SearchSourceBuilder searchQuery = new SearchSourceBuilder();
  // 建立要評估的搜索查詢
  searchQuery.query(QueryBuilders.matchQuery(field, content));
  // 將前三部分合併為 RatedRequest
  RatedRequest ratedRequest = new RatedRequest("content_query", ratedDocs,
      searchQuery);
  List<RatedRequest> ratedRequests = Arrays.asList(ratedRequest);
  // 建立排序評估標準
  RankEvalSpec specification = new RankEvalSpec(ratedRequests, metric);
  // 建立排序評估請求
  RankEvalRequest request = new RankEvalRequest(specification, new String[]
      {index});
  return request;
  }
// 以非同步方式執行 RankEvalRequest
  public void executeRankEvalRequestAsync(String index, String documentId,
      String field,
    String content) {
  // 建置 RankEvalRequest
  RankEvalRequest request = buildRankEvalRequest(index, documentId, field,
      content);
  // 建置監聽器
  ActionListener<RankEvalResponse> listener = new ActionListener
      <RankEvalResponse>() {
   @Override
   public void onResponse(RankEvalResponse response) {
   }
```

```
    @Override
    public void onFailure(Exception e) {
    }
  };
  try {
    restClient.rankEvalAsync(request, RequestOptions.DEFAULT, listener);
  } catch (Exception e) {
    e.printStackTrace();
  } finally {
    // 關閉 Elasticsearch 連接
    closeEs();
  }
}
```

當然，在非同步請求執行過程中可能會出現例外，例外的處理與同步方式執行情況相同。

3 解析排序評估請求的回應結果

不論同步方式，還是非同步方式，在排序評估請求執行後，用戶端均需要對回應結果 RankEvalResponse 進行處理和解析。RankEvalResponse 中不僅包含了整體評估分數和查詢每個搜索請求的分數，還包含了有關搜索命中的詳細資訊及每個結果度量計算的詳細資訊。

RankEvalResponse 的解析程式共分為三層，以同步方式為例，分別是 Controller 層、Service 層和 ServiceImpl 實現層。

其中，在 Controller 層中新增 RankEvalController 類別，用於撰寫排序評估請求對應介面，程式如下所示：

```
package com.niudong.esdemo.controller;
import org.elasticsearch.common.Strings;
import org.springframework.beans.factory.annotation.Autowired;
import org.springframework.web.bind.annotation.RequestMapping;
import org.springframework.web.bind.annotation.RestController;
import com.niudong.esdemo.service.RankEvalService;
@RestController
```

```
@RequestMapping("/springboot/es/ranksearch")
public class RankEvalController {
  @Autowired
  private RankEvalService rankEvalService;

// 以同步方式執行 RankEvalResponse
 @RequestMapping("/sr")
 public String executeRankEvalRequest(String indexName, String document,
     String field, String content) {
   // 參數驗證
   if (Strings.isNullOrEmpty(indexName) || Strings.isNullOrEmpty(document) ||
       Strings.isNullOrEmpty(field) || Strings.isNullOrEmpty(content)) {
     return "Parameters are wrong!";
   }

   rankEvalService.executeRankEvalRequest(indexName, document, field, content);

   return "Execute RankEvalRequest success!";
 }
}
```

在 Service 層新增 RankEvalService 介面類別，用於撰寫 Service 層中的業務介面定義，程式如下所示：

```
package com.niudong.esdemo.service;
public interface RankEvalService {
  // 以同步方式執行 RankEvalRequest
  public void executeRankEvalRequest(String index, String documentId, String
      field, String content);
}
```

在 ServiceImpl 層中新增 RankEvalServiceImpl 介面實現類別，用於撰寫 Service 層中業務介面的邏輯實現，程式如下所示：

```
package com.niudong.esdemo.service.impl;
import java.util.ArrayList;
import java.util.Arrays;
import java.util.List;
```

```java
import java.util.Map;
import javax.annotation.PostConstruct;
import org.apache.commons.logging.Log;
import org.apache.commons.logging.LogFactory;
import org.apache.http.HttpHost;
import org.elasticsearch.action.ActionListener;
import org.elasticsearch.client.RequestOptions;
import org.elasticsearch.client.RestClient;
import org.elasticsearch.client.RestHighLevelClient;
import org.elasticsearch.index.query.QueryBuilders;
import org.elasticsearch.index.rankeval.EvalQueryQuality;
import org.elasticsearch.index.rankeval.EvaluationMetric;
import org.elasticsearch.index.rankeval.MetricDetail;
import org.elasticsearch.index.rankeval.PrecisionAtK;
import org.elasticsearch.index.rankeval.RankEvalRequest;
import org.elasticsearch.index.rankeval.RankEvalResponse;
import org.elasticsearch.index.rankeval.RankEvalSpec;
import org.elasticsearch.index.rankeval.RatedDocument;
import org.elasticsearch.index.rankeval.RatedRequest;
import org.elasticsearch.index.rankeval.RatedSearchHit;
import org.elasticsearch.search.builder.SearchSourceBuilder;
import org.springframework.stereotype.Service;
import com.niudong.esdemo.service.RankEvalService;
@Service
public class RankEvalServiceImpl implements RankEvalService {
  private static Log log = LogFactory.getLog(RankEvalServiceImpl.class);
  private RestHighLevelClient restClient;
  // 初始化連接
  @PostConstruct
  public void initEs() {
    restClient = new RestHighLevelClient(RestClient.builder(new HttpHost
        ("localhost", 9200, "http"),
        new HttpHost("localhost", 9201, "http")));
    log.info("ElasticSearch init in service.");
  }
  // 關閉連接
  public void closeEs() {
```

```
    try {
      restClient.close();
    } catch (Exception e) {
      e.printStackTrace();
    }
  }


// 建置 RankEvalRequest
public RankEvalRequest buildRankEvalRequest(String index, String documentId,
    String field,
    String content) {
  EvaluationMetric metric = new PrecisionAtK();
  List<RatedDocument> ratedDocs = new ArrayList<>();
  // 增加按索引名稱、ID 和分級指定的分級文件
  ratedDocs.add(new RatedDocument(index, documentId, 1));
  SearchSourceBuilder searchQuery = new SearchSourceBuilder();
  // 建立要評估的搜索查詢
  searchQuery.query(QueryBuilders.matchQuery(field, content));
  // 將前三個部分合併為 RatedRequest
  RatedRequest ratedRequest = new RatedRequest("content_query", ratedDocs,
      searchQuery);
  List<RatedRequest> ratedRequests = Arrays.asList(ratedRequest);
  // 建立排序評估標準
  RankEvalSpec specification = new RankEvalSpec(ratedRequests, metric);
  // 建立排序評估請求
  RankEvalRequest request = new RankEvalRequest(specification, new String[]
      {index});
  return request;
}
// 以同步方式執行 RankEvalRequest
public void executeRankEvalRequest(String index, String documentId, String
    field,
    String content) {
  // 建置 RankEvalRequest
  RankEvalRequest request = buildRankEvalRequest(index, documentId, field,
      content);
  try {
```

```java
    RankEvalResponse response = restClient.rankEval(request, RequestOptions.
        DEFAULT);
    // 處理 RankEvalResponse
    processRankEvalResponse(response);
  } catch (Exception e) {
    e.printStackTrace();
  } finally {
    // 關閉 Elasticsearch 連接
    closeEs();
  }
}
// 處理 RankEvalResponse
private void processRankEvalResponse(RankEvalResponse response) {
  // 整體評價結果
  double evaluationResult = response.getMetricScore();
  log.info("evaluationResult is " + evaluationResult);
  Map<String, EvalQueryQuality> partialResults = response. getPartialResults();
  // 取得關鍵字 content_query 對應的評估結果
  EvalQueryQuality evalQuality = partialResults.get("content_query");
  log.info("content_query id is " + evalQuality.getId());
  // 每部分結果的度量分數
  double qualityLevel = evalQuality.metricScore();
  log.info("qualityLevel is " + qualityLevel);
  List<RatedSearchHit> hitsAndRatings = evalQuality.getHitsAndRatings();
  RatedSearchHit ratedSearchHit = hitsAndRatings.get(2);
  // 在分級搜索命中裡包含完全成熟的搜索命中 SearchHit
  log.info("SearchHit id is " + ratedSearchHit.getSearchHit().getId());
      // 分級搜索命中還包含一個可選的 <integer> 分級 Optional<Integer> ，如果文件
      // 在請求中未獲得分級，則該分級不存在
  log.info("rate's isPresent is " + ratedSearchHit.getRating().isPresent());
  MetricDetail metricDetails = evalQuality.getMetricDetails();
  String metricName = metricDetails.getMetricName();
  // 度量詳細資訊，以請求中使用的度量命名
  log.info("metricName is " + metricName);

  PrecisionAtK.Detail detail = (PrecisionAtK.Detail) metricDetails;
  // 在轉換到請求中使用的度量之後，度量詳細資訊提供了對度量計算部分的深入了解
```

```
    log.info("detail's relevantRetrieved is " + detail.getRelevantRetrieved
        ());
    log.info("detail's retrieved is " + detail.getRetrieved());
  }
}
```

隨後編輯專案，在專案根目錄下輸入以下指令：

```
mvn clean package
```

透過以下指令啟動專案服務：

```
java -jar ./target/esdemo-0.0.1-SNAPSHOT.jar
```

在專案服務啟動後，在瀏覽器中呼叫以下介面檢視排序評估的情況：

```
http://localhost:8080/springboot/es/ranksearch/sr?indexName=ultraman&document
=1&field=content&content=美聯儲
```

請求執行後，在伺服器中輸出如下所示內容：

```
2019-09-03 15:04:38.632  INFO 39048 --- [nio-8080-exec-2] c.n.e.service.impl.RankEvalServiceImpl   : evaluationResult is
 0.3333333333333333
2019-09-03 15:04:38.634  INFO 39048 --- [nio-8080-exec-2] c.n.e.service.impl.RankEvalServiceImpl   : content_query id is
 content_query
2019-09-03 15:04:38.645  INFO 39048 --- [nio-8080-exec-2] c.n.e.service.impl.RankEvalServiceImpl   : qualityLevel is 0.3
333333333333333
2019-09-03 15:04:38.647  INFO 39048 --- [nio-8080-exec-2] c.n.e.service.impl.RankEvalServiceImpl   : SearchHit id is 1
2019-09-03 15:04:38.650  INFO 39048 --- [nio-8080-exec-2] c.n.e.service.impl.RankEvalServiceImpl   : rate's isPresent is
 true
2019-09-03 15:04:38.651  INFO 39048 --- [nio-8080-exec-2] c.n.e.service.impl.RankEvalServiceImpl   : metricName is preci
sion
2019-09-03 15:04:38.662  INFO 39048 --- [nio-8080-exec-2] c.n.e.service.impl.RankEvalServiceImpl   : detail's relevantRe
trieved is 1
2019-09-03 15:04:38.662  INFO 39048 --- [nio-8080-exec-2] c.n.e.service.impl.RankEvalServiceImpl   : detail's retrieved
is 3
```

此時，如果在呼叫介面的瀏覽器中輸出以下內容，則表示介面呼叫成功：

```
Execute RankEvalRequest success!
```

7.6 搜索結果解釋

除排序評估介面外，Elasticsearch 還提供了搜索結果解釋 API，即 Explain API。Explain API 用於為查詢請求和相關的文件計算解釋性的分數。無論文件是否比對這個查詢請求，Elasticsearch 伺服器都可以給使用者提供一些有用的回饋。

1 建置搜索結果解釋請求

在發起搜索結果解釋請求前，需要先建置搜索結果解釋請求，即 ExplainRequest。ExplainRequest 有兩個必選參數，即索引名稱和文件 ID，同時需要透過 QueryBuilder 來建置查詢運算式。

建置搜索結果解釋請求的程式如下所示：

```
// 建置 ExplainRequest
  public ExplainRequest buildExplainRequest(String indexName, String
      document, String field, String content) {
    ExplainRequest request = new ExplainRequest(indexName, document);
    request.query(QueryBuilders.termQuery(field, content));

    return request;
}
```

同理，ExplainRequest 的建置也有可選參數供使用者進行設定。ExplainRequest 的可選參數主要有路由、搜索偏好、儲存欄位的控制、是否搜索文件來源等。

配製 ExplainRequest 的可選參數的程式如下所示：

```
// 建置 ExplainRequest
  public ExplainRequest buildExplainRequest(String indexName, String document,
String field, String content) {
    ExplainRequest request = new ExplainRequest(indexName, document);
    request.query(QueryBuilders.termQuery(field, content));
    // 設定路由
    request.routing("routing");
    // 使用首選參數，例如執行搜索以首選本地碎片。預設值是在分片之間隨機進行
    request.preference("_local");
    // 設定為 " 真 "，以檢索解釋的文件來源。還可以透過使用 " 包含原始程式碼 " 和 " 排除
    // 原始程式碼 " 來檢索部分文件
    request.fetchSourceContext(new FetchSourceContext(true, new String[]
        {field}, null));
    // 允許控制一部分的儲存欄位 ( 要求在對映中單獨儲存該欄位 )，並將其傳回作為說明文件
    request.storedFields(new String[] {field});
```

```
    return request;
  }
```

❷ 執行搜索結果解釋請求

在 ExplainRequest 建置後，即可執行搜索結果解釋請求。與文件索引請求類似，搜索結果解釋請求也有同步和非同步兩種執行方式。

◿ 同步方式

當以同步方式執行搜索結果解釋請求時，用戶端會等待 Elasticsearch 伺服器傳回的查詢結果 ExplainResponse。在收到 ExplainResponse 後，用戶端會繼續執行相關的邏輯程式。以同步方式執行的核心程式如下所示：

```
ExplainResponse response = client.explain(request, RequestOptions.DEFAULT);
```

以同步方式執行的完整程式如下所示：

```
// 建置 ExplainRequest
  public ExplainRequest buildExplainRequest(String indexName, String
      document, String field, String content) {
    ExplainRequest request = new ExplainRequest(indexName, document);
    request.query(QueryBuilders.termQuery(field, content));

    return request;
}
// 以同步方式執行 ExplainRequest
  public void executeExplainRequest(String indexName, String document,
      String field,
      String content) {
    // 建置 ExplainRequest
    ExplainRequest request = buildExplainRequest(indexName, document, field,
        content);
    // 執行請求，接收傳回結果
    try {
      ExplainResponse response = restClient.explain(request, RequestOptions.
          DEFAULT);
    } catch (Exception e) {
```

```
      e.printStackTrace();
    } finally {
      // 關閉 Elasticsearch 的連接
      closeEs();
    }
  }
```

◪ 非同步方式

當以非同步方式執行搜索結果解釋請求時,進階用戶端不必同步等待請求結果的傳回,可以直接向介面呼叫方傳回非同步介面執行成功的結果。

為了處理非同步傳回的回應資訊或處理在請求執行過程中引發的例外資訊,使用者需要指定監聽器。以非同步方式執行的核心程式如下所示:

```
client.explainAsync(request, RequestOptions.DEFAULT, listener);
```

其中,listener 是監聽器。

在非同步請求處理後,如果請求執行成功,則呼叫 ActionListener 類別中的 onResponse 方法進行相關邏輯的處理;如果請求執行失敗,則呼叫 ActionListener 類別中的 onFailure 方法進行相關邏輯的處理。

以非同步方式執行的完整程式如下所示:

```
// 建置 ExplainRequest
  public ExplainRequest buildExplainRequest(String indexName, String
      document, String field, String content) {
    ExplainRequest request = new ExplainRequest(indexName, document);
    request.query(QueryBuilders.termQuery(field, content));

    return request;
}
// 以非同步方式執行 ExplainRequest
  public void executeExplainRequestAsync(String indexName, String document,
      String field,
      String content) {
    // 建置 ExplainRequest
```

```
    ExplainRequest request = buildExplainRequest(indexName, document, field,
        content);
    // 建置監聽器
    ActionListener<ExplainResponse> listener = new ActionListener
<ExplainResponse>
        () {
      @Override
      public void onResponse(ExplainResponse explainResponse) {
      }
      @Override
      public void onFailure(Exception e) {
      }
    };

    // 執行請求，接收傳回結果
    try {
      restClient.explainAsync(request, RequestOptions.DEFAULT, listener);
    } catch (Exception e) {
      e.printStackTrace();
    } finally {
      // 關閉 Elasticsearch 連接
      closeEs();
    }
  }
```

當然，在非同步請求執行過程中可能會出現例外，例外的處理與同步方式執
行情況相同。

3 解析搜索結果解釋請求的回應結果

不論同步方式，還是非同步方式，在搜索合理解釋請求執行後，用戶端均需
要對請求的回應結果 ExplainResponse 進行處理和解析。以同步方式為例，程
式共分為三層，分別是 Controller 層、Service 層和 ServiceImpl 實現層。

其中，在 Controller 層中增加新類別 ExplainController，用來撰寫與搜索結果
解釋請求相關的介面，程式如下所示：

```java
package com.niudong.esdemo.controller;
import org.elasticsearch.common.Strings;
import org.springframework.beans.factory.annotation.Autowired;
import org.springframework.web.bind.annotation.RequestMapping;
import org.springframework.web.bind.annotation.RestController;
import com.niudong.esdemo.service.ExplainService;
/**
 *
@RestController
@RequestMapping("/springboot/es/explainsearch")
public class ExplainController {
  @Autowired
  private ExplainService explainService;
  // 以同步方式執行 ExplainRequest
  @RequestMapping("/sr")
  public String executeExplainRequest(String indexName, String document,
      String field,
      String content) {
    // 參數驗證
    if (Strings.isNullOrEmpty(indexName) || Strings.isNullOrEmpty(document)
        || Strings.isNullOrEmpty(field) || Strings.isNullOrEmpty(content)) {
      return "Parameters are wrong!";
    }
    explainService.executeExplainRequest(indexName, document, field, content);
    return "Execute ExplainRequest success!";
  }
}
```

在 Service 層中增加新類別 ExplainService，用來撰寫 Service 層中的搜索結果解釋方法宣告，程式如下所示：

```java
package com.niudong.esdemo.service;
public interface ExplainService {
  // 以同步方式執行 ExplainRequest
  public void executeExplainRequest(String indexName, String document,
      String field,
      String content);
}
```

在 ServiceImpl 實現層中新增 ExplainServiceImpl 類別，用來撰寫 Service 層中
的搜索結果解釋方法的實際邏輯實現，程式如下所示：

```java
package com.niudong.esdemo.service.impl;
import java.util.Map;
import javax.annotation.PostConstruct;
import org.apache.commons.logging.Log;
import org.apache.commons.logging.LogFactory;
import org.apache.http.HttpHost;
import org.apache.lucene.search.Explanation;
import org.elasticsearch.action.ActionListener;
import org.elasticsearch.action.explain.ExplainRequest;
import org.elasticsearch.action.explain.ExplainResponse;
import org.elasticsearch.client.RequestOptions;
import org.elasticsearch.client.RestClient;
import org.elasticsearch.client.RestHighLevelClient;
import org.elasticsearch.common.document.DocumentField;
import org.elasticsearch.index.get.GetResult;
import org.elasticsearch.index.query.QueryBuilders;
import org.elasticsearch.search.fetch.subphase.FetchSourceContext;
import org.springframework.stereotype.Service;
import com.niudong.esdemo.service.ExplainService;
@Service
public class ExplainServiceImpl implements ExplainService {
  private static Log log = LogFactory.getLog(ExplainServiceImpl.class);
  private RestHighLevelClient restClient;
  // 初始化連接
  @PostConstruct
  public void initEs() {
    restClient = new RestHighLevelClient(RestClient.builder(new HttpHost
        ("localhost", 9200, "http"),
        new HttpHost("localhost", 9201, "http")));
    log.info("ElasticSearch init in service.");
  }
  // 關閉連接
  public void closeEs() {
    try {
```

```
      restClient.close();
    } catch (Exception e) {
      e.printStackTrace();
    }
}
// 建置 ExplainRequest
public ExplainRequest buildExplainRequest(String indexName, String
    document, String field,
    String content) {
  ExplainRequest request = new ExplainRequest(indexName, document);
  request.query(QueryBuilders.termQuery(field, content));
  return request;
}
// 以同步方式執行 ExplainRequest
public void executeExplainRequest(String indexName, String document,
    String field,
    String content) {
  // 建置 ExplainRequest
  ExplainRequest request = buildExplainRequest(indexName, document, field,
      content);
  // 執行請求，接收傳回結果
  try {
    ExplainResponse response = restClient.explain(request,
        RequestOptions.DEFAULT);
    // 解析 ExplainResponse
    processExplainResponse(response);
  } catch (Exception e) {
    e.printStackTrace();
  } finally {
    // 關閉 Elasticsearch 連接
    closeEs();
  }
}
// 解析 ExplainResponse
private void processExplainResponse(ExplainResponse response) {
  // 解釋文件的索引名稱
  String index = response.getIndex();
```

```java
// 解釋文件的 ID
String id = response.getId();
// 檢視解釋的文件是否存在
boolean exists = response.isExists();
log.info("index is " + index + ";id is " + id + ";exists is " + exists);
```
// 解釋的文件與提供的查詢之間是否比對（比對是從後台的 Lucene 解釋中檢索的，如果
// Lucene 解釋建模比對，則傳回 true，否則傳回 false）
```java
boolean match = response.isMatch();
// 檢視是否存在此請求的 Lucene 解釋
boolean hasExplanation = response.hasExplanation();
log.info("match is " + match + ";hasExplanation is " + hasExplanation);
// 取得 Lucene 解釋物件（如果存在）
Explanation explanation = response.getExplanation();
if (explanation != null) {
  log.info("explanation is " + explanation.toString());
}
// 如果檢索到來源或儲存欄位，則取得 getresult 物件
GetResult getResult = response.getGetResult();
if (getResult == null) {
  return;
}
// getresult 內部包含兩個對映，用於儲存分析的來源欄位和儲存的欄位
// 以 Map 形式檢索來源
Map<String, Object> source = getResult.getSource();
if (source == null) {
  return;
}
for (String str : source.keySet()) {
  log.info("str key is " + str);
}
// 以對映形式檢索指定的儲存欄位
Map<String, DocumentField> fields = getResult.getFields();
if (fields == null) {
  return;
}
for (String str : fields.keySet()) {
  log.info("field str key is " + str);
```

```
    }
  }
}
```

隨後編輯專案，在專案根目錄下輸入以下指令：

```
mvn clean package
```

透過以下指令啟動專案服務：

```
java -jar ./target/esdemo-0.0.1-SNAPSHOT.jar
```

在專案服務啟動後，在瀏覽器中呼叫以下介面檢視搜索結果解釋請求的執行情況：

```
http://localhost:8080/springboot/es/explainsearch/sr?indexName=
ultraman&document=1&field=content&content=美聯儲
```

請求執行後，在伺服器中輸出如下所示內容：

```
2019-09-04 14:58:34.216  INFO 48744 --- [nio-8080-exec-8] c.n.e.service.impl.ExplainServiceImpl    : index is ultraman;i
d is 1;exists is true
2019-09-04 14:58:34.216  INFO 48744 --- [nio-8080-exec-8] c.n.e.service.impl.ExplainServiceImpl    : match is false;hasE
xplanation is true
2019-09-04 14:58:34.219  INFO 48744 --- [nio-8080-exec-8] c.n.e.service.impl.ExplainServiceImpl    : explanation is 0.0
= no matching term
```

此時，如果在執行請求的瀏覽器中輸出以下內容，則表示介面呼叫成功：

```
Execute ExplainRequest success!
```

7.7 統計

除前文提及的排序評估介面、搜索解釋介面外，Elasticsearch 還提供了統計 API，即 Count API。統計介面用於執行查詢請求，並傳回與請求符合的統計結果。

1 建置統計請求

在執行統計請求前，需要先建置統計請求，即 CountRequest。CountRequest 的用法與 SearchRequest 類似，且二者都是基於 SearchSourceBuilder 實例。

建置 CountRequest 的程式如下所示：

```
// 建置 CountRequest
  public CountRequest buildCountRequest() {
  // 建立 CountRequest。如果沒有參數，則對所有索引執行
  CountRequest countRequest = new CountRequest();

  // 大多數搜索參數都需要增加到 SearchSourceBuilder 中
  SearchSourceBuilder searchSourceBuilder = new SearchSourceBuilder();

  // 在 SearchSourceBuilder 中增加 " 全部比對 " 查詢
  searchSourceBuilder.query(QueryBuilders.matchAllQuery());

  // 將 SearchSourceBuilder 增加到 CountRequest 中
  countRequest.source(searchSourceBuilder);
  return countRequest;
 }
```

同理，CountRequest 還提供了可選參數列表供使用者進行設定。CountRequest 的參數清單主要有索引設定項目、路由設定項目、IndicesOptions 控制項、首選參數設定項目等。

在 CountRequest 中設定可選參數的程式如下所示：

```
// 建置 CountRequest
  public CountRequest buildCountRequest(String indexName,String routeName) {
      // 將請求限制為特定名稱的索引
  CountRequest countRequest = new CountRequest(indexName).
      // 設定路由參數
      routing(routeName)
      // 設定 IndiceOptions，控制如何解析不可用索引及如何展開萬用字元運算式
      .indicesOptions(IndicesOptions.lenientExpandOpen())
      // 使用首選參數，例如執行搜索以首選本地分片。預設值是在分片之間隨機選擇
      .preference("_local");

  return countRequest;
 }
```

正如前文所提及的，在建置 CountRequest 時使用了 SearchSourceBuilder 實例。在統計介面的呼叫中，大多數控制搜索行為的選項都可以在 SearchSourceBuilder 實例中進行設定。設定常見選項的程式如下所示：

```
// 建置 CountRequest
public CountRequest buildCountRequest(String indexName, String routeName,
    String field,
    String content) {
  // 將請求限制為特定名稱的索引
  CountRequest countRequest = new CountRequest(indexName).
    // 設定路由參數
    routing(routeName)
    // 設定 IndicesOptions，控制如何解析不可用索引及如何展開萬用字元運算式
    .indicesOptions(IndicesOptions.lenientExpandOpen())
    // 使用首選參數，例如執行搜索以首選本地分片。預設值是在分片之間隨機選擇
    .preference("_local");
  // 使用預設選項建立 SearchSourceBuilder
  SearchSourceBuilder sourceBuilder = new SearchSourceBuilder();
  // 設定查詢可以是任意類型的 QueryBuilder
  sourceBuilder.query(QueryBuilders.termQuery(field, content));
  // 將 SearchSourceBuilder 增加到 CountRequest 中
  countRequest.source(sourceBuilder);
  return countRequest;
}
```

2 執行統計請求

在 CountRequest 建置後，即可執行統計請求。與文件索引請求類似，統計請求也有同步和非同步兩種執行方式。

同步方式

當以同步方式執行統計請求時，用戶端會等待 Elasticsearch 伺服器傳回的查詢結果 CountResponse。在收到 CountResponse 後，用戶端會繼續執行相關的邏輯程式。以同步方式執行的核心程式如下所示：

```
CountResponse countResponse = client.count(countRequest, RequestOptions.DEFAULT);
```

以同步方式執行的全部程式如下所示：

```
// 建置 CountRequest
  public CountRequest buildCountRequest() {
    // 建立 CountRequest。如果沒有參數，則對所有索引執行
    CountRequest countRequest = new CountRequest();
    // 大多數搜索參數都需要增加到 SearchSourceBuilder 中
    SearchSourceBuilder searchSourceBuilder = new SearchSourceBuilder();
    // 在 SearchSourceBuilder 中增加 " 全部比對 " 查詢
    searchSourceBuilder.query(QueryBuilders.matchAllQuery());
    // 把 SearchSourceBuilder 增加到 CountRequest 中
    countRequest.source(searchSourceBuilder);
    return countRequest;
  }
// 以同步方式執行 CountRequest
  public void executeCountRequest() {
    CountRequest countRequest = buildCountRequest();
    try {
      CountResponse countResponse = restClient.count(countRequest,
RequestOptions.DEFAULT);
    } catch (Exception e) {
      e.printStackTrace();
    } finally {
      // 關閉 Elasticsearch 連接
      closeEs();
    }
  }
```

▨ 非同步方式

當以非同步方式執行統計請求時，進階用戶端不必同步等待請求結果的傳回，可以直接向介面呼叫方傳回非同步介面執行成功的結果。

為了處理非同步傳回的回應資訊或處理在請求執行過程中引發的例外資訊，使用者需要指定監聽器。以非同步方式執行的核心程式如下所示：

```
client.countAsync(countRequest, RequestOptions.DEFAULT, listener);
```

其中，listener 為監聽器物件。

在非同步請求處理後，如果請求執行成功，則呼叫 ActionListener 類別中的 onResponse 方法進行相關邏輯的處理；如果請求執行失敗，則呼叫 ActionListener 類別中的 onFailure 方法進行相關邏輯的處理。

以非同步方式執行的全部程式如下所示：

```
// 以非同步方式執行 CountRequest
public void executeCountRequestAsync() {
  // 建置 CountRequest
  CountRequest countRequest = buildCountRequest();
  // 建置監聽器
  ActionListener<CountResponse> listener = new ActionListener
      <CountResponse>() {
   @Override
   public void onResponse(CountResponse countResponse) {
   }
   @Override
   public void onFailure(Exception e) {
   }
  };
  try {
   restClient.countAsync(countRequest, RequestOptions.DEFAULT, listener);
  } catch (Exception e) {
   e.printStackTrace();
  } finally {
   // 關閉 Elasticsearch 的連接
   closeEs();
  }
}
```

當然，在非同步請求執行過程中可能會出現例外，例外的處理與同步方式執行情況相同。

3 解析統計請求的回應結果

不論同步方式，還是非同步方式，在統計請求執行後，用戶端均需要對請求的回應結果 CountResponse 進行處理和解析。

CountResponse 提供了統計請求對應的結果命中總數和統計執行本身的詳細資訊，如 HTTP 狀態碼、請求是否提前終止等。

CountResponse 還提供了與統計請求對應的分片總數、成功執行與失敗執行的分片的統計資訊，以及有關分片等級執行的資訊等。使用者可以透過檢查 ShardSearchFailures 陣列來處理可能的失敗資訊。

以同步方式執行為例，CountResponse 的解析程式如下所示。程式共分為三層，分別是 Controller 層、Service 層和 ServiceImpl 實現層。

其中，在 Controller 層中新增 CountController 類別，用來撰寫與統計相關的介面，程式如下所示：

```
package com.niudong.esdemo.controller;
import org.springframework.beans.factory.annotation.Autowired;
import org.springframework.web.bind.annotation.RequestMapping;
import org.springframework.web.bind.annotation.RestController;
import com.niudong.esdemo.service.CountService;
@RestController
@RequestMapping("/springboot/es/countsearch")
public class CountController {
  @Autowired
  private CountService countService;
  // 以同步方式執行 CountRequest
  @RequestMapping("/sr")
  public String executeCount() {
    countService.executeCountRequest();
    return "Execute CountRequest success!";
  }
}
```

在 Service 層中新增 CountService 類別，用來撰寫 Service 層中介面方法的宣告，程式如下所示：

```
package com.niudong.esdemo.service;
public interface CountService {
  // 以同步方式執行 CountRequest
```

```
  public void executeCountRequest();
}
```

在 ServiceImpl 實現層中新增 CountServiceImpl 類別，用來撰寫實際的業務邏輯實現，程式如下所示：

```
package com.niudong.esdemo.service.impl;
import javax.annotation.PostConstruct;
import org.apache.commons.logging.Log;
import org.apache.commons.logging.LogFactory;
import org.apache.http.HttpHost;
import org.elasticsearch.action.ActionListener;
import org.elasticsearch.action.search.ShardSearchFailure;
import org.elasticsearch.action.support.IndicesOptions;
import org.elasticsearch.client.RequestOptions;
import org.elasticsearch.client.RestClient;
import org.elasticsearch.client.RestHighLevelClient;
import org.elasticsearch.client.core.CountRequest;
import org.elasticsearch.client.core.CountResponse;
import org.elasticsearch.index.query.QueryBuilders;
import org.elasticsearch.rest.RestStatus;
import org.elasticsearch.search.builder.SearchSourceBuilder;
import org.springframework.stereotype.Service;
import com.niudong.esdemo.service.CountService;
@Service
public class CountServiceImpl implements CountService {
  private static Log log = LogFactory.getLog(CountServiceImpl.class);
  private RestHighLevelClient restClient;
  // 初始化連接
  @PostConstruct
  public void initEs() {
    restClient = new RestHighLevelClient(RestClient.builder(new HttpHost
        ("localhost", 9200, "http"),
        new HttpHost("localhost", 9201, "http")));
    log.info("ElasticSearch init in service.");
  }
  // 關閉連接
  public void closeEs() {
```

```
    try {
      restClient.close();
    } catch (Exception e) {
      e.printStackTrace();
    }
  }
  // 建置 CountRequest
  public CountRequest buildCountRequest() {
    // 建立 CountRequest。如果沒有參數，則將針對所有索引執行
    CountRequest countRequest = new CountRequest();
    // 大多數搜索參數都需要增加到 SearchSourceBuilder 中
    SearchSourceBuilder searchSourceBuilder = new SearchSourceBuilder();
    // 在 SearchSourceBuilder 中增加 " 全部比對 " 查詢
    searchSourceBuilder.query(QueryBuilders.matchAllQuery());
    // 將 SearchSourceBuilder 增加到 CountRequest 中
    countRequest.source(searchSourceBuilder);
    return countRequest;
  }
  // 以同步方式執行 CountRequest
  public void executeCountRequest() {
    CountRequest countRequest = buildCountRequest();
    try {
      CountResponse countResponse = restClient.count(countRequest,
RequestOptions.DEFAULT);
      // 解析 CountResponse
      processCountResponse(countResponse);
    } catch (Exception e) {
      e.printStackTrace();
    } finally {
      // 關閉 Elasticsearch 連接
      closeEs();
    }
  }
  // 解析 CountResponse
  private void processCountResponse(CountResponse countResponse) {
    // 統計請求對應的結果命中總數
    long count = countResponse.getCount();
```

```java
    // HTTP 狀態碼
    RestStatus status = countResponse.status();
    // 請求是否提前終止
    Boolean terminatedEarly = countResponse.isTerminatedEarly();
    log.info("count is " + count + ";status is " + status.getStatus() +
        ";terminatedEarly is "
      + terminatedEarly);
    // 與統計請求對應的分片總數
    int totalShards = countResponse.getTotalShards();
    // 執行統計請求跳過的分片數量
    int skippedShards = countResponse.getSkippedShards();
    // 執行統計請求成功的分片數量
    int successfulShards = countResponse.getSuccessfulShards();
    // 執行統計請求失敗的分片數量
    int failedShards = countResponse.getFailedShards();
    log.info("totalShards is " + totalShards + ";skippedShards is " +
        skippedShards
      + ";successfulShards is " + successfulShards + ";failedShards is " +
      failedShards);
    // 透過檢查 ShardSearchFailures 陣列來處理可能的失敗資訊
    if (countResponse.getShardFailures() == null) {
      return;
    }
    for (ShardSearchFailure failure : countResponse.getShardFailures()) {
      log.info("fail index is " + failure.index());
    }
  }
}
```

隨後編輯專案，在專案根目錄下輸入以下指令：

```
mvn clean package
```

透過以下指令啟動專案服務：

```
java -jar ./target/esdemo-0.0.1-SNAPSHOT.jar
```

在專案服務啟動後，在瀏覽器中呼叫以下介面檢視統計請求的情況：

```
http://localhost:8080/springboot/es/countsearch/sr
```

介面執行後，在伺服器中輸出如下所示內容：

```
2019-09-04 20:13:02.119  INFO 54400 --- [nio-8080-exec-1] c.n.e.service.impl.CountServiceImpl      : count is 6;status i
s 200;terminatedEarly is null
2019-09-04 20:13:02.119  INFO 54400 --- [nio-8080-exec-1] c.n.e.service.impl.CountServiceImpl      : totalShards is 2;sk
ippedShards is 0;successfulShards is 2;failedShards is 0
```

目前，在 Elasticsearch 中 存 在 2 個 索 引，索 引 名 稱 分 別 是 ultraman 和 ultraman1。每個索引各有 3 筆資料，因此輸出的 6 個資料與預期相符；而傳回碼 200 表示介面執行成功。

此時，如果在執行請求的瀏覽器中顯示以下內容，則表示介面執行成功：

```
Execute CountRequest success!
```

7.8 搜索過程解析

下面介紹 Elasticsearch 內部如何搜索文件。

7.8.1 對已知文件的搜索

如果被搜索的文件（不論是單一文件，還是批次文件）能夠從主分片或任意一個備份分片中被檢索到，則與索引文件過程相同，對已知文件的搜索也會用到路由演算法，Elasticsearch 中的路由演算法如下所示：

```
shard = hash(routing) % number_of_primary_shards
```

下面以圖 7-3 為例，展示在主分片搜索一個文件的必要步驟。

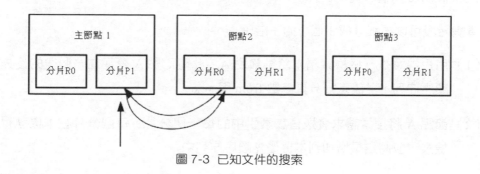

圖 7-3 已知文件的搜索

（1）用戶端給主節點 1 發送文件的 Get 請求，此時主節點 1 就成為協作節點。主節點使用路由演算法算出文件所在的主分片；隨後協作節點將請求轉發給主分片所在的節點 2，當然，也可以基於輪詢演算法轉發給備份分片。

（2）如圖 7-3 所示，主節點 1 根據文件的 ID 確定文件屬於分片 R0。分片 R0 對應的備份分片在三個節點上都有。此時，主節點 1 轉發請求到節點 2。

（3）節點 2 在本機分片進行搜索，並將目的文件資訊作為結果返給主節點 1。

對於讀取請求，為了在各節點間負載平衡，請求節點一般會為每個請求選擇不同的分片，一般採用輪詢演算法循環在所有備份分片中進行請求。

7.8.2 對未知文件的搜索

除對已知文件的搜索外，大部分請求實際上是不知道查詢準則會命中哪些文件的。這些被查詢準則命中的文件可能位於 Elasticsearch 叢集中的任意位置上。因此，搜索請求的執行不得不去詢問每個索引中的每一個分片。

在 Elasticsearch 中，搜索過程分為查詢階段（Query Phase）和取得階段（Fetch Phase）。

在查詢階段，查詢請求會廣播到索引中的每一個主分片和備份中，每一個分片都會在本機執行檢索，並在本機各建立一個優先順序佇列（Priority Queue）。該優先順序佇列是一份根據文件相關度指標進行排序的列表，列表的長度由 from 和 size 兩個分頁參數決定。

查詢階段可以再細分成 3 個小的子階段：

（1）用戶端發送一個檢索請求給某節點 A，此時節點 A 會建立一個空的優先順序佇列，並設定好分頁參數 from 與 size。

（2）節點 A 將搜索請求發送替該索引中的每一個分片，每個分片在本機執行檢索，並將結果增加到本機優先順序佇列中。

（3）每個分片傳回本機優先順序序列中所記錄的 ID 與 sort 值，平行處理送給
　　節點 A。節點 A 將這些值合併到自己的本機的優先順序佇列中，並做出
　　全域的排序。

在取得階段，主要是以上一階段找到所要搜索文件資料為基礎的實際位置，
將文件資料內容取回並傳回給用戶端。

在 Elasticsearch 中，預設的搜索類型就是上面介紹的 Query then Fetch。上述
描述運作方式就是 Query then fetch。Query then Fetch 有可能會出現評分偏離
的情形，幸好，Elasticsearch 還提供了一個稱為 "DFS Query then Fetch" 的搜
索方式，它和 Query then Fetch 大致相同，但是它會執行一個預查詢來計算整
體文件的 frequency。其處理過程如下所示：

（1）預查詢每個分片，詢問 Term 和 Document Frequency 等資訊。

（2）發送查詢請求到每個分片。

（3）找到各個分片中所有符合的文件，並使用全域的 Term/Document
　　Frequency 資訊進行評分。在執行過程中依然需要對結果建置一個優先佇
　　列，如排序等。

（4）傳回關於結果的中繼資料到請求節點。需要指出的是，此時實際文件還
　　沒有發送到請求節點，發送的只是分數。

（5）請求節點將來自所有分片的分數合併起來，並在請求節點上進行排序，
　　文件被按照查詢要求進行選擇。最後，實際文件從它們各自所在的獨立
　　的分片被檢索出來，結果被傳回給讀者。

7.8.3 對詞條的搜索

前面介紹了 Elasticsearch 是如何對已知文件和未知文件進行搜索，那麼實際到
一個分片，Elasticsearch 是如何按照詞條進行搜索的呢？

前文中已提及，Elasticsearch 分別為每個文件中的欄位建立了一個倒排索引。
倒排索引示意圖如圖 2-6 所示。

當要搜索中文詞 1 時，我們透過倒排索引可以知道，與待搜索中文詞 1 的文件是中文網頁 1、中文網頁 2 和中文網頁 3。

當詞筆數量較少時，我們可以順序檢查詞條取得結果，但如果詞條有成千上萬個呢？

Elasticsearch 為了能快速找到某個詞條，它對所有的詞條都進行了排序，隨後使用二分法尋找詞條，其尋找效率為 log(N)。這個過程就像查字典一樣，因此排序詞條的集合也稱為 Term Dictionary。

為了加強查詢效能，Elasticsearch 直接透過記憶體尋找詞條，而非從磁碟中讀取。但當詞條太多時，顯然 Term Dictionary 也會很大，此時全部放在記憶體有些不現實，於是引用了 Term Index。

Term Index 就像字典中的索引頁，其中的內容如字母 A 開頭的有哪些詞條，這些詞條分別在哪頁。透過 Term Index，Elasticsearch 可以快速地定位到 Term Dictionary 的某個 OffSet（位置偏移），然後從這個位置再往後順序尋找。

前面提及了單一詞條的搜索方法，而在實際應用中，更常見的常常是多個詞條連接成的「聯集查詢」，那麼 Elasticsearch 是如何以倒排索引實現快速查詢為基礎的呢？

核心思想是利用跳表快速做「與」運算，還有一種方法是利用 BitSet（點陣圖）逐位元「與」運算。

先來看利用跳表快速做「與」運算的方式，首先介紹跳表，如圖 7-4 所示。

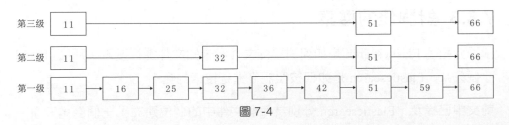

圖 7-4

跳表由多級鏈結串列組成，上一級是下一級元素的子集，因此上一級常常資料較少。在尋找時，資料從上級向下級逐級尋找。如尋找資料 36，從第三級中找到元素 11，從第二級中到資料 32，從第三級中的資料 32 之後尋找到 36，總共用了 3 次尋找。

那麼跳表是如何用在多詞條索引中的呢？

透過倒排索引，每個詞條都會有一個命中的文件 ID 列表（如果能命中的話），此時，我們可以找到這些文件 ID 列表中最短的那一個。接下來，用最短的文件 ID 列表中的 ID 一個一個在其他文件 ID 列表中進行尋找，都能找到的 ID 即為多個詞條的交集結果。

再來看 BitSet 逐位元「與」運算的方式。

BitSet 中的每一位只有兩個可能設定值，即 0 或 1。如果資料存在，則對應的標記置為 1，否則置為 0。

因此，將詞條的文件 ID 列表轉化為點陣圖後，將多個詞條對應的點陣圖取「與」運算，即可獲得交集結果。

7.9 基礎知識連結

在資料儲存領域，游標與捲動搜索的思想一脈相承。在資料庫領域，游標是一個十分重要的方法。

與使用捲動搜索類似，在使用游標時一般也有 4 個步驟：

（1）宣告游標，並把游標與資料庫查詢結果集關聯起來。

（2）開啟游標。

（3）使用游標操作資料，每次都需要指定下一次游標的 NEXT 值，NEXT 為預設的游標分析選項。

（4）關閉游標。

游標和 C 語言中的檔案控制代碼類似，一旦開啟檔案，該檔案控制代碼就可代表該檔案。同理，游標亦然。

7.10 小結

本章主要介紹了搜索 API 的使用，主要有關搜索、捲動搜索、批次搜索、跨索引欄位搜索、搜索結果的排序評估、搜索結果解釋和統計等介面。

索引實戰

爰著目錄
略述鴻烈

本章介紹索引相關 API 的使用。在介紹這些 API 的使用之前，我們先簡單介紹什麼是索引。

儲存資料的行為就叫作索引。在 Elasticsearch 中，文件會歸屬於一種類型，這些類型會存在於索引中。

Elasticsearch 叢集和資料庫中核心概念的對應關係如表 8-1 所示。

表 8-1

Elasticsearch 叢集	關聯式資料庫
索引	資料庫
類型	表
文件	行資料
欄位	列資料

需要指出的是，「索引」一詞在 Elasticsearch 中具有不同的屬性和含義。

（1）索引作為名詞時。一個索引有如傳統關聯式資料庫中的資料庫，它是儲存相關文件的地方。

（2）索引作為動詞時。「索引文件」指的是將一個文件儲存到索引（這裡是名詞）裡，以便該文件可以被檢索或查詢。這類似關聯式資料庫 SQL 敘述中的 INSERT 關鍵字，不同的是，如果文件已經存在，則新的文件將覆蓋舊的文件。

（3）指代倒排索引。在預設情況下，文件中的所有欄位都會被索引，即欄位都有一個倒排索引，也就是説，所有欄位都可被搜索。

進階用戶端支援以下索引相關的 API：

（1）Analyze API。 （10）Flush API。

（2）Create Index API。 （11）Flush Synced API。

（3）Delete Index API。 （12）Clear Cache API。

（4）Indices Exists API。 （13）Force Merge API。

（5）Open Index API。 （14）Rollover Index API。

（6）Close Index API。 （15）Get Index API。

（7）Shrink Index API。 （16）Index Aliases API。

（8）Split Index API。 （17）Exists Alias API。

（9）Refresh API。 （18）Get Alias API。

這些索引相關的 API 可以分為三種，索引的增刪改查、索引的更新及合併 / 拆分、索引別名的使用。下面一一介紹。

8.1 欄位索引分析

前文提及，在 Elasticsearch 中，當文件被索引到索引檔案時會進行斷詞操作，在使用者查詢時，也會基於斷詞進行檢索。

但有的時候，明明被索引的文件包含了下一步要搜索的關鍵字，但為什麼搜索不出來呢？由於索引過程對使用者來説是黑盒操作，因此使用者十分好奇文件索引後的斷詞結果是什麼？

對於這些場景，我們都可以使用分析介面，即 Analyze API，來分析欄位如何建立索引。

1 建置分析請求

在執行分析請求之前，需要建置分析請求，即 AnalyzeRequest。AnalyzeRequest 中一般包含要分析的文字，並指定如何執行分析。

建置 AnalyzeRequest 的程式增加在 IndexServiceImpl 類別中，如下所示：

```
// 建置 AnalyzeRequest
    public void buildAnalyzeRequest() {
    AnalyzeRequest request = new AnalyzeRequest();
    // 要包含的文字。多個字串被視為多值欄位
    request.text(" 中國天眼系統第一次探測到宇宙深處的神秘射頻訊號 ");
    // 內建分析器
request.analyzer("standard");
}
```

如果是對英文字串進行分析，則程式如下所示，詳見 buildAnalyzeRequest 方法：

```
// 要包含的文字。多個字串被視為多值欄位
    request.text("Some text to analyze", "Some more text to analyze");
    // 內建分析器
    request.analyzer("english");
```

上述對中、英文字串的分析採用了內建分析器，Elasticsearch 還支援使用者自訂分析器。自訂分析器的程式如下所示，詳見 buildAnalyzeRequest 方法：

```
// 自訂分析器 1
    request.text("<b>Some text to analyze</b>");
    // 設定字元篩選器
    request.addCharFilter("html_strip");
    // 設定標記器
    request.tokenizer("standard");
    // 增加內建標記篩選器
    request.addTokenFilter("lowercase");

    // 自訂分析器 2
    Map < String, Object > stopFilter = new HashMap<>();
    stopFilter.put("type", "stop");
```

```
// 自訂權杖篩選器 tokenfilter 的設定
stopFilter.put("stopwords", new String[] {
    "to"
});
// 增加自訂標記篩選器
request.addTokenFilter(stopFilter);
```

同理，AnalyzeRequest 也有可選參數供使用者進行設定，其可選參數主要有 explain 欄位和屬性設定，程式如下所示，詳見 buildAnalyzeRequest 方法：

```
// 可選參數
// 將 explain 設定為 true，為回應增加更多詳細資訊
request.explain(true);
// 設定屬性，允許只傳回使用者有興趣的權杖屬性
request.attributes("keyword", "type");
```

2 執行分析請求

在 AnalyzeRequest 建置後，即可執行分析請求。與文件索引請求類似，分析請求也有同步和非同步兩種執行方式。

同步方式

當以同步方式執行分析請求時，用戶端會等待 Elasticsearch 伺服器傳回的查詢結果 AnalyzeResponse。在收到 AnalyzeResponse 後，用戶端會繼續執行相關的邏輯程式。以同步方式執行的核心程式如下所示：

```
AnalyzeResponse response = client.indices().analyze(request, RequestOptions.
DEFAULT);
```

以同步方式執行的全部程式增加在 IndexServiceImpl 類別中，如下所示：

```
// 建置 AnalyzeRequest
  public AnalyzeRequest buildAnalyzeRequest(String text) {
    AnalyzeRequest request = new AnalyzeRequest();
    // 要包含的文字。多個字串被視為多值欄位
    request.text(text);
    // 內建分析器
    request.analyzer("standard");
```

```
   return request;
 }
 // 以同步方式執行 AnalyzeRequest
 public void executeAnalyzeRequest(String text) {
   // 建置 AnalyzeRequest
   AnalyzeRequest request = buildAnalyzeRequest(text);
   try {
     AnalyzeResponse response = restClient.indices().analyze(request,
RequestOptions.DEFAULT);
   } catch (Exception e) {
     e.printStackTrace();
   } finally {
     // 關閉 Elasticsearch 的連接
     closeEs();
   }
 }
```

非同步方式

當以非同步方式執行 AnalyzeRequest 時，進階用戶端不必同步等待請求結果的傳回，可以直接向介面呼叫方傳回非同步介面執行成功的結果。

為了處理非同步傳回的回應資訊或處理在請求執行過程中引發的例外資訊，使用者需要指定監聽器。以非同步方式執行的核心程式如下所示：

```
client.indices().analyzeAsync(request, RequestOptions.DEFAULT, listener);
```

其中，listener 為監聽器。

在非同步請求處理後，如果請求執行成功，則呼叫 ActionListener 類別中的 onResponse 方法進行相關邏輯的處理；如果請求執行失敗，則呼叫 ActionListener 類別中的 onFailure 方法進行相關邏輯的處理。

以非同步方式執行的全部程式增加在 IndexServiceImpl 類別中，如下所示：

```
// 以非同步方式執行 AnalyzeRequest
   public void executeAnalyzeRequestAsync(String text) {
   // 建置 AnalyzeRequest
```

```
AnalyzeRequest request = buildAnalyzeRequest(text);
// 建置監聽器
ActionListener<AnalyzeResponse> listener = new ActionListener
    <AnalyzeResponse>() {
  @Override
  public void onResponse(AnalyzeResponse analyzeTokens) {
  }
  @Override
  public void onFailure(Exception e) {
  }
};

try {
  restClient.indices().analyzeAsync(request, RequestOptions.DEFAULT,
      listener);
} catch (Exception e) {
  e.printStackTrace();
} finally {
  // 關閉 Elasticsearch 連接
  closeEs();
}
}
```

當然，在非同步請求執行過程中可能會出現例外，例外的處理與同步方式執行情況相同。

3 解析分析請求的回應結果

不論同步方式，還是非同步方式，用戶端均需要對 AnalyzeResponse 進行處理和解析。以同步方式為例，AnalyzeResponse 的解析程式共分為三層，分別是 Controller 層、Service 層和 ServiceImpl 實現層。

在 Controller 層中新增 IndexController 類別，用來撰寫索引操作相關的介面，程式如下所示：

```
package com.niudong.esdemo.controller;
import org.elasticsearch.common.Strings;
```

```
import org.springframework.beans.factory.annotation.Autowired;
import org.springframework.web.bind.annotation.RequestMapping;
import org.springframework.web.bind.annotation.RestController;
import com.niudong.esdemo.service.IndexService;
@RestController
@RequestMapping("/springboot/es/indexsearch")
public class IndexController {
  @Autowired
  private IndexService indexService;
  // 以同步方式執行 IndexRequest
  @RequestMapping("/sr")
  public String executeIndex(String text) {
    // 參數驗證
    if (Strings.isNullOrEmpty(text)) {
      return " Parameters are wrong!";
    }
    indexService.executeAnalyzeRequest(text);
    return "Execute IndexRequest success!";
  }
}
```

在 Service 層中新增 IndexService 類別，用來撰寫 Service 層中索引操作相關方法的宣告，程式如下所示：

```
package com.niudong.esdemo.service;
public interface IndexService {
  // 以同步方式執行 AnalyzeRequest
  public void executeAnalyzeRequest(String text);
}
```

在 ServiceImpl 實現層中新增 IndexServiceImpl 類別，用來撰寫 Service 層中索引操作相關方法的實作方式，程式如下所示：

```
package com.niudong.esdemo.service.impl;
import java.util.HashMap;
import java.util.List;
import java.util.Map;
import javax.annotation.PostConstruct;
```

```java
import org.apache.commons.logging.Log;
import org.apache.commons.logging.LogFactory;
import org.apache.http.HttpHost;
import org.elasticsearch.action.ActionListener;
import org.elasticsearch.action.admin.indices.analyze.AnalyzeRequest;
import org.elasticsearch.action.admin.indices.analyze.AnalyzeResponse;
import org.elasticsearch.action.admin.indices.analyze.DetailAnalyzeResponse;
import org.elasticsearch.client.RequestOptions;
import org.elasticsearch.client.RestClient;
import org.elasticsearch.client.RestHighLevelClient;
import org.springframework.stereotype.Service;
import com.niudong.esdemo.service.IndexService;
@Service
public class IndexServiceImpl implements IndexService {
  private static Log log = LogFactory.getLog(IndexServiceImpl.class);
  private RestHighLevelClient restClient;
  // 初始化連接
  @PostConstruct
  public void initEs() {
    restClient = new RestHighLevelClient(RestClient.builder(new HttpHost
        ("localhost", 9200, "http"),
      new HttpHost("localhost", 9201, "http")));
    log.info("ElasticSearch init in service.");
  }
  // 關閉連接
  public void closeEs() {
    try {
      restClient.close();
    } catch (Exception e) {
      e.printStackTrace();
    }
  }
  // 建置 AnalyzeRequest
  public AnalyzeRequest buildAnalyzeRequest(String text) {
    AnalyzeRequest request = new AnalyzeRequest();
    // 要包含的文字。多個字串被視為多值欄位
    request.text(text);
```

```
  // 內建分析器
  request.analyzer("standard");
  return request;
}
// 以同步方式執行 AnalyzeRequest
public void executeAnalyzeRequest(String text) {
  // 建置 AnalyzeRequest
  AnalyzeRequest request = buildAnalyzeRequest(text);
  try {
    AnalyzeResponse response = restClient.indices().analyze(request,
        RequestOptions.DEFAULT);
    // 解析 AnalyzeResponse
    processAnalyzeResponse(response);
  } catch (Exception e) {
    e.printStackTrace();
  } finally {
    // 關閉 Elasticsearch 連接
    closeEs();
  }
}
// 解析 AnalyzeResponse
private void processAnalyzeResponse(AnalyzeResponse response) {
  // AnalyzeToken 儲存了有關分析產生的單一權杖的資訊
  List<AnalyzeResponse.AnalyzeToken> tokens = response.getTokens();
  if (tokens == null) {
    return;
  }
  for (AnalyzeResponse.AnalyzeToken token : tokens) {
    log.info(token.getTerm() + " start offset is " + token.getStartOffset()
            + ";end offset is "
        + token.getEndOffset() + ";position is" + token.getPosition());
  }
  // 如果把 explain 設定為 true，則透過 detail 方法傳回資訊
  // DetailAnalyzeResponse 包含有關分析鏈中不同子步驟產生的權杖的更詳細的資訊
  DetailAnalyzeResponse detail = response.detail();
  if (detail == null) {
    return;
```

```
    }
    log.info("detail is " + detail.toString());
  }
}
```

隨後編輯專案，在專案根目錄下輸入以下指令：

```
mvn clean package
```

透過以下指令啟動專案服務：

```
java -jar ./target/esdemo-0.0.1-SNAPSHOT.jar
```

在專案服務啟動後，在瀏覽器中呼叫以下介面檢視欄位索引分析的情況：

```
http://localhost:8080/springboot/es/indexsearch/sr?text= 中國天眼系統第一次探測
到宇宙深處的神秘射頻訊號
```

此時，在伺服器中輸出內容如下所示：

```
2019-09-05 20:17:26.790  INFO 54068 --- [nio-8080-exec-1] c.n.e.service.impl.IndexServiceImpl      : 中 start offset is
0;end offset is 1;position is0
2019-09-05 20:17:26.790  INFO 54068 --- [nio-8080-exec-1] c.n.e.service.impl.IndexServiceImpl      : 国 start offset is
1;end offset is 2;position is1
2019-09-05 20:17:26.803  INFO 54068 --- [nio-8080-exec-1] c.n.e.service.impl.IndexServiceImpl      : 天 start offset is
2;end offset is 3;position is2
2019-09-05 20:17:26.806  INFO 54068 --- [nio-8080-exec-1] c.n.e.service.impl.IndexServiceImpl      : 眼 start offset is
3;end offset is 4;position is3
2019-09-05 20:17:26.807  INFO 54068 --- [nio-8080-exec-1] c.n.e.service.impl.IndexServiceImpl      : 系 start offset is
4;end offset is 5;position is4
2019-09-05 20:17:26.828  INFO 54068 --- [nio-8080-exec-1] c.n.e.service.impl.IndexServiceImpl      : 统 start offset is
5;end offset is 6;position is5
2019-09-05 20:17:26.829  INFO 54068 --- [nio-8080-exec-1] c.n.e.service.impl.IndexServiceImpl      : 首 start offset is
6;end offset is 7;position is6
2019-09-05 20:17:26.831  INFO 54068 --- [nio-8080-exec-1] c.n.e.service.impl.IndexServiceImpl      : 次 start offset is
7;end offset is 8;position is7
2019-09-05 20:17:26.833  INFO 54068 --- [nio-8080-exec-1] c.n.e.service.impl.IndexServiceImpl      : 探 start offset is
8;end offset is 9;position is8
2019-09-05 20:17:26.834  INFO 54068 --- [nio-8080-exec-1] c.n.e.service.impl.IndexServiceImpl      : 測 start offset is
9;end offset is 10;position is9
2019-09-05 20:17:26.840  INFO 54068 --- [nio-8080-exec-1] c.n.e.service.impl.IndexServiceImpl      : 到 start offset is
10;end offset is 11;position is10
2019-09-05 20:17:26.879  INFO 54068 --- [nio-8080-exec-1] c.n.e.service.impl.IndexServiceImpl      : 宇 start offset is
11;end offset is 12;position is11
2019-09-05 20:17:26.884  INFO 54068 --- [nio-8080-exec-1] c.n.e.service.impl.IndexServiceImpl      : 宙 start offset is
12;end offset is 13;position is12
2019-09-05 20:17:26.884  INFO 54068 --- [nio-8080-exec-1] c.n.e.service.impl.IndexServiceImpl      : 深 start offset is
13;end offset is 14;position is13
2019-09-05 20:17:26.885  INFO 54068 --- [nio-8080-exec-1] c.n.e.service.impl.IndexServiceImpl      : 处 start offset is
14;end offset is 15;position is14
2019-09-05 20:17:26.885  INFO 54068 --- [nio-8080-exec-1] c.n.e.service.impl.IndexServiceImpl      : 的 start offset is
15;end offset is 16;position is15
2019-09-05 20:17:26.886  INFO 54068 --- [nio-8080-exec-1] c.n.e.service.impl.IndexServiceImpl      : 神 start offset is
16;end offset is 17;position is16
2019-09-05 20:17:26.925  INFO 54068 --- [nio-8080-exec-1] c.n.e.service.impl.IndexServiceImpl      : 秘 start offset is
17;end offset is 18;position is17
2019-09-05 20:17:26.930  INFO 54068 --- [nio-8080-exec-1] c.n.e.service.impl.IndexServiceImpl      : 射 start offset is
18;end offset is 19;position is18
2019-09-05 20:17:26.931  INFO 54068 --- [nio-8080-exec-1] c.n.e.service.impl.IndexServiceImpl      : 电 start offset is
19;end offset is 20;position is19
2019-09-05 20:17:26.932  INFO 54068 --- [nio-8080-exec-1] c.n.e.service.impl.IndexServiceImpl      : 信 start offset is
20;end offset is 21;position is20
2019-09-05 20:17:26.961  INFO 54068 --- [nio-8080-exec-1] c.n.e.service.impl.IndexServiceImpl      : 号 start offset is
21;end offset is 22;position is21
```

8.2 建立索引

透過建立索引介面，建立使用者所需的索引。

1 建置建立索引請求

在執行建立索引請求前，需要建置索引請求，即 CreateIndexRequest。在建置
CreateIndexRequest 時，主要參數有索引名稱，以及與其連結的特定設定，如
分片數量和備份數量。程式增加在 IndexServiceImpl 類別中，如下所示：

```
// 建立索引
public void buildIndexRequest(String index, int shardsNumber, int
     replicasNumber) {
 CreateIndexRequest request = new CreateIndexRequest(index);
 // 設定分片數量和備份數量
 request.settings(Settings.builder().put("index.number_of_shards",
     shardsNumber)
   .put("index.number_of_replicas", replicasNumber));
}
```

在建置 CreateIndexRequest 時，還可以使用文件類型的 Map 對映來建立索
引。Elasticsearch 提供了字串、Map 對映、XContentBuilder 物件等不同的方式
提供對映來源，在上述 buildIndexRequest 方法中新增以下程式：

```
// 設定對映來源
  // 以字串方式提供對映來源
  request.mapping("{\n" + "  \"properties\": {\n" + "    \"message\": {\n"
     + "      \"type\": \"text\"\n" + "    }\n" + "  }\n" + "}",
             XContentType.JSON);
  // 以 Map 方式提供對映來源
  Map<String, Object> message = new HashMap<>();
  message.put("type", "text");
  Map<String, Object> properties = new HashMap<>();
  properties.put("message", message);
  Map<String, Object> mapping = new HashMap<>();
  mapping.put("properties", properties);
```

```
request.mapping(mapping);
// 以 XContentBuilder 方式提供對映來源
try {
  XContentBuilder builder = XContentFactory.jsonBuilder();
  builder.startObject();
  {
    builder.startObject("properties");
    {
      builder.startObject("message");
      {
        builder.field("type", "text");
      }
      builder.endObject();
    }
    builder.endObject();
  }
  builder.endObject();
  request.mapping(builder);
} catch (Exception e) {
  e.printStackTrace();
} finally {
}
```

在建立索引的過程中，還可以為索引設定別名。設定別名的方式有兩種，一種是在建立索引時設定，另一種是透過提供整個索引來源來設定。兩種設定索引別名的程式增加在上述 **buildIndexRequest** 方法中，如下所示：

```
// 設定別名：在索引建立時設定
request.alias(new Alias(index + "_alias").filter(QueryBuilders.termQuery
            ("user", "niudong")));
// 設定別名：透過提供整個索引來源
request.source("{\n" +
    "    \"settings\" : {\n" +
    "        \"number_of_shards\" : 1,\n" +
    "        \"number_of_replicas\" : 0\n" +
    "    },\n" +
    "    \"mappings\" : {\n" +
```

```
    "          \"properties\" : {\n" +
    "              \"message\" : { \"type\" : \"text\" }\n" +
    "          }\n" +
    "      },\n" +
    "      \"aliases\" : {\n" +
    "          \"niudong_alias\" : {}\n" +
    "      }\n" +
    "}", XContentType.JSON);
```

在建置 CreateIndexRequest 時，Elasticsearch 還提供了可選參數供使用者進行設定。可選參數主要有等待所有節點確認建立索引的逾時、從節點連接到主節點的逾時、在請求回應傳回前活動狀態的分片數量和拷貝數量。相關程式增加在 buildIndexRequest 方法中，如下所示：

```
// 可選參數設定
    // 等待所有節點確認建立索引的逾時
    request.setTimeout(TimeValue.timeValueMinutes(2));
    // 從節點連接到主節點的逾時
    request.setMasterTimeout(TimeValue.timeValueMinutes(1));

    // 在請求回應傳回前活動狀態的分片數量
    request.waitForActiveShards(ActiveShardCount.from(2));
    // 在請求回應傳回前活動狀態的拷貝數量
    request.waitForActiveShards(ActiveShardCount.DEFAULT);
```

2 執行建立索引請求

在 CreateIndexRequest 建置後，即可執行建立索引請求。與文件索引請求類似，建立索引請求也有同步和非同步兩種執行方式。

◿ 同步方式

當以同步方式執行建立索引請求時，用戶端會等待 Elasticsearch 伺服器傳回的查詢結果 CreateIndexResponse。在收到 CreateIndexResponse 後，用戶端會繼續執行相關的邏輯程式。以同步方式執行的程式增加在 IndexServiceImpl 類別中，如下所示：

```
// 建立索引請求
public CreateIndexRequest buildIndexRequest(String index) {
  CreateIndexRequest request = new CreateIndexRequest(index);
  // 設定預設分片數量和備份數量
  request.settings(
      Settings.builder().put("index.number_of_shards",
          3).put("index.number_of_replicas", 2));
  return request;
}
// 以同步方式執行建立索引請求
public void executeIndexRequest(String index) {
  // 建立索引請求
  CreateIndexRequest request = buildIndexRequest(index);
  try {
    CreateIndexResponse createIndexResponse =
        restClient.indices().create(request, RequestOptions.DEFAULT);
  } catch (Exception e) {
    e.printStackTrace();
  } finally {
    // 關閉 Elasticsearch 連接
    closeEs();
  }
}
```

非同步方式

當以非同步方式執行建立索引請求時，進階用戶端不必同步等待請求結果的傳回，可以直接向介面呼叫方傳回非同步介面執行成功的結果。

為了處理非同步傳回的回應資訊或處理在請求執行過程中引發的例外資訊，使用者需要指定監聽器。以非同步方式執行的核心程式如下所示：

```
client.indices().createAsync(request, RequestOptions.DEFAULT, listener);
```

其中，listener 是監聽器。

在非同步請求處理後，如果請求執行成功，則呼叫 ActionListener 類別中的 onResponse 方法進行相關邏輯的處理；如果請求執行失敗，則呼叫 ActionListener 類別中的 onFailure 方法進行相關邏輯的處理。

以非同步方式執行的全部程式增加在 IndexServiceImpl 類別中，如下所示：

```
// 以非同步方式執行建立索引請求
public void executeIndexRequestAsync(String index) {
  // 建立索引請求
  CreateIndexRequest request = buildIndexRequest(index);
  // 建立監聽器
  ActionListener<CreateIndexResponse> listener = new ActionListener
      <CreateIndexResponse>() {
    @Override
    public void onResponse(CreateIndexResponse createIndexResponse) {
    }
    @Override
    public void onFailure(Exception e) {
    }
  };
  // 非同步執行
  try {
    restClient.indices().createAsync(request, RequestOptions.DEFAULT,
        listener);
  } catch (Exception e) {
    e.printStackTrace();
  } finally {
    // 關閉 Elasticsearch 連接
    closeEs();
  }
}
```

當然，在非同步請求執行過程中可能會出現例外，例外的處理與同步方式執行情況相同。

❸ 解析建立索引請求的回應結果

不論同步方式，還是非同步方式，在建立索引請求執行後，用戶端均需要對回應結果 CreateIndexResponse 進行處理和解析。CreateIndexResponse 包含有關已執行操作的資訊，解析 CreateIndexResponse 的程式共分為三層，分別是 Controller 層、Service 層和 ServiceImpl 實現層。

在 Controller 層的 IndexController 類別中新增以下程式：

```
// 以同步方式執行 CreateIndexRequest
@RequestMapping("/create/sr")
public String executeCreateIndexRequest(String indexName) {
  // 參數驗證
  if (Strings.isNullOrEmpty(indexName)) {
    return "Parameters are wrong!";
  }
  indexService.executeIndexRequest(indexName);
  return "Execute CreateIndexRequest success!";
}
```

在 Service 層的 IndexService 類別中新增以下程式：

```
// 以同步方式執行建立索引請求
public void executeIndexRequest(String index);
```

在 ServiceImpl 實現層的 IndexServiceImpl 類別中新增以下程式：

```
// 以同步方式執行建立索引請求
public void executeIndexRequest(String index) {
  // 建立索引請求
  CreateIndexRequest request = buildIndexRequest(index);
  try {
    CreateIndexResponse createIndexResponse =
        restClient.indices().create(request, RequestOptions.DEFAULT);
    // 解析 CreateIndexResponse
    processCreateIndexResponse(createIndexResponse);
  } catch (Exception e) {
    e.printStackTrace();
  } finally {
    // 關閉 Elasticsearch 連接
    closeEs();
  }
}
// 解析 CreateIndexResponse
private void processCreateIndexResponse(CreateIndexResponse
  createIndexResponse) {
```

```
  // 所有節點是否已確認請求
  boolean acknowledged = createIndexResponse.isAcknowledged();
  // 是否在逾時前為索引中的每個分片啟動了所需數量的分片備份
  boolean shardsAcknowledged = createIndexResponse.isShardsAcknowledged();
  log.info("acknowledged is " + acknowledged + ";shardsAcknowledged is " +
      shardsAcknowledged);
}
```

隨後編輯專案，在專案根目錄下輸入以下指令：

```
mvn clean package
```

透過以下指令啟動專案服務：

```
java -jar ./target/esdemo-0.0.1-SNAPSHOT.jar
```

在專案服務啟動後，在瀏覽器中呼叫以下介面檢視索引建立情況：

```
http://localhost:8080/springboot/es/indexsearch/create/sr?indexName=ultraman2
```

請求執行後，如果在伺服器中輸出如下所示內容，則表示教材索引已經建立成功：

```
2019-09-06 10:29:05.758  INFO 78164 --- [nio-8080-exec-1] c.n.e.service.impl.IndexServiceImpl      : acknowledged is true;shardsAcknowledged is true
```

8.3 取得索引

在索引建立後，使用者可以透過取得索引請求檢視索引的建立情況。

1 建置取得索引請求

在取得索引請求前，需要建置取得索引請求，即 GetIndexRequest。GetIndexRequest 需要一個或多個索引參數，程式增加在 IndexServiceImpl 類別中，如下所示：

```
// 建置取得索引請求
  public GetIndexRequest buildGetIndexRequest(String index) {
    GetIndexRequest request = new GetIndexRequest(index);
```

```
    return request;
 }
```

　GetIndexRequest 提供了可選參數列表供使用者進行設定，可選參數主要有 IndicesOptions 和 includeDefaults。IndicesOptions 用於控制解析不可用索引及展開萬用字元運算式。includeDefaults 的值如果設定為 true，則對於未在索引上顯性設定的內容，將傳回預設值。程式增加在 IndexServiceImpl 類別中的 buildGetIndexRequest 方法內，如下所示：

```
// 如果設定為 true，則對於未在索引上顯性設定的內容，將傳回預設值
   request.includeDefaults(true);
   // 控制解析不可用索引及展開萬用字元運算式
   request.indicesOptions(IndicesOptions.lenientExpandOpen());
```

2 執行取得索引請求

在 GetIndexRequest 建置後，即可執行取得索引請求。與建立索引請求類似，取得索引請求也有同步和非同步兩種執行方式。

▨ 同步方式

當以同步方式執行取得索引請求時，用戶端會等待 Elasticsearch 伺服器傳回的查詢結果 GetIndexResponse。在收到 GetIndexResponse 後，用戶端會繼續執行相關的邏輯程式。以同步方式執行的程式增加在 IndexServiceImpl 類別中，如下所示：

```
// 以同步方式執行 GetIndexRequest
  public void excuteGetIndexRequest(String index) {
  GetIndexRequest request = buildGetIndexRequest(index);

  try {
    GetIndexResponse getIndexResponse = restClient.indices().get(request,
        RequestOptions.DEFAULT);
  } catch (Exception e) {
    e.printStackTrace();
  } finally {
    // 關閉 Elasticsearch 連接
```

```
    closeEs();
  }
}
```

非同步方式

當以非同步方式執行取得索引請求時,進階用戶端不必同步等待請求結果的傳回,可以直接向介面呼叫方傳回非同步介面執行成功的結果。

為了處理非同步傳回的回應資訊或處理在請求執行過程中引發的例外資訊,使用者需要指定監聽器。以非同步方式執行的核心程式如下所示:

```
client.indices().getAsync(request, RequestOptions.DEFAULT, listener);
```

其中,listener 為監聽器。

在非同步請求處理後,如果請求執行成功,則呼叫 ActionListener 類別中的 onResponse 方法進行相關邏輯的處理;如果請求執行失敗,則呼叫 ActionListener 類別中的 onFailure 方法進行相關邏輯的處理。

以非同步方式執行的全部程式增加在 IndexServiceImpl 類別中,如下所示:

```
// 以非同步方式執行 GetIndexRequest
public void excuteGetIndexRequestAsync(String index) {
  GetIndexRequest request = buildGetIndexRequest(index);
  // 建置監聽器
  ActionListener<GetIndexResponse> listener = new ActionListener
      <GetIndexResponse>() {
    @Override
    public void onResponse(GetIndexResponse getIndexResponse) {
    }
    @Override
    public void onFailure(Exception e) {
    }
  };
  try {
    restClient.indices().getAsync(request, RequestOptions.DEFAULT, listener);
  } catch (Exception e) {
```

```
    e.printStackTrace();
  } finally {
    // 關閉 Elasticsearch 連接
    closeEs();
  }
}
```

當然,在非同步請求執行過程中可能會出現例外,例外的處理與同步方式執行情況相同。

3 解析取得索引請求的回應結果

不論同步方式,還是非同步方式,在取得索引請求執行後,用戶端均需要對回應結果 GetIndexResponse 進行處理和解析。GetIndexResponse 中包含了有關已執行操作的資訊。解析程式共分為三層,分別是 Controller 層、Service 層和 ServiceImpl 實現層。

其中,在 Controller 層的 IndexController 類別中新增以下程式:

```
// 以同步方式執行 GetIndexRequest
@RequestMapping("/get/sr")
public String executeGetIndexRequest(String indexName) {
  // 參數驗證
  if (Strings.isNullOrEmpty(indexName)) {
    return "Parameters are wrong!";
  }
  indexService.excuteGetIndexRequest(indexName);
  return "Execute GetIndexRequest success!";
}
```

在 Service 層的 IndexService 類別中新增以下程式:

```
// 以同步方式執行 GetIndexRequest
public void excuteGetIndexRequest(String index);
```

在 ServiceImpl 實現層的 IndexServiceImpl 類別中新增以下程式:

```
// 以同步方式執行 GetIndexRequest
public void excuteGetIndexRequest(String index) {
```

```
    GetIndexRequest request = buildGetIndexRequest(index);
    try {
      GetIndexResponse getIndexResponse = restClient.indices().get(request,
          RequestOptions.DEFAULT);
      // 解析 GetIndexResponse
      processGetIndexResponse(getIndexResponse, index);
    } catch (Exception e) {
      e.printStackTrace();
    } finally {
      // 關閉 Elasticsearch 連接
      closeEs();
    }
}
// 解析 GetIndexResponse
private void processGetIndexResponse(GetIndexResponse getIndexResponse,
    String index) {
  // 檢索不同類型的對映到索引的對映中繼資料 MappingMetadata
  MappingMetaData indexMappings = getIndexResponse.getMappings().get(index);
  if (indexMappings == null) {
    return;
  }
  // 檢索文件類型和文件屬性的對映
  Map<String, Object> indexTypeMappings = indexMappings.getSourceAsMap();
  for (String str : indexTypeMappings.keySet()) {
    log.info("key is " + str);
  }
  // 取得索引的別名清單
  List<AliasMetaData> indexAliases = getIndexResponse.getAliases().get
      (index);
  if (indexAliases == null) {
    return;
  }
  log.info("indexAliases is " + indexAliases.size());
  // 取得為索引設定字串 index.number_shards 的值。該設定是預設設定的一部分
  // (includeDefault 為 true)，如果未顯性指定設定，則將檢索預設設定
  String numberOfShardsString = getIndexResponse.getSetting(index, "index.
      number_of_shards");
```

```
    // 檢索索引的所有設定
    Settings indexSettings = getIndexResponse.getSettings().get(index);
    // 設定物件提供更多的靈活性。在這裡,它被用來分析作為整數的碎片設定 index.number
    Integer numberOfShards = indexSettings.getAsInt("index.number_of_shards",
        null);
    // 取得預設設定 index.refresh_interval (includeDefault 預設設定為 true,如果
    // includeDefault 設定為 false,則 getIndexResponse.defaultSettings() 將傳回
    // 空對映
    TimeValue time = getIndexResponse.getDefaultSettings().get(index).
        getAsTime("index.refresh_interval", null);
    log.info("numberOfShardsString is " + numberOfShardsString
        +";indexSettings is "
        + indexSettings.toString() + ";numberOfShards is " + numberOfShards.
        intValue() + ";time is "

        + time.getMillis());
}
```

隨後編輯專案,在專案根目錄下輸入以下指令:

```
mvn clean package
```

透過以下指令啟動專案服務:

```
java -jar ./target/esdemo-0.0.1-SNAPSHOT.jar
```

在專案服務啟動後,在瀏覽器中呼叫以下介面檢視索引取得情況:

```
http://localhost:8080/springboot/es/indexsearch/get/sr?indexName=ultraman2
```

請求執行後,在伺服器中輸出如下所示內容:

```
2019-09-06 11:09:41.832  INFO 30192 --- [nio-8080-exec-1] c.n.e.service.impl.IndexServiceImpl      : indexAliases is 0
2019-09-06 11:09:41.835  INFO 30192 --- [nio-8080-exec-1] c.n.e.service.impl.IndexServiceImpl      : numberOfShardsStrin
g is 3;indexSettings is {"index.creation_date":"1567736944366","index.number_of_replicas":"2","index.number_of_shards":"
3","index.provided_name":"ultraman2","index.uuid":"3MJyBU9yQ-m0xyjN1NQDCQ","index.version.created":"7020099"};numberOfSh
ards is 3;time is 1000
```

從中可以看出,我們在建立索引時沒有設定別名,因此別名列表長度為 0;分片數量與設定相同,其他參數為預設值。

8.4 刪除索引

在 Elasticsearch 中，索引既可以建立，也可以刪除。

1 建置刪除索引請求

在執行刪除索引請求前，需要建置刪除索引請求，即 DeleteIndexRequest。

DeleteIndexRequest 需要以索引名稱作為參數，程式增加在 IndexServiceImpl 類別中，如下所示：

```
//建置刪除索引請求
public DeleteIndexRequest  buildDeleteIndexRequest(String index) {
  DeleteIndexRequest request = new DeleteIndexRequest(index);
  return request;
}
```

DeleteIndexRequest 提供了可選參數供使用者進行設定。在 DeleteIndexRequest 中，使用者可以設定所有節點刪除索引的確認等待逾時、從節點連接到主節點的逾時和 IndicesOptions。 其中，IndicesOptions 用於控制解析不可用索引及展開萬用字元運算式。程式增加在 IndexServiceImpl 類別的 buildDeleteIndexRequest 方法中，如下所示：

```
// 設定可選參數
    // 等待所有節點刪除索引的確認逾時
    request.timeout(TimeValue.timeValueMinutes(2));
    // 等待所有節點刪除索引的確認的逾時
    request.timeout("2m");
    // 從節點連接到主節點的逾時
    request.masterNodeTimeout(TimeValue.timeValueMinutes(1));
    // 從節點連接到主節點的逾時
    request.masterNodeTimeout("1m");
    // 設定 IndicesOptions，控制解析不可用索引及展開萬用字元運算式
    request.indicesOptions(IndicesOptions.lenientExpandOpen());
```

2 執行刪除索引請求

在 DeleteIndexRequest 建置後,即可執行刪除索引請求。與建立索引請求類似,刪除索引請求也有同步和非同步兩種執行方式。

✎ 同步方式

當以同步方式執行刪除索引請求時,用戶端會等待 Elasticsearch 伺服器傳回的查詢結果 DeleteIndexResponse。在收到 DeleteIndexResponse 後,用戶端會繼續執行相關的邏輯程式。以同步方式執行的程式增加在 IndexServiceImpl 類別中,如下所示:

```
// 以同步方式執行 DeleteIndexRequest
  public void executeDeleteIndexRequest(String index) {
    DeleteIndexRequest request = buildDeleteIndexRequest(index);
    try {
      AcknowledgedResponse deleteIndexResponse =
          restClient.indices().delete(request, RequestOptions.DEFAULT);
    } catch (Exception e) {
      e.printStackTrace();
    } finally {
    // 關閉 Elasticsearch 連接
      closeEs();
    }
  }
```

✎ 非同步方式

當以非同步方式執行刪除索引請求時,進階用戶端不必同步等待請求結果的傳回,可以直接向介面呼叫方傳回非同步介面執行成功的結果。

為了處理非同步傳回的回應資訊或處理在請求執行過程中引發的例外資訊,使用者需要指定監聽器。以非同步方式執行的核心程式如下所示:

```
client.indices().deleteAsync(request, RequestOptions.DEFAULT, listener);
```

其中,listener 為監聽器。

在非同步請求處理後，如果請求執行成功，則呼叫 ActionListener 類別中的 onResponse 方法進行相關邏輯的處理；如果請求執行失敗，則呼叫 ActionListener 類別中的 onFailure 方法進行相關邏輯的處理。

以非同步方式執行的全部程式增加在 IndexServiceImpl 類別中，如下所示：

```
// 以非同步方式執行 DeleteIndexRequest
public void executeDeleteIndexRequestAsync(String index) {
  DeleteIndexRequest request = buildDeleteIndexRequest(index);
  // 建置監聽器
  ActionListener<AcknowledgedResponse> listener = new ActionListener
      <AcknowledgedResponse>() {
    @Override
    public void onResponse(AcknowledgedResponse deleteIndexResponse) {
    }
    @Override
    public void onFailure(Exception e) {
    }
  };
  try {
    restClient.indices().deleteAsync(request, RequestOptions.DEFAULT,
        listener);
  } catch (Exception e) {
    e.printStackTrace();
  } finally {
    // 關閉 Elasticsearch 連接
    closeEs();
  }
}
```

當然，在非同步請求執行過程中可能會出現例外，例外的處理與同步方式執行情況相同。

3 解析刪除索引請求的回應結果

不論同步方式，還是非同步方式，在刪除索引請求執行後，用戶端均需要對回應結果 AcknowledgedResponse 進行解析和處理。程式增加在 IndexServiceImpl 類別中，共分為三層，分別是 Controller 層、Service 層和 ServiceImpl 實現層。

在 Controller 層的 IndexController 類別中新增以下程式：

```
// 以同步方式執行 DeleteIndexRequest
@RequestMapping("/delete/sr")
public String executeDeleteIndexRequest(String indexName) {
  // 參數驗證
  if (Strings.isNullOrEmpty(indexName)) {
    return "Parameters are wrong!";
  }
  indexService.executeDeleteIndexRequest(indexName);
  return "Execute DeleteIndexRequest success!";
}
```

在 Service 層的 IndexService 類別中新增以下程式：

```
// 以同步方式執行 DeleteIndexRequest
public void executeDeleteIndexRequest(String index);
```

在 ServiceImpl 實現層的 IndexServiceImpl 類別中新增以下程式：

```
// 以同步方式執行 DeleteIndexRequest
public void executeDeleteIndexRequest(String index) {
  DeleteIndexRequest request = buildDeleteIndexRequest(index);
  try {
    AcknowledgedResponse deleteIndexResponse =
        restClient.indices().delete(request, RequestOptions.DEFAULT);
    // 解析 AcknowledgedResponse
    processAcknowledgedResponse(deleteIndexResponse);
  } catch (Exception e) {
    e.printStackTrace();
  } finally {
    // 關閉 Elasticsearch 連接
    closeEs();
  }
}
// 解析 AcknowledgedResponse
private void processAcknowledgedResponse(AcknowledgedResponse
    deleteIndexResponse) {
  // 所有節點是否已確認請求
```

```
    boolean acknowledged = deleteIndexResponse.isAcknowledged();
    log.info("acknowledged is " + acknowledged);
  }
```

隨後編輯專案，在專案根目錄下輸入以下指令：

```
mvn clean package
```

透過以下指令啟動專案服務：

```
java -jar ./target/esdemo-0.0.1-SNAPSHOT.jar
```

在專案服務啟動後，在瀏覽器中呼叫以下介面刪除索引情況：

```
http://localhost:8080/springboot/es/indexsearch/delete/sr?indexName=ultraman3
```

請求執行成功後，在伺服器主控台輸出如下所示內容：

```
2019-09-06 11:35:43.309  INFO 77484 --- [io-8080-exec-10] c.n.e.service.impl.IndexServiceImpl       : acknowledged is tru
e
```

8.5 索引存在驗證

想要檢視目標索引是否已經建立，使用者需要使用索引存在驗證介面，即 Exists API。

1 建置索引存在驗證請求

在進階用戶端中，使用者稱需要基於 GetIndexRequest 才能發起索引存在驗證請求。GetIndexRequest 的必選參數是索引名稱，程式增加在 IndexServiceImpl 類別中，如下所示：

```
// 建置索引存在驗證請求
  public GetIndexRequest buildExistsIndexRequest(String index) {
    GetIndexRequest request = new GetIndexRequest(index);
    return request;
  }
```

在索引存在驗證請求中，GetIndexRequest 的可選參數有是否從主節點傳回本機資訊或檢索狀態、是否回歸到適合人類的格式、是否傳回每個索引的所有預設設定和 IndicesOptions。其中，IndicesOptions 用於解析不可用的索引及展開萬用字元運算式，程式增加在 IndexServiceImpl 類別中，如下所示：

```
// 從主節點傳回本機資訊或檢索狀態
    request.local(false);
    // 回歸到適合人類的格式
    request.humanReadable(true);
    // 是否傳回每個索引的所有預設設定
    request.includeDefaults(false);
    // 控制如何解析不可用索引及如何展開萬用字元運算式
    request.indicesOptions(IndicesOptions.lenientExpandOpen());
```

2 執行索引存在驗證請求

在 GetIndexRequest 建置完成後，即可執行索引存在驗證請求。與文件索引請求類似，索引存在驗證請求也有同步和非同步兩種執行方式。

▨ 同步方式

當以同步方式執行索引存在驗證請求時，用戶端會等待 Elasticsearch 伺服器傳回的布林型查詢結果。以同步方式執行的程式增加在 IndexServiceImpl 類別中，如下所示：

```
// 以同步方式執行索引存在驗證請求
  public void executeExistsIndexRequest(String index) {
    GetIndexRequest request = buildExistsIndexRequest(index);
    try {
      boolean exists = restClient.indices().exists(request, RequestOptions.
          DEFAULT);
      log.info("exists is " + exists);
    } catch (Exception e) {
      e.printStackTrace();
    } finally {
      // 關閉 Elasticsearch 連接
      closeEs();
    }
  }
```

✍ 非同步方式

當以非同步方式執行索引存在驗證請求時，進階用戶端不必同步等待請求結果的傳回，可以直接向介面呼叫方傳回非同步介面執行成功的結果。

為了處理非同步傳回的回應資訊或處理在請求執行過程中引發的例外資訊，使用者需要指定監聽器。以非同步方式執行的核心程式如下所示：

```
client.indices().existsAsync(request, RequestOptions.DEFAULT, listener);
```

其中，listener 為監聽器。

在非同步請求處理完成後，如果請求執行成功，則呼叫 ActionListener 類別中的 onResponse 方法進行相關邏輯的處理；如果請求執行失敗，則呼叫 ActionListener 類別中的 onFailure 方法進行相關邏輯的處理。

以非同步方式執行的全部程式增加在 IndexServiceImpl 類別中，如下所示：

```
// 以非同步方式執行索引存在驗證請求
  public void executeExistsIndexRequestAsync(String index) {
    GetIndexRequest request = buildExistsIndexRequest(index);
    // 建置監聽器
    ActionListener<Boolean> listener = new ActionListener<Boolean>() {
      @Override
      public void onResponse(Boolean exists) {
      }
      @Override
      public void onFailure(Exception e) {
      }
    };
    try {
      restClient.indices().existsAsync(request, RequestOptions.DEFAULT,
          listener);
    } catch (Exception e) {
      e.printStackTrace();
    } finally {
      // 關閉 Elasticsearch 連接
      closeEs();
    }
  }
```

當然，在非同步請求執行過程中可能會出現例外，例外的處理與同步方式執行情況相同。

❸ 解析索引存在驗證請求的回應結果

不論同步方式，還是非同步方式，在索引存在驗證請求執行後，用戶端均會獲得索引是否存在的結果。程式增加在 IndexServiceImpl 類別中，共分為三層，分別是 Controller 層、Service 層和 ServiceImpl 實現層。

在 Controller 層的 IndexController 類別中新增以下程式：

```
// 以同步方式執行 ExistsIndexRequest
@RequestMapping("/exists/sr")
public String executeExistsIndexRequest(String indexName) {
  // 參數驗證
  if (Strings.isNullOrEmpty(indexName)) {
    return "Parameters are wrong!";
  }
  indexService.executeExistsIndexRequest(indexName);
  return "Execute ExistsIndexRequest success!";
}
```

在 Service 層的 IndexService 類別中新增以下程式：

```
// 以同步方式執行索引存在驗證請求
public void executeExistsIndexRequest(String index);
```

在 ServiceImpl 實現層的 IndexServiceImpl 類別中的程式和「同步方式」部分一致，不再贅述。

隨後編輯專案，在專案根目錄下輸入以下指令：

```
mvn clean package
```

透過以下指令啟動專案服務：

```
java -jar ./target/esdemo-0.0.1-SNAPSHOT.jar
```

在專案服務啟動後，在瀏覽器中呼叫以下介面檢視索引 ultraman 的存在情況：

```
http://localhost:8080/springboot/es/indexsearch/exists/sr?indexName=ultraman
```

請求執行後,在伺服器中輸出如下所示內容,與預期相符:

```
2019-09-06 14:05:14.685  INFO 78932 --- [nio-8080-exec-4] c.n.e.service.impl.IndexServiceImpl     : exists is true
```

8.6 開啟索引

在索引建立後,即可透過開啟索引介面開啟索引。

1 建置開啟索引請求

在執行開啟索引請求前,需要建置開啟索引請求,即 OpenIndexRequest。
OpenIndexRequest 需要以索引名稱作為參數,程式增加在 IndexServiceImpl 類
別中,如下所示:

```
// 建置 OpenIndexRequest
  public OpenIndexRequest buildOpenIndexRequest(String index) {
    OpenIndexRequest request = new OpenIndexRequest(index);
    return request;
  }
```

在建置開啟索引請求時,OpenIndexRequest 提供了可選參數供使用者進行設
定。OpenIndexRequest 的可選參數有所有節點確認索引開啟的逾時、從節點
連接到主節點的逾時、請求傳回回應前活躍的分片數量和 IndicesOptions。
IndicesOptions 用於控制解析不可用索引及展開萬用字元運算式,程式增加在
IndexServiceImpl 類別中的 buildOpenIndexRequest() 方法內,如下所示:

```
// 設定可選參數
    // 所有節點確認索引開啟的逾時
    request.timeout(TimeValue.timeValueMinutes(2));
    request.timeout("2m");
    // 從節點連接到主節點的逾時
    request.masterNodeTimeout(TimeValue.timeValueMinutes(1));
    request.masterNodeTimeout("1m");
    // 請求傳回回應前活躍的分片數量
    request.waitForActiveShards(2);
    request.waitForActiveShards(ActiveShardCount.DEFAULT);
```

```
   // 設定 IndicesOptions
   request.indicesOptions(IndicesOptions.strictExpandOpen());
```

2 執行開啟索引請求

在 OpenIndexRequest 建置後，即可執行開啟索引請求。與建立索引請求類似，開啟索引請求也有同步和非同步兩種執行方式。

◩ 同步方式

當以同步方式執行開啟索引請求時，用戶端會等待 Elasticsearch 伺服器傳回的查詢結果 OpenIndexResponse。在收到 OpenIndexResponse 後，用戶端會繼續執行相關的邏輯程式。以同步方式執行的程式增加在 IndexServiceImpl 類別中，如下所示：

```
// 以同步方式執行 OpenIndexRequest
  public void executeOpenIndexRequest(String index) {
    OpenIndexRequest request = buildOpenIndexRequest(index);
    try {
      OpenIndexResponse openIndexResponse =
          restClient.indices().open(request, RequestOptions.DEFAULT);
    } catch (Exception e) {
      e.printStackTrace();
    } finally {
      // 關閉 Elasticsearch 連接
      closeEs();
    }
  }
```

◩ 非同步方式

當以非同步方式執行開啟索引請求時，進階用戶端不必同步等待請求結果的傳回，可以直接向介面呼叫方傳回非同步介面執行成功的結果。

為了處理非同步傳回的回應資訊或處理在請求執行過程中引發的例外資訊，使用者需要指定監聽器。以非同步方式執行的核心程式如下所示：

```
client.indices().openAsync(request, RequestOptions.DEFAULT, listener);
```

其中，listener 為監聽器。

在非同步請求處理後，如果請求執行成功，則呼叫 ActionListener 類別中的 onResponse 方法進行相關邏輯的處理；如果請求執行失敗，則呼叫 ActionListener 類別中的 onFailure 方法進行相關邏輯的處理。

以非同步方式執行的全部程式增加在 IndexServiceImpl 類別中，如下所示：

```
// 以非同步方式執行 OpenIndexRequest
public void executeOpenIndexRequestAsync(String index) {
  OpenIndexRequest request = buildOpenIndexRequest(index);
  // 建置監聽器
  ActionListener<OpenIndexResponse> listener = new ActionListener
      <OpenIndexResponse>() {
    @Override
    public void onResponse(OpenIndexResponse openIndexResponse) {
    }
    @Override
    public void onFailure(Exception e) {
    }
  };
  try {
    restClient.indices().openAsync(request, RequestOptions.DEFAULT, listener);
  } catch (Exception e) {
    e.printStackTrace();
  } finally {
    // 關閉 Elasticsearch 連接
    closeEs();
  }
}
```

當然，在非同步請求執行過程中可能會出現例外，例外的處理與同步方式執行情況相同。

3 解析開啟索引請求的回應結果

不論同步方式，還是非同步方式，在開啟索引請求執行後，用戶端均需要對回應結果 OpenIndexResponse 進行解析和處理。程式增加在 IndexServiceImpl

類別中，共分為三層，分別是 Controller 層、Service 層和 ServiceImpl 實現層。

在 Controller 層的 IndexController 類別中新增以下程式：

```
// 以同步方式執行 OpenIndexRequest
@RequestMapping("/open/sr")
public String executeOpenIndexRequest(String indexName) {
  // 參數驗證
  if (Strings.isNullOrEmpty(indexName)) {
    return "Parameters are wrong!";
  }
  indexService.executeOpenIndexRequest(indexName);
  return "Execute OpenIndexRequest success!";
}
```

在 Service 層的 IndexService 類別中新增以下程式：

```
// 以同步方式執行 OpenIndexRequest
public void executeOpenIndexRequest(String index);
```

在 ServiceImpl 實現層的 IndexServiceImpl 類別中新增以下程式：

```
// 以同步方式執行 OpenIndexRequest
public void executeOpenIndexRequest(String index) {
  OpenIndexRequest request = buildOpenIndexRequest(index);
  try {
    OpenIndexResponse openIndexResponse =
        restClient.indices().open(request, RequestOptions.DEFAULT);
    // 解析 OpenIndexResponse
    processOpenIndexResponse(openIndexResponse);
  } catch (Exception e) {
    e.printStackTrace();
  } finally {
    // 關閉 Elasticsearch 連接
    closeEs();
  }
}
// 解析 OpenIndexResponse
private void processOpenIndexResponse(OpenIndexResponse openIndexResponse)
```

```
{
    // 所有節點是否已確認請求
    boolean acknowledged = openIndexResponse.isAcknowledged();
    // 是否在逾時前為索引中的每個分片啟動了所需數量的分片備份
    boolean shardsAcked = openIndexResponse.isShardsAcknowledged();
    log.info("acknowledged is " + acknowledged + ";shardsAcked is " +
        shardsAcked);
}
```

隨後編輯專案，在專案根目錄下輸入以下指令：

```
mvn clean package
```

透過以下指令啟動專案服務：

```
java -jar ./target/esdemo-0.0.1-SNAPSHOT.jar
```

在專案服務啟動後，在瀏覽器中呼叫以下介面檢視索引 ultraman 的開啟情況：

```
http://localhost:8080/springboot/es/indexsearch/open/sr?indexName=ultraman
```

在請求執行後，伺服器輸出如下所示內容：

```
2019-09-06 14:19:45.202  INFO 76468 --- [nio-8080-exec-1] c.n.e.service.impl.IndexServiceImpl      : acknowledged is tru
e;shardsAcked is true
```

8.7 關閉索引

用完索引後，還需要關閉索引，這時就需要使用關閉索引介面。

1 建置關閉索引請求

在執行關閉索引請求前，需要建置關閉索引請求，即 CloseIndexRequest。
CloseIndexRequest 必須以索引名稱作為參數，程式增加在 IndexServiceImpl 類
別中，如下所示：

```
// 建置 CloseIndexRequest
 public CloseIndexRequest buildCloseIndexRequest(String index) {
   CloseIndexRequest request = new CloseIndexRequest(index);
```

```
    return request;
  }
```

在建置關閉索引請求時，CloseIndexRequest 提供了可選參數列表供使用者進行設定。CloseIndexRequest 支援的可選參數有所有節點確認索引關閉的逾時、從節點連接到主節點的逾時和 IndicesOptions。IndicesOptions 用於控制解析不可用索引及展開萬用字元運算式。程式增加在 IndexServiceImpl 類別中的 buildCloseIndexRequest 方法內，如下所示：

```
// 設定可選參數
    // 所有節點確認索引關閉的逾時
    request.timeout(TimeValue.timeValueMinutes(2));
    request.timeout("2m");
    // 從節點連接到主節點的逾時
    request.masterNodeTimeout(TimeValue.timeValueMinutes(1));
    request.masterNodeTimeout("1m");
    // 用於控制解析不可用索引及展開萬用字元運算式
    request.indicesOptions(IndicesOptions.lenientExpandOpen());
```

2 執行關閉索引請求

在 CloseIndexRequest 建置後，即可執行關閉索引請求。與建立索引請求類似，關閉索引請求也有同步和非同步兩種執行方式。

◿ 同步方式

當以同步方式執行關閉索引請求時，用戶端會等待 Elasticsearch 伺服器傳回的查詢結果 AcknowledgedResponse。在收到 AcknowledgedResponse 後，用戶端會繼續執行相關的邏輯程式。以同步方式執行的程式增加在 IndexServiceImpl 類別中，如下所示：

```
// 以同步方式執行關閉索引請求
  public void executeCloseIndexRequest(String index) {
    CloseIndexRequest request = buildCloseIndexRequest(index);
    try {
      AcknowledgedResponse closeIndexResponse =
          restClient.indices().close(request, RequestOptions.DEFAULT);
```

```
    // 所有節點是否已確認請求
    boolean acknowledged = closeIndexResponse.isAcknowledged();
    log.info(index + " acknowledged is " + acknowledged);
  } catch (Exception e) {
    e.printStackTrace();
  } finally {
    // 關閉Elasticsearch連接
    closeEs();
  }
}
```

🗹 非同步方式

當以非同步方式執行關閉索引請求時，進階用戶端不必同步等待請求結果的傳回，可以直接向介面呼叫方傳回非同步介面執行成功的結果。

為了處理非同步傳回的回應資訊或處理在請求執行過程中引發的例外資訊，使用者需要指定監聽器。以非同步方式執行的核心程式如下所示：

```
client.indices().closeAsync(request, RequestOptions.DEFAULT, listener);
```

其中，listener 為監聽器。

在非同步請求處理後，如果請求執行成功，則呼叫 ActionListener 類別中的 onResponse 方法進行相關邏輯的處理；如果請求執行失敗，則呼叫 ActionListener 類別中的 onFailure 方法進行相關邏輯的處理。

以非同步方式執行的全部程式增加在 IndexServiceImpl 類別中，如下所示：

```
// 以非同步方式執行關閉索引請求
  public void executeCloseIndexRequestAsync(String index) {
    CloseIndexRequest request = buildCloseIndexRequest(index);
    // 建置監聽器
    ActionListener<AcknowledgedResponse> listener = new ActionListener
        <AcknowledgedResponse>() {
      @Override
      public void onResponse(AcknowledgedResponse closeIndexResponse) {
      }
```

```
    @Override
    public void onFailure(Exception e) {
    }
  };
  try {
    restClient.indices().closeAsync(request, RequestOptions.DEFAULT,
listener);
  } catch (Exception e) {
    e.printStackTrace();
  } finally {
    // 關閉 Elasticsearch 連接
    closeEs();
  }
}
```

當然，在非同步請求執行過程中可能會出現例外，例外的處理與同步方式執行情況相同。

❸ 解析關閉索引請求的回應結果

以同步方式執行的程式如下所示，AcknowledgedResponse 的 isAcknowledged 方法會傳回非同步布林值，表示索引是否已經關閉。

程式增加在 IndexServiceImpl 類別中，共分為三層，分別是 Controller 層、Service 層和 ServiceImpl 實現層。

在 Controller 層的 IndexController 類別中新增以下程式：

```
// 以同步方式執行 CloseIndexRequest
@RequestMapping("/close/sr")
public String executeCloseIndexRequest(String indexName) {
  // 參數驗證
  if (Strings.isNullOrEmpty(indexName)) {
    return "Parameters are wrong!";
  }
  indexService.executeCloseIndexRequest(indexName);
  return "Execute CloseIndexRequest success!";
}
```

在 Service 層的 IndexService 類別中新增以下程式：

```
// 以同步方式執行 CloseIndexRequest
  public void executeCloseIndexRequest(String index);
```

在 ServiceImpl 實現層的 IndexServiceImpl 類別中的程式與「同步方式」中的
程式一致，不再贅述。

隨後編輯專案，在專案根目錄下輸入以下指令：

```
mvn clean package
```

透過以下指令啟動專案服務：

```
java -jar ./target/esdemo-0.0.1-SNAPSHOT.jar
```

在專案服務啟動後，在瀏覽器中呼叫以下介面檢視索引 ultraman 的關閉情況：

```
http://localhost:8080/springboot/es/indexsearch/close/sr?indexName=ultraman
```

請求執行後，在伺服器中輸出如下所示內容：

```
2019-09-06 14:41:57.947  INFO 82424 --- [nio-8080-exec-1] c.n.e.service.impl.IndexServiceImpl      : ultraman acknowledg
ed is true
```

8.8 縮小索引

如果要縮小索引，則需要使用縮小索引介面，即 Shrink API。

1 建置縮小索引大小請求

在執行調整索引大小請求前，需要建置調整索引大小請求，即 ResizeRequest。
ResizeRequest 的必選參數是兩個字串參數，分別是來源索引名稱和目標索引
名稱。程式增加在 IndexServiceImpl 類別中，如下所示：

```
// 建置 ResizeRequest
  public ResizeRequest buildResizeRequest(String sourceIndex, String
      targetIndex) {
    ResizeRequest request = new ResizeRequest(targetIndex, sourceIndex);
    return request;
  }
```

在執行調整索引大小請求時，ResizeRequest 提供了可選參數列表供使用者進行設定。ResizeRequest 支援的可選參數有所有節點確認索引開啟的逾時、從節點連接到主節點的逾時、請求傳回前需要等待的活躍狀態的分片數量、在縮小索引上的目標索引中的分片數、刪除從來源索引複製的分配要求、與目標索引連結的別名等。程式增加在 IndexServiceImpl 類別中的 buildResizeRequest 方法內，如下所示：

```
// 設定可選參數
    // 所有節點確認索引開啟的逾時
    request.timeout(TimeValue.timeValueMinutes(2));
    request.timeout("2m");
    // 從節點連接到主節點的逾時
    request.masterNodeTimeout(TimeValue.timeValueMinutes(1));
    request.masterNodeTimeout("1m");
    // 請求傳回前需要等待的活躍狀態的分片數量
    request.setWaitForActiveShards(2);
    request.setWaitForActiveShards(ActiveShardCount.DEFAULT);
    // 在縮小索引上的目標索引中的分片數、刪除從來源索引複製的分配要求
    request.getTargetIndexRequest().settings(Settings.builder().put
    ("index.number_of_shards",2).putNull("index.routing.allocation.require._
    name"));
    // 與目標索引連結的別名
    request.getTargetIndexRequest().alias(new Alias(targetIndex + "_alias"));
```

2 執行縮小索引請求

在建置 ResizeRequest 之後，即可執行縮小索引請求。與建立索引請求類似，縮小索引請求也有同步和非同步兩種執行方式。

同步方式

當以同步方式執行縮小索引請求時，用戶端會等待 Elasticsearch 伺服器傳回的查詢結果 ResizeResponse。在收到 ResizeResponse 後，用戶端會繼續執行相關的邏輯程式。以同步方式執行的程式增加在 IndexServiceImpl 類別中，如下所示：

```
// 以同步方式執行 ResizeRequest
  public void executeResizeRequest(String sourceIndex, String targetIndex) {
    ResizeRequest request = buildResizeRequest(sourceIndex, targetIndex);
    try {
      ResizeResponse resizeResponse = restClient.indices().shrink(request,
          RequestOptions.DEFAULT);
      // 解析 ResizeResponse
      processResizeResponse(resizeResponse);
    } catch (Exception e) {
      e.printStackTrace();
    } finally {
      // 關閉 Elasticsearch 連接
      closeEs();
    }
  }
```

📝 非同步方式

當以非同步方式執行縮小索引請求時，進階用戶端不必同步等待請求結果的傳回，可以直接向介面呼叫方傳回非同步介面執行成功的結果。

為了處理非同步傳回的回應資訊或處理在請求執行過程中引發的例外資訊，使用者需要指定監聽器。以非同步方式執行的核心程式如下所示：

```
client.indices().shrinkAsync(request, RequestOptions.DEFAULT, listener);
```

其中，listener 為監聽器。

在非同步請求處理後，如果請求執行成功，則呼叫 ActionListener 類別中的 onResponse 方法進行相關邏輯的處理；如果請求執行失敗，則呼叫 ActionListener 類別中的 onFailure 方法進行相關邏輯的處理。

以非同步方式執行的全部程式如下所示：

```
// 以非同步方式執行 ResizeRequest
  public void executeResizeRequestAsync(String sourceIndex, String
      targetIndex) {
    ResizeRequest request = buildResizeRequest(sourceIndex, targetIndex);
```

```
  // 建置監聽器
  ActionListener<ResizeResponse> listener = new ActionListener
      <ResizeResponse>() {
   @Override
   public void onResponse(ResizeResponse resizeResponse) {
   }
   @Override
   public void onFailure(Exception e) {
   }
  };

  try {
    restClient.indices().shrinkAsync(request, RequestOptions.DEFAULT,
listener);
  } catch (Exception e) {
    e.printStackTrace();
  } finally {
    // 關閉 Elasticsearch 連接
    closeEs();
  }
 }
```

當然，在非同步請求執行過程中可能會出現例外，例外的處理與同步方式執行情況相同。

3 解析縮小索引請求的回應結果

不論同步方式，還是非同步方式，在縮小索引請求執行後，用戶端均需要對回應結果 ResizeResponse 進行處理和解析。

程式增加在 IndexServiceImpl 類別中，共分為三層，分別是 Controller 層、Service 層和 ServiceImpl 實現層。

在 Controller 層的 IndexController 類別中新增以下程式：

```
// 以同步方式執行 ResizeRequest
  @RequestMapping("/resize/sr")
  public String executeResizeRequest(String sourceIndexName, String
```

```
    targetIndexName) {
  // 參數驗證
  if (Strings.isNullOrEmpty(sourceIndexName) || Strings.isNullOrEmpty
     (targetIndexName)) {
    return "Parameters are wrong!";
  }
  indexService.executeResizeRequest(sourceIndexName, targetIndexName);
  return "Execute ResizeRequest success!";
}
```

在 Service 層的 IndexService 類別中新增以下程式：

```
// 以同步方式執行 ResizeRequest
  public void executeResizeRequest(String sourceIndex, String targetIndex);
```

在 ServiceImpl 實現層的 IndexServiceImpl 類別中新增以下程式：

```
// 解析 ResizeResponse
  private void processResizeResponse(ResizeResponse resizeResponse) {
    // 所有節點是否已確認請求
    boolean acknowledged = resizeResponse.isAcknowledged();
    // 是否在逾時前為索引中的每個分片啟動了所需數量的分片備份
    boolean shardsAcked = resizeResponse.isShardsAcknowledged();
    log.info("acknowledged is " + acknowledged + ";shardsAcked is " +
        shardsAcked);
  }
```

隨後編輯專案，在專案根目錄下輸入以下指令：

```
mvn clean package
```

透過以下指令啟動專案服務：

```
java -jar ./target/esdemo-0.0.1-SNAPSHOT.jar
```

在專案服務啟動後，在瀏覽器中呼叫以下介面檢視索引 ultraman6 縮小的情況：

```
http://localhost:8080/springboot/es/indexsearch/resize/sr?sourceIndexName=
ultraman6&targetIndexName=ultraman7
```

請求執行後，在伺服器中輸出如下所示內容，可見縮小索引請求執行成功：

```
2019-09-06 15:22:06.474  INFO 83236 --- [nio-8080-exec-2] c.n.e.service.impl.IndexServiceImpl      : acknowledged is true;shardsAcked is true
```

8.9 拆分索引

如果要拆分索引，則需要呼叫拆分索引介面，即 Split API。

◼ 建置拆分索引請求

在拆分索引請求前和樣需要建置 ResizeRequest。在 8.8 節中，曾建置 ResizeRequest 程式，兩個請求介面的差別在於，在拆分索引請求中，需要將「調整大小」的類型設定為「拆分」。程式增加在 IndexServiceImpl 類別中，如下所示：

```
// 建置 ResizeRequest
  public ResizeRequest buildSplitRequest(String sourceIndex, String
      targetIndex) {
    ResizeRequest request = new ResizeRequest(targetIndex, sourceIndex);
    // 把 " 調整大小 " 的類型設定為 " 拆分 "
    request.setResizeType(ResizeType.SPLIT);
    return request;
  }
```

在拆分索引請求建置過程中，ResizeRequest 可設定的可選參數清單與縮小索引請求建置過程中的相同，不再贅述。

◼ 執行拆分索引請求

在 ResizeRequest 建置後，即可執行拆分索引請求。與建立索引請求類似，拆分索引請求也有同步和非同步兩種執行方式。

◩ 同步方式

當以同步方式執行拆分索引請求時，用戶端會等待 Elasticsearch 伺服器傳回的

查詢結果 ResizeResponse。在收到 ResizeResponse 後，用戶端繼續執行相關的
邏輯程式。以同步方式執行的核心程式增加在 IndexServiceImpl 類別中，如下
所示：

```
// 以同步方式執行 ResizeRequest
  public void executeSplitRequest(String sourceIndex, String targetIndex) {
    ResizeRequest request = buildSplitRequest(sourceIndex, targetIndex);
    try {
      ResizeResponse resizeResponse = restClient.indices().split(request,
          RequestOptions.DEFAULT);

      // 解析 ResizeResponse
      processResizeResponse(resizeResponse);
    } catch (Exception e) {
      e.printStackTrace();
    } finally {
    // 關閉 Elasticsearch 連接
      closeEs();
    }
  }
```

☑ 非同步方式

當以非同步方式執行拆分索引請求時，進階用戶端不必同步等待請求結果的
傳回，可以直接向介面呼叫方傳回非同步介面執行成功的結果。

為了處理非同步傳回的回應資訊或處理在請求執行過程中引發的例外資訊，
使用者需要指定監聽器。以非同步方式執行的核心程式如下所示：

```
client.indices().splitAsync(request, RequestOptions.DEFAULT,listener);
```

其中，listener 為監聽器。

在非同步請求處理後，如果請求執行成功，則呼叫 ActionListener 類別
中的 onResponse 方法進行相關邏輯的處理；如果請求執行失敗，則呼叫
ActionListener 類別中的 onFailure 方法進行相關邏輯的處理。

以非同步方式執行的全部程式增加在 IndexServiceImpl 類別中，如下所示：

```
// 以非同步方式執行ResizeRequest
public void executeSplitRequestAsync(String sourceIndex, String
    targetIndex) {
  ResizeRequest request = buildSplitRequest(sourceIndex, targetIndex);
  // 建置監聽器
  ActionListener<ResizeResponse> listener = new ActionListener
      <ResizeResponse>() {
    @Override
    public void onResponse(ResizeResponse resizeResponse) {

    }
    @Override
    public void onFailure(Exception e) {

    }
  };
  try {
    restClient.indices().splitAsync(request, RequestOptions.DEFAULT,
        listener);
  } catch (Exception e) {
    e.printStackTrace();
  } finally {
    // 關閉Elasticsearch連接
    closeEs();
  }
}
```

當然，在非同步請求執行過程中可能會出現例外，例外的處理與同步方式執行情況相同。

3 解析拆分索引請求的回應結果

不論同步方式，還是非同步方式，在拆分索引請求執行後，用戶端均需要對 ResizeResponse 進行處理和解析。

程式增加在 IndexServiceImpl 類別中，共分為三層，分別是 Controller 層、Service 層和 ServiceImpl 實現層。

在 Controller 層的 IndexController 類別中新增以下程式：

```
// 以同步方式執行 ResizeRequest
@RequestMapping("/split/sr")
public String executeSplitRequest(String sourceIndexName, String
    targetIndexName) {
  // 參數驗證
  if (Strings.isNullOrEmpty(sourceIndexName) || Strings.isNullOrEmpty
      (targetIndexName)) {
    return "Parameters are wrong!";
  }
  indexService.executeSplitRequest(sourceIndexName, targetIndexName);
  return "Execute SplitResizeRequest success!";
}
```

在 Service 層的 IndexService 類別中新增以下程式：

```
// 以同步方式執行 ResizeRequest
public void executeSplitRequest(String sourceIndex, String targetIndex);
```

在 ServiceImpl 實現層 IndexServiceImpl 類別中的程式與本節的「同步方式」
中的程式相同，不再贅述。

隨後編輯專案，在專案根目錄下輸入以下指令：

```
mvn clean package
```

透過以下指令啟動專案服務：

```
java -jar ./target/esdemo-0.0.1-SNAPSHOT.jar
```

當專案服務啟動後，在瀏覽器中呼叫以下介面檢視索引 ultraman6 的拆分情
況：

```
http://localhost:8080/springboot/es/indexsearch/split/sr?sourceIndexName=
ultraman6&targetIndexName=ultraman10
```

請求執行後，在伺服器中輸出的內容如下所示，可見相關節點均已完成索引
ultraman6 的拆分：

```
2019-09-06 16:17:13.897  INFO 83936 --- [nio-8080-exec-1] c.n.e.service.impl.IndexServiceImpl      : acknowledged is tru
e;shardsAcked is true
```

8.10 更新索引

更新索引介面可以應用於單一或多個索引，甚至可以應用於全部索引。

1 建置更新索引請求

在執行更新索引請求前，需要建置更新索引請求，即 RefreshRequest。建置 RefreshRequest 的方法有多種，可以更新單一索引、多個索引或全部索引。程式增加在 IndexServiceImpl 類別中，如下所示：

```
// 建置更新索引請求
public RefreshRequest buildRefreshRequest(String index) {
    // 更新單一索引
    RefreshRequest request = new RefreshRequest(index);
    // 更新多個索引
    RefreshRequest requestMultiple = new RefreshRequest(index, index);
    // 更新全部索引
    RefreshRequest requestAll = new RefreshRequest();
    return request;
}
```

在執行 RefreshRequest 時，RefreshRequest 提供了可選參數供使用者進行設定，其中最主要的是 IndicesOptions，用於解析不可用索引及展開萬用字元運算式。程式增加在 IndexServiceImpl 類別中的 buildRefreshRequest 方法內，如下所示：

```
// 解析不可用索引及展開萬用字元運算式
request.indicesOptions(IndicesOptions.lenientExpandOpen());
```

2 執行更新索引請求

在 RefreshRequest 建置後，即可執行更新索引請求。與建立索引請求類似，更新索引請求也有同步和非同步兩種執行方式。

☑ 同步方式

當以同步方式執行更新索引請求時，用戶端會等待 Elasticsearch 伺服器傳回的

查詢結果 RefreshResponse。在收到 RefreshResponse 後，用戶端會繼續執行相
關的邏輯程式。以同步方式執行的程式增加在 IndexServiceImpl 類別中，如下
所示：

```
// 以同步方式執行 RefreshRequest
public void executeRefreshRequest(String index) {
  RefreshRequest request = buildRefreshRequest(index);
  try {
    RefreshResponse refreshResponse =
        restClient.indices().refresh(request, RequestOptions.DEFAULT);
    // 解析 RefreshResponse
    processRefreshResponse(refreshResponse);
  } catch (Exception e) {
    e.printStackTrace();
  } finally {
    // 關閉 Elasticsearch 連接
    closeEs();
  }
}
```

非同步方式

當以非同步方式執行更新索引請求時，進階用戶端不必同步等待請求結果的
傳回，可以直接向介面呼叫方傳回非同步介面執行成功的結果。

為了處理非同步傳回的回應資訊或處理在請求執行過程中引發的例外資訊，
使用者需要指定監聽器。以非同步方式執行的核心程式如下所示：

```
client.indices().refreshAsync(request, RequestOptions.DEFAULT, listener);
```

其中，listener 為監聽器。

在非同步請求處理後，如果請求執行成功，則呼叫 ActionListener 類別
中的 onResponse 方法進行相關邏輯的處理；如果請求執行失敗，則呼叫
ActionListener 類別中的 onFailure 方法進行相關邏輯的處理。

以非同步方式執行的全部程式如下所示：

```
// 以非同步方式執行 RefreshRequest
  public void executeRefreshRequestAsync(String index) {
    RefreshRequest request = buildRefreshRequest(index);
    // 建置監聽器
    ActionListener<RefreshResponse> listener = new ActionListener
        <RefreshResponse>() {
      @Override
      public void onResponse(RefreshResponse refreshResponse) {
      }
      @Override
      public void onFailure(Exception e) {
      }
    };
    try {
      restClient.indices().refreshAsync(request, RequestOptions.DEFAULT,
listener);
    } catch (Exception e) {
      e.printStackTrace();
    } finally {
      // 關閉 Elasticsearch 連接
      closeEs();
    }
  }
```

當然，在非同步請求執行過程中可能會出現例外，例外的處理與同步方式執行情況相同。

3 解析更新索引請求的回應結果

不論同步方式，還是非同步方式，在更新索引請求執行後，用戶端均需要對 RefreshResponse 進行處理和解析。

程式增加在 IndexServiceImpl 類別中，共分為三層，分別是 Controller 層、Service 層和 ServiceImpl 實現層。

在 Controller 層的 IndexController 類別中新增以下程式：

```
// 以同步方式執行 RefreshRequest
@RequestMapping("/refresh/sr")
public String executeRefreshRequest(String indexName) {
  // 參數驗證
  if (Strings.isNullOrEmpty(indexName)) {
    return "Parameters are wrong!";
  }
  indexService.executeRefreshRequest(indexName);
  return "Execute RefreshRequest success!";
}
```

在 Service 層的 IndexService 類別中新增以下程式：

```
// 以同步方式執行 RefreshRequest
public void executeRefreshRequest(String index);
```

在 ServiceImpl 實現層的 IndexServiceImpl 類別中新增以下程式：

```
// 解析 RefreshResponse
private void processRefreshResponse(RefreshResponse refreshResponse) {
  // 更新請求命中的分片總數
  int totalShards = refreshResponse.getTotalShards();
  // 更新成功的分片數
  int successfulShards = refreshResponse.getSuccessfulShards();
  // 更新失敗的分片數
  int failedShards = refreshResponse.getFailedShards();
  // 在一個或多個分片更新失敗時的失敗列表
  DefaultShardOperationFailedException[] failures = refreshResponse.
getShardFailures();
  log.info("totalShards is " + totalShards + ";successfulShards is " +
successfulShards
      + ";failedShards is " + failedShards + "; failures is "
      + (failures == null ? 0 : failures.length));
}
```

隨後編輯專案，在專案根目錄下輸入以下指令：

```
mvn clean package
```

透過以下指令啟動專案服務：

```
java -jar ./target/esdemo-0.0.1-SNAPSHOT.jar
```

在專案服務啟動後，在瀏覽器中呼叫以下介面檢視索引 ultraman8 的更新情況：

```
http://localhost:8080/springboot/es/indexsearch/refresh/sr?indexName=ultraman8
```

請求執行後，在伺服器中輸出如下所示內容，表示更新請求執行成功：

```
2019-09-06 16:37:50.659  INFO 82996 --- [nio-8080-exec-1] c.n.e.service.impl.IndexServiceImpl      : totalShards is 12;s
uccessfulShards is 6;failedShards is 0; failures is 0
```

8.11 Flush 更新

除更新索引介面外，Elasticsearch 還提供了 Flush 更新介面。Flush 更新介面可以應用於單一或多個索引，甚至可以應用於全部索引。

索引的更新過程是將資料更新到索引儲存，然後清除內部交易記錄檔釋放索引記憶體。在預設情況下，Elasticsearch 使用記憶體啟發式方式根據需要自動觸發更新操作，以清理記憶體。

1 建置 Flush 更新請求

在執行 Flush 更新請求前，需要建置 Flush 更新請求，即 FlushRequest。建置 FlushRequest 的方法有多種，可以更新單一索引、多個索引或全部索引。程式增加在 IndexServiceImpl 類別中，如下所示：

```
// 建置更新索引請求物件
 public FlushRequest  buildFlushRequest(String index) {
   // 更新單一索引
   FlushRequest  request = new FlushRequest (index);
   // 更新全部索引
   FlushRequest  requestMultiple = new FlushRequest (index, index);
   // 更新所有索引
   FlushRequest  requestAll = new FlushRequest ();
   return request;
 }
```

在執行 FlushRequest 時，FlushRequest 還提供了可選參數供使用者進行設定，其中最主要的是 IndicesOptions，用於解析不可用索引及展開萬用字元運算式。程式增加在 IndexServiceImpl 類別中的 buildFlushRequest 方法內，如下所示：

```
// 控制解析不可用索引及展開萬用字元運算式
request.indicesOptions(IndicesOptions.lenientExpandOpen());
```

2 執行 Flush 更新請求

在建置 FlushRequest 之後，即可執行 Flush 更新請求。與建立索引請求類似，Flush 更新請求也有同步和非同步兩種執行方式。

同步方式

當以同步方式執行 Flush 更新請求時，用戶端會等待 Elasticsearch 伺服器傳回的查詢結果 FlushResponse。在收到 FlushResponse 後，用戶端會繼續執行相關的邏輯程式。以同步方式執行的程式增加在 IndexServiceImpl 類別中，如下所示：

```
// 以同步方式執行 FlushRequest
  public void executeFlushRequest(String index) {
    FlushRequest request = buildFlushRequest(index);
    try {
      FlushResponse flushResponse = restClient.indices().flush(request,
RequestOptions.DEFAULT);

      // 解析 FlushResponse
      processFlushResponse(flushResponse);
    } catch (Exception e) {
      e.printStackTrace();
    } finally {
      // 關閉 Elasticsearch 連接
      closeEs();
    }
  }
```

非同步方式

當以非同步方式執行 Flush 更新索引請求時,進階用戶端不必同步等待請求結果的傳回,可以直接向介面呼叫方傳回非同步介面執行成功的結果。

為了處理非同步傳回的回應資訊或處理在請求執行過程中引發的例外資訊,使用者需要指定監聽器。以非同步方式執行的核心程式如下所示:

```
client.indices().flushAsync(request, RequestOptions.DEFAULT, listener);
```

其中,listener 為監聽器。

在非同步請求處理後,如果請求執行成功,則呼叫 ActionListener 類別中的 onResponse 方法進行相關邏輯的處理;如果請求執行失敗,則呼叫 ActionListener 類別中的 onFailure 方法進行相關邏輯的處理。

以非同步方式執行的全部程式如下所示:

```
// 以非同步方式執行 FlushRequest
  public void executeFlushRequestAsync(String index) {
    FlushRequest request = buildFlushRequest(index);
    // 建置監聽器
    ActionListener<FlushResponse> listener = new ActionListener <FlushResponse>
    () {
      @Override
      public void onResponse(FlushResponse refreshResponse) {
      }
      @Override
      public void onFailure(Exception e) {
      }
    };
    try {
      restClient.indices().flushAsync(request, RequestOptions.DEFAULT,
listener);
    } catch (Exception e) {
      e.printStackTrace();
    } finally {
      // 關閉 Elasticsearch 連接
```

```
    closeEs();
  }
}
```

當然，在非同步請求執行過程中可能會出現例外，例外的處理與同步方式執行情況相同。

3 解析 Flush 更新請求的回應結果

不論同步方式，還是非同步方式，在 Flush 更新索引請求執行後，用戶端均需要對 FlushResponse 進行處理和解析。

程式增加在 IndexServiceImpl 類別中，共分為三層，分別是 Controller 層、Service 層和 ServiceImpl 實現層。

在 Controller 層的 IndexController 類別中新增以下程式：

```
// 以同步方式執行 FlushRequest
@RequestMapping("/flush/sr")
public String executeFlushRequest(String indexName) {
  // 參數驗證
  if (Strings.isNullOrEmpty(indexName)) {
    return "Parameters are wrong!";
  }

  indexService.executeFlushRequest(indexName);
  return "Execute FlushRequest success!";
}
```

在 Service 層的 IndexService 類別中新增以下程式：

```
// 以同步方式執行 FlushRequest
public void executeFlushRequest(String index);
```

在 ServiceImpl 實現層的 IndexServiceImpl 類別中新增以下程式：

```
// 解析 FlushResponse
private void processFlushResponse(FlushResponse flushResponse) {
  // 更新請求命中的分片總數
```

```
    int totalShards = flushResponse.getTotalShards();
    // 更新成功的分片數
    int successfulShards = flushResponse.getSuccessfulShards();
    // 更新失敗的分片數
    int failedShards = flushResponse.getFailedShards();
    // 在一個或多個分片更新失敗時的失敗列表
    DefaultShardOperationFailedException[] failures = flushResponse.
        getShardFailures();
    log.info("totalShards is " + totalShards + ";successfulShards is " +
        successfulShards
        + ";failedShards is " + failedShards + "; failures is "
        + (failures == null ? 0 : failures.length));
}
```

隨後編輯專案，在專案根目錄下輸入以下指令：

```
mvn clean package
```

透過以下指令啟動專案服務：

```
java -jar ./target/esdemo-0.0.1-SNAPSHOT.jar
```

在專案服務啟動後，在瀏覽器中呼叫以下介面檢視索引 ultraman8 的 Flush 更新情況：

```
http://localhost:8080/springboot/es/indexsearch/flush/sr?indexName=ultraman8
```

請求執行後，如果在伺服器中輸出如下所示內容，則表示 Flush 更新成功：

```
2019-09-06 17:10:01.470  INFO 85664 --- [nio-8080-exec-1] c.n.e.service.impl.IndexServiceImpl     : totalShards is 12;s
uccessfulShards is 6;failedShards is 0; failures is 0
```

8.12 同步 Flush 更新

Elasticsearch 不僅提供了 Flush 更新方式，還提供了同步 Flush 更新方式。與普通 Flush 更新方式一樣，同步 Flush 更新也可以應用於單一或多個索引，甚至可以應用於全部索引。

什麼是同步 Flush 更新呢？

Elasticsearch 會追蹤每個分片的索引活動，在 5 分鐘內未收到任何索引操作的
分片會自動標記為非活動狀態，這樣 Elasticsearch 就可以減少分片資源。

1 建置同步 Flush 更新請求

在執行同步 Flush 更新請求前，需要建置同步 Flush 更新請求，即
SyncedFlushRequest。 與 FlushRequest 類似， 建置 SyncedFlushRequest 的
方法有多種，可以更新單一索引、多個索引或全部索引。程式增加在
IndexServiceImpl 類別中，如下所示：

```
// 建置同步 Flush 更新索引請求
  public SyncedFlushRequest buildSyncedFlushRequest(String index) {
    // 更新單一索引
    SyncedFlushRequest request = new SyncedFlushRequest(index);
    // 更新多個索引
    SyncedFlushRequest requestMultiple = new SyncedFlushRequest(index, index);
    // 更新全部索引
    SyncedFlushRequest requestAll = new SyncedFlushRequest();
    return request;
  }
```

在執行 SyncedFlushRequest 時，SyncedFlushRequest 還提供了可選參數供使用
者進行設定，其中主要的是 IndicesOptions，用於解析不可用索引及展開萬用
字元運算式。程式增加在 IndexServiceImpl 類別中的 buildSyncedFlushRequest
方法內，如下所示：

```
// 控制解析不可用索引及展開萬用字元運算式
request.indicesOptions(IndicesOptions.lenientExpandOpen());
```

2 執行同步 Flush 更新請求

在 SyncedFlushRequest 建置後，即可執行同步 Flush 更新請求。與建立索引請
求類似，同步 Flush 更新請求也有同步和非同步兩種執行方式。

☑ 同步方式

當以同步方式執行同步 Flush 更新請求時，用戶端會等待 Elasticsearch 伺服器

傳回的查詢結果 SyncedFlushResponse。在收到 SyncedFlushResponse 後，用戶端會繼續執行相關的邏輯程式。

以同步方式執行的程式增加在 IndexServiceImpl 類別中，如下所示：

```java
// 以同步方式執行 SyncedFlushRequest
  public void executeSyncedFlushRequest(String index) {
    SyncedFlushRequest request = buildSyncedFlushRequest(index);
    try {
      SyncedFlushResponse flushResponse = restClient.indices().flushSynced
          (request, RequestOptions.DEFAULT);
      // 解析 SyncedFlushResponse
      processSyncedFlushResponse(flushResponse);
    } catch (Exception e) {
      e.printStackTrace();
    } finally {
      // 關閉 Elasticsearch 連接
      closeEs();
    }
  }
```

☑ 非同步方式

當以非同步方式執行同步 Flush 更新請求時，進階用戶端不必同步等待請求結果的傳回，可以直接向介面呼叫方傳回非同步介面執行成功的結果。

為了處理非同步傳回的回應資訊或處理在請求執行過程中引發的例外資訊，使用者需要指定監聽器。以非同步方式執行的核心程式如下所示：

```java
client.indices().flushSyncedAsync(request, RequestOptions.DEFAULT, listener);
```

其中，listener 為監聽器。

在非同步請求處理後，如果請求執行成功，則呼叫 ActionListener 類別中的 onResponse 方法進行相關邏輯的處理；如果請求執行失敗，則呼叫 ActionListener 類別中的 onFailure 方法進行相關邏輯的處理。

以非同步方式執行的全部程式如下所示：

```
// 非同步方式執行 SyncedFlushRequest
 public void executeSyncedFlushRequestAsync(String index) {
   SyncedFlushRequest request = buildSyncedFlushRequest(index);
   // 建置監聽器
   ActionListener<SyncedFlushResponse> listener = new ActionListener
       <SyncedFlushResponse>() {
     @Override
     public void onResponse(SyncedFlushResponse refreshResponse) {
     }
     @Override
     public void onFailure(Exception e) {
     }
   };
   try {
     restClient.indices().flushSyncedAsync(request, RequestOptions.DEFAULT,
         listener);
   } catch (Exception e) {
     e.printStackTrace();
   } finally {
     // 關閉 Elasticsearch 連接
     closeEs();
   }
 }
```

當然，在非同步請求執行過程中可能會出現例外，例外的處理與同步方式執行情況相同。

❸ 解析同步 Flush 更新請求的回應結果

不論同步方式，還是非同步方式，在同步 Flush 更新索引請求執行後，用戶端均需要對 SyncedFlushResponse 進行處理和解析。

程式增加在 IndexServiceImpl 類別中，共分為三層，分別是 Controller 層、Service 層和 ServiceImpl 實現層。

在 Controller 層的 IndexController 類別中新增以下程式：

```
// 以同步方式執行 SyncedFlushRequest
@RequestMapping("/syncedflush/sr")
public String executeSyncedFlushRequest(String indexName) {
  // 參數驗證
  if (Strings.isNullOrEmpty(indexName)) {
    return "Parameters are wrong!";
  }
  indexService.executeSyncedFlushRequest(indexName);
  return "Execute SyncedFlushRequest success!";
}
```

在 Service 層的 IndexService 類別中新增以下程式：

```
// 以同步方式執行 SyncedFlushRequest
public void executeSyncedFlushRequest(String index);
```

在 ServiceImpl 實現層的 IndexServiceImpl 類別中新增以下程式：

```
// 解析 SyncedFlushResponse
private void processSyncedFlushResponse(SyncedFlushResponse flushResponse)
{
  // 更新請求命中的分片總數
  int totalShards = flushResponse.totalShards();
  // 更新成功的分片數
  int successfulShards = flushResponse.successfulShards();
  // 更新失敗的分片數
  int failedShards = flushResponse.failedShards();
  log.info("totalShards is " + totalShards + ";successfulShards is " +
      successfulShards
    + ";failedShards is " + failedShards);
}
```

隨後編輯專案，在專案根目錄下輸入以下指令：

```
mvn clean package
```

透過以下指令啟動專案服務：

```
java -jar ./target/esdemo-0.0.1-SNAPSHOT.jar
```

在專案服務啟動後，在瀏覽器中呼叫以下介面檢視索引 ultraman8 的同步
Flush 更新情況：

```
http://localhost:8080/springboot/es/indexsearch/syncedflush/sr?indexName=
ultraman8
```

請求執行後，在伺服器主控台輸出如下所示內容，表示同步 Flush 更新成功。

```
2019-09-06 17:35:12.267  INFO 73596 --- [nio-8080-exec-9] c.n.e.service.impl.IndexServiceImpl      : totalShards is 12;s
uccessfulShards is 6;failedShards is 0
```

8.13　清除索引快取

清除索引快取介面允許使用者清除與單一或多個與索引相連結的所有快取和
特定快取。

1　建置清除索引快取請求

在 執 行 清 除 索 引 快 取 請 求 前，需 要 建 置 清 除 索 引 快 取 請 求，即
ClearIndicesCacheRequest。與 FlushRequest 類似，建置 ClearIndicesCacheRequest
的方法有多種，可以消除單一索引、多個索引或全部索引。程式增加在
IndexServiceImpl 類別中，如下所示：

```
// 建置清除索引快取請求
public ClearIndicesCacheRequest buildClearIndicesCacheRequest(String index)
{
    // 清除單一索引
    ClearIndicesCacheRequest request = new ClearIndicesCacheRequest(index);
    // 清除多個索引
    ClearIndicesCacheRequest requestMultiple = new ClearIndicesCacheRequest
    (index, index);
    // 清除全部索引
    ClearIndicesCacheRequest requestAll = new ClearIndicesCacheRequest();
    return request;
}
```

在建置 ClearIndicesCacheRequest 時可以設定可選參數,其中最主要的是 IndicesOptions,用於解析不可用索引及展開萬用字元運算式。程式增加在 IndexServiceImpl 類別中的 buildClearIndicesCacheRequest 方法內,如下所示:

```
// 用於解析不可用索引及展開萬用字元運算式
request.indicesOptions(IndicesOptions.lenientExpandOpen());
```

在預設情況下,清除索引快取 API 會清除所有快取,但是使用者可以透過設定 query、fieldData 或 request 來顯性清除特定快取記憶體。程式增加在 IndexServiceImpl 類別中的 buildClearIndicesCacheRequest 方法內,如下所示:

```
// 將查詢標示設定為 true
  request.queryCache(true);
  // 將 FieldData 標示設定為 true
  request.fieldDataCache(true);
  // 將請求標示設定為 true
  request.requestCache(true);
  // 設定欄位參數
  request.fields("field1", "field2", "field3");
```

2 執行清除索引快取請求

在 ClearIndicesCacheRequest 建置後,即可執行清除索引快取請求。與文件索引請求類似,清除索引快取請求也有同步和非同步兩種執行方式。

同步方式

當以同步方式執行清除索引快取請求時,用戶端會等待 Elasticsearch 伺服器傳回的查詢結果 ClearIndicesCacheResponse。在收到 ClearIndicesCacheResponse 後,用戶端會繼續執行相關的邏輯程式。以同步方式執行的程式增加在 IndexServiceImpl 類別中,如下所示:

```
// 以同步方式執行 ClearIndicesCacheRequest
  public void executeClearIndicesCacheRequest(String index) {
    ClearIndicesCacheRequest request = buildClearIndicesCacheRequest(index);
    try {
      ClearIndicesCacheResponse clearCacheResponse = restClient.indices().
```

```
      clearCache(request, RequestOptions.DEFAULT);
      // 解析 ClearIndicesCacheRequest
      processClearIndicesCacheRequest (clearCacheResponse);
  } catch (Exception e) {
      e.printStackTrace();
  } finally {
      // 關閉 Elasticsearch 連接
      closeEs();
  }
}
```

☑ 非同步方式

當以非同步方式執行清除索引快取請求時,進階用戶端不必同步等待請求結果的傳回,可以直接向介面呼叫方傳回非同步介面執行成功的結果。

為了處理非同步傳回的回應資訊或處理在請求執行過程中引發的例外資訊,使用者需要指定監聽器。以非同步方式執行的核心程式如下所示:

```
client.indices().clearCacheAsync(request, RequestOptions.DEFAULT, listener);
```

其中,listener 為監聽器。

在非同步請求處理後,如果請求執行成功,則呼叫 ActionListener 類別中的 onResponse 方法進行相關邏輯的處理;如果請求執行失敗,則呼叫 ActionListener 類別中的 onFailure 方法進行相關邏輯的處理。

以非同步方式執行的全部程式如下所示:

```
// 以非同步方式執行 ClearIndicesCacheRequest
  public void executeClearIndicesCacheRequestAsync(String index) {
    ClearIndicesCacheRequest request = buildClearIndicesCacheRequest(index);
    // 建置監聽器
    ActionListener<ClearIndicesCacheResponse> listener =
      new ActionListener<ClearIndicesCacheResponse>() {
        @Override
        public void onResponse(ClearIndicesCacheResponse refreshResponse) {
      }
```

```
        @Override
        public void onFailure(Exception e) {
      }
      };
  try {
    restClient.indices().clearCacheAsync(request, RequestOptions.DEFAULT,
        listener);
  } catch (Exception e) {
    e.printStackTrace();
  } finally {
    // 關閉 Elasticsearch 連接
    closeEs();
  }
}
```

當然，在非同步請求執行過程中可能會出現例外，例外的處理與同步方式執行情況相同。

3 解析清除索引快取請求的回應結果

不論同步方式，還是非同步方式，在清除索引快取請求執行後，用戶端均需要對 ClearIndicesCacheResponse 進行處理和解析。

程式增加在 IndexServiceImpl 類別中，共分為三層，分別是 Controller 層、Service 層和 ServiceImpl 實現層。

在 Controller 層的 IndexController 類別中新增以下程式：

```
// 以同步方式執行 ClearIndicesCacheRequest
@RequestMapping("/clearcache/sr")
public String executeClearIndicesCacheRequest(String indexName) {
  // 參數驗證
  if (Strings.isNullOrEmpty(indexName)) {
    return "Parameters are wrong!";
  }
  indexService.executeClearIndicesCacheRequest(indexName);
  return "Execute ClearIndicesCacheRequest success!";
}
```

在 Service 層的 IndexService 類別中新增以下程式：

```
// 以同步方式執行 ClearIndicesCacheRequest
  public void executeClearIndicesCacheRequest(String index);
```

在 ServiceImpl 實現層的 IndexServiceImpl 類別中新增以下程式：

```
// 解析 ClearIndicesCacheRequest
  private void processClearIndicesCacheRequest(ClearIndicesCacheResponse
      clearCacheResponse) {
    // 清除索引快取請求命中的分片總數
    int totalShards = clearCacheResponse.getTotalShards();
    // 清除成功的分片數
    int successfulShards = clearCacheResponse.getSuccessfulShards();
    // 清除失敗的分片數
    int failedShards = clearCacheResponse.getFailedShards();
    log.info("totalShards is " + totalShards + ";successfulShards is " +
        successfulShards
      + ";failedShards is " + failedShards);
  }
```

隨後編輯專案，在專案根目錄下輸入以下指令：

```
mvn clean package
```

透過以下指令啟動專案服務：

```
java -jar ./target/esdemo-0.0.1-SNAPSHOT.jar
```

在專案服務啟動後，在瀏覽器中呼叫以下介面檢視索引 ultraman8 的清除快取後的情況：

```
http://localhost:8080/springboot/es/indexsearch/clearcache/sr?indexName=
ultraman8
```

請求執行後，如果在伺服器中輸出如下所示內容，則表明清除成功：

```
2019-09-06 18:12:25.020  INFO 84892 --- [nio-8080-exec-1] c.n.e.service.impl.IndexServiceImpl    : totalShards is 12;s
uccessfulShards is 6;failedShards is 0
```

8.14 強制合併索引

Elasticsearch 提供了強制合併索引介面，當使用者需要合併單一、多個或全部索引時，該介面會合併依賴於 Lucene 索引在每個分片中儲存的分段數。強制合併操作透過合併分段來減少分段數量。如果 HTTP 連接遺失，則請求將在後台繼續執行，並且任何新的請求都會被阻塞，直到前面的強制合併完成。

■ 建置強制合併索引請求

在執行強制合併索引請求前，需要建置強制合併索引請求，即 ForceMergeRequest。建置 ForceMergeRequest 的方法有多種，可以強制合併單一索引、多個索引或全部索引。程式增加在 IndexServiceImpl 類別中，如下所示：

```
// 建置 ForceMergeRequest
  public ForceMergeRequest buildForceMergeRequest(String index) {
    // 強制合併單一索引
    ForceMergeRequest request = new ForceMergeRequest(index);
    // 強制合併多個索引
    ForceMergeRequest requestMultiple = new ForceMergeRequest(index + "1",
        index + "2");
    // 強制合併全部索引
    ForceMergeRequest requestAll = new ForceMergeRequest();
    return request;
  }
```

在建置 ForceMergeRequest 時，使用者可以設定可選參數。可選參數有 flush 標識、唯一刪除標識、合併的段數和 IndicesOptions。其中，IndicesOptions 用於解析不可用索引及展開萬用字元運算式。程式增加在 IndexServiceImpl 類別中的 buildForceMergeRequest 方法內，如下所示：

```
// 設定可選參數
    // 設定 IndicesOptions，用於解析不可用索引及展開萬用字元運算式
    request.indicesOptions(IndicesOptions.lenientExpandOpen());
    // 設定 max_num_segments，以控制合併後的段數
```

```
request.maxNumSegments(1);
// 將唯一刪除標識設定為 true
request.onlyExpungeDeletes(true);
// 將 flush 標識設定為 true
request.flush(true);
```

2 執行強制合併索引請求

在 ForceMergeRequest 建置後，即可執行強制合併索引請求。與建立索引請求類似，強制合併索引請求也有同步和非同步兩種執行方式。

▨ 同步方式

當以同步方式執行強制合併索引請求時，用戶端會等待 Elasticsearch 伺服器傳回的查詢結果 ForceMergeResponse。在收到 ForceMergeResponse 後，用戶端會繼續執行相關的邏輯程式。以同步方式執行的程式增加在 IndexServiceImpl 類別中，如下所示：

```
// 以同步方式執行 ForceMergeRequest
  public void executeForceMergeRequest(String index) {
    ForceMergeRequest request = buildForceMergeRequest(index);
    try {
      ForceMergeResponse forceMergeResponse =
          restClient.indices().forcemerge(request, RequestOptions.DEFAULT);
    } catch (Exception e) {
      e.printStackTrace();
    } finally {
      // 關閉 Elasticsearch 連接
      closeEs();
    }
}
```

▨ 非同步方式

當以非同步方式執行強制合併索引請求時，Java 進階用戶端不必同步等待請求結果的傳回，可以直接向介面呼叫方傳回非同步介面執行成功的結果。

為了處理非同步傳回的回應資訊或處理在請求執行過程中引發的例外資訊，

使用者需要指定監聽器。以非同步方式執行的核心程式如下所示：

```
client.indices().forcemergeAsync(request, RequestOptions.DEFAULT, listener);
```

其中，listener 為監聽器。

在非同步請求處理後，如果請求執行成功，則呼叫 ActionListener 類別中的 onResponse 方法進行相關邏輯的處理；如果請求執行失敗，則呼叫 ActionListener 類別中的 onFailure 方法進行相關邏輯的處理。

以非同步方式執行的全部程式增加在 IndexServiceImpl 類別中，如下所示：

```
// 以非同步方式執行 ForceMergeRequest
public void executeForceMergeRequestAsync(String index) {
  ForceMergeRequest request = buildForceMergeRequest(index);
  // 建置監聽器
  ActionListener<ForceMergeResponse> listener = new ActionListener
      <ForceMergeResponse>() {
    @Override
    public void onResponse(ForceMergeResponse forceMergeResponse) {
    }
    @Override
    public void onFailure(Exception e) {
    }
  };
  try {
    restClient.indices().forcemergeAsync(request, RequestOptions.DEFAULT,
        listener);
  } catch (Exception e) {
    e.printStackTrace();
  } finally {
    // 關閉 Elasticsearch 連接
    closeEs();
  }
}
```

當然，在非同步請求執行過程中可能會出現例外，例外的處理與同步方式執行情況相同。

3 解析強制合併索引請求的回應結果

不論同步方式，還是非同步方式，在強制合併索引請求執行後，用戶端均需要對請求的回應結果 ForceMergeResponse 進行處理和解析。

程式共分為三層，分別是 Controller 層、Service 層和 ServiceImpl 實現層。

在 Controller 層的 IndexController 類別中新增以下程式：

```
// 以同步方式執行 ForceMergeRequest
@RequestMapping("/merge/sr")
public String executeForceMergeRequest(String indexName) {
    // 參數驗證
    if (Strings.isNullOrEmpty(indexName)) {
        return "Parameters are wrong!";
    }
    indexService.executeForceMergeRequest(indexName);
    return "Execute ForceMergeRequest success!";
}
```

在 Service 層的 IndexService 類別中新增以下程式：

```
// 以同步方式執行 ForceMergeRequest
public void executeForceMergeRequest(String index);
```

在 ServiceImpl 實現層的 IndexServiceImpl 類別中新增以下程式：

```
// 以同步方式執行 ForceMergeRequest
public void executeForceMergeRequest(String index) {
    ForceMergeRequest request = buildForceMergeRequest(index);
    try {
        ForceMergeResponse forceMergeResponse =
            restClient.indices().forcemerge(request, RequestOptions.DEFAULT);
        // 解析 ForceMergeResponse
        processForceMergeResponse(forceMergeResponse);
    } catch (Exception e) {
        e.printStackTrace();
    } finally {
        // 關閉 Elasticsearch 連接
        closeEs();
```

```
    }
}
// 解析 ForceMergeResponse
private void processForceMergeResponse(ForceMergeResponse forceMergeResponse)
{
    // 強制合併索引請求命中的分片總數
    int totalShards = forceMergeResponse.getTotalShards();
    // 強制合併成功的分片數
    int successfulShards = forceMergeResponse.getSuccessfulShards();
    // 強制合併失敗的分片數
    int failedShards = forceMergeResponse.getFailedShards();
    // 在一個或多個分片強制合併失敗的失敗列表
    DefaultShardOperationFailedException[] failures = forceMergeResponse.
        getShardFailures();
    log.info("totalShards is " + totalShards + ";successfulShards is " +
        successfulShards
        + ";failedShards is " + failedShards + ";failures size is "
        + (failures == null ? 0 : failures.length));
}
```

隨後編輯專案，在專案根目錄下輸入以下指令：

```
mvn clean package
```

透過以下指令啟動專案服務：

```
java -jar ./target/esdemo-0.0.1-SNAPSHOT.jar
```

在專案服務啟動後，在瀏覽器中呼叫以下介面檢視索引 ultraman8 的合併情況：

```
http://localhost:8080/springboot/es/indexsearch/merge/sr?indexName=ultraman8
```

請求執行後，如果在伺服器中輸出如下所示內容，則表示索引合併成功：

```
2019-09-07 13:59:47.318  INFO 26636 --- [nio-8080-exec-1] c.n.e.service.impl.IndexServiceImpl      : totalShards is 12;s
uccessfulShards is 6;failedShards is 0;failures size is 0
```

8.15 捲動索引

當索引較大或資料很老舊時，可以使用 Elasticsearch 提供的捲動索引 API 將別名捲動到新的索引。

1 建置捲動索引請求

在執行捲動索引請求前，需要建置捲動索引請求，即 RolloverRequest。RolloverRequest 的必選參數是兩個字串參數，以及一個或多個條件參數。其中，字串參數為索引別名和新索引名稱，條件參數用於確定何時回覆索引。新索引名稱對應的索引會在條件參數滿足時建立，並將別名指向新索引。

程式增加在 IndexServiceImpl 類別中，如下所示：

```
// 建置 RolloverRequest
public RolloverRequest buildRolloverRequest(String index) {
    // 指向要捲動的索引別名 (第一個參數)，以及執行捲動操作時的新索引名稱。new index
    // 參數是可選的，可以設定為空
    RolloverRequest request = new RolloverRequest(index, index + "-2");
    // 指數年齡
    request.addMaxIndexAgeCondition(new TimeValue(7, TimeUnit.DAYS));
    // 索引中的文件數
    request.addMaxIndexDocsCondition(1000);
    // 索引的大小
    request.addMaxIndexSizeCondition(new ByteSizeValue(5, ByteSizeUnit.GB));
    return request;
}
```

在建置 RolloverRequest 時，使用者還可以設定可選參數。可選參數主要有是否執行捲動、所有節點確認索引開啟的逾時、從節點連接到主節點的逾時、請求傳回前等待的活躍分片數量，以及新索引相關的設定。程式增加在 IndexServiceImpl 類別中的 buildRolloverRequest 方法內，如下所示：

```
// 設定可選參數
    // 是否執行捲動 (預設為 true)
    request.dryRun(true);
```

```
      // 所有節點確認索引開啟的逾時
      request.setTimeout(TimeValue.timeValueMinutes(2));
      // 從節點連接到主節點的逾時
      request.setMasterTimeout(TimeValue.timeValueMinutes(1));
      // 請求傳回前等待的活躍分片數量
      request.getCreateIndexRequest().waitForActiveShards(ActiveShardCount.
from(2));
      // 請求傳回前等待的活躍分片數量，重置為預設值
      request.getCreateIndexRequest().waitForActiveShards(ActiveShardCount.
DEFAULT);
      // 增加應用於新索引的設定，其中包含要為其建立的分片數
      request.getCreateIndexRequest().settings(Settings.builder().put("index.
number_of_shards", 4));
      // 增加與新索引連結的對映
      String mappings = "{\"properties\":{\"field\":{\"type\":\"content\"}}}";
      request.getCreateIndexRequest().mapping(mappings, XContentType.JSON);
      // 增加與新索引連結的別名
      request.getCreateIndexRequest().alias(new Alias(index + "-2_alias"));
```

▌2 執行捲動索引請求

在 RolloverRequest 建置後，即可執行捲動索引請求。與建立索引請求類似，
捲動索引請求也有同步和非同步兩種執行方式。

▨ 同步方式

當以同步方式執行捲動索引請求時，用戶端會等待 Elasticsearch 伺服器傳回的
查詢結果 RolloverResponse。在收到 RolloverResponse 後，用戶端會繼續執行
相關的邏輯程式。以同步方式執行的程式增加在 IndexServiceImpl 類別中，如
下所示：

```
// 以同步方式執行 RolloverRequest
  public void executeRolloverRequest(String index) {
    RolloverRequest request = buildRolloverRequest(index);
    try {
      RolloverResponse rolloverResponse =
          restClient.indices().rollover(request, RequestOptions.DEFAULT);
```

```
    // 解析 RolloverResponse
    processRolloverResponse(rolloverResponse);
  } catch (Exception e) {
    e.printStackTrace();
  } finally {
    // 關閉 Elasticsearch 連接
    closeEs();
  }
}
```

非同步方式

當以非同步方式執行捲動索引請求時，進階用戶端不必同步等待請求結果的傳回，可以直接向介面呼叫方傳回非同步介面執行成功的結果。

為了處理非同步傳回的回應資訊或處理在請求執行過程中引發的例外資訊，使用者需要指定監聽器。以非同步方式執行的核心程式如下所示：

```
client.indices().rolloverAsync(request, RequestOptions.DEFAULT, listener);
```

其中，listener 為監聽器。

在非同步請求處理後，如果請求執行成功，則呼叫 ActionListener 類別中的 onResponse 方法進行相關邏輯的處理；如果請求執行失敗，則呼叫 ActionListener 類別中的 onFailure 方法進行相關邏輯的處理。

以非同步方式執行的全部程式如下所示：

```
// 以非同步方式執行 RolloverRequest
  public void executeRolloverRequestAsync(String index) {
    RolloverRequest request = buildRolloverRequest(index);
    // 建置 RolloverRequest
    ActionListener<RolloverResponse> listener = new ActionListener
        <RolloverResponse>() {
      @Override
      public void onResponse(RolloverResponse rolloverResponse) {
      }
      @Override
```

```
    public void onFailure(Exception e) {
    }
};

try {
    restClient.indices().rolloverAsync(request, RequestOptions.DEFAULT,
        listener);
} catch (Exception e) {
    e.printStackTrace();
} finally {
    // 關閉 Elastcsearch 連接
    closeEs();
}
}
```

當然，在非同步請求執行過程中可能會出現例外，例外的處理與同步方式執行情況相同。

❸ 解析捲動索引請求的回應結果

不論同步方式，還是非同步方式，在捲動索引請求執行後，用戶端均需要對請求的回應結果 RolloverResponse 進行處理和解析。

程式共分為三層，分別是 Controller 層、Service 層和 ServiceImpl 實現層。

在 Controller 層的 IndexController 類別中新增以下程式：

```
// 以同步方式執行 RolloverRequest
@RequestMapping("/rollover/sr")
public String executeRolloverRequest(String indexName) {
    // 參數驗證
    if (Strings.isNullOrEmpty(indexName)) {
        return "Parameters are wrong!";
    }
    indexService.executeRolloverRequest(indexName);
    return "Execute RolloverRequest success!";
}
```

在 Service 層的 IndexService 類別中新增以下程式：

```
// 以同步方式執行 RolloverRequest
  public void executeRolloverRequest(String index);
```

在 ServiceImpl 實現層的 IndexServiceImpl 中新增以下程式：

```
// 解析 RolloverResponse
  private void processRolloverResponse(RolloverResponse rolloverResponse) {
      // 所有節點是否已確認請求
      boolean acknowledged = rolloverResponse.isAcknowledged();
      // 是否在逾時前為索引中的每個分片啟動了所需數量的分片備份
      boolean shardsAcked = rolloverResponse.isShardsAcknowledged();
      // 舊索引名稱，最後被捲動
      String oldIndex = rolloverResponse.getOldIndex();
      // 新索引名稱
      String newIndex = rolloverResponse.getNewIndex();
      // 索引是否已回覆
      boolean isRolledOver = rolloverResponse.isRolledOver();
      boolean isDryRun = rolloverResponse.isDryRun();
      // 不同的條件，是否比對
      Map<String, Boolean> conditionStatus = rolloverResponse. getConditionStatus
                                                    ();
      log.info("acknowledged is " + acknowledged + ";shardsAcked is " +
          shardsAcked + ";oldIndex is "
        + oldIndex + ";newIndex is " + newIndex + ";isRolledOver is " +
          isRolledOver
        + ";isDryRun is " + isDryRun + ";conditionStatus size is "
        + (conditionStatus == null ? 0 : conditionStatus.size()));
  }
```

隨後編輯專案，在專案根目錄下輸入以下指令：

```
mvn clean package
```

透過以下指令啟動專案服務：

```
java -jar ./target/esdemo-0.0.1-SNAPSHOT.jar
```

在專案服務啟動後，在瀏覽器中呼叫以下介面檢視捲動索引的執行情況：

```
http://localhost:8080/springboot/es/indexsearch/rollover/sr?indexName=
ultraman7_alias
```

請求執行後，如果在伺服器中輸出如下所示內容，則表示新索引已經建置成功：

```
2019-09-07 14:31:36.421  INFO 70892 --- [nio-8080-exec-1] c.n.e.service.impl.IndexServiceImpl      : acknowledged is fal
se;shardsAcked is false;oldIndex is ultraman7;newIndex is ultraman7_alias-2;isRolledOver is false;isDryRun is true;condi
tionStatus size is 3
```

8.16 索引別名

在 Elasticsearch 中，有專門的介面為索引別名進行命名，即索引別名介面。當透過索引別名呼叫索引時，所有的介面都將自動轉為索引實際名稱。

1 建置索引別名請求

在執行索引別名請求前，需要建置索引別名請求，即 IndicatesAliasesRequest。IndicatesAliasesRequest 至少含有一個別名命名操作。一個別名命名操作必須包含目前索引名稱和將要命名的別名，對應的程式增加在 IndexServiceImpl 類別中，如下所示：

```
// 建置 IndicatesAliasesRequest
  public IndicesAliasesRequest buildIndicatesAliasesRequest(String index,
      String indexAlias) {
    // 建立 IndicatesAliasesRequest
    IndicesAliasesRequest request = new IndicesAliasesRequest();
    // 建立別名操作，將索引的別名設為 indexAlias
    AliasActions aliasAction =new AliasActions(AliasActions.Type.ADD).index
    (index).alias(indexAlias);
    // 將別名操作增加到請求中
    request.addAliasAction(aliasAction);
    return request;
  }
```

其中，AliasActions 支援的操作類型有新增別名（AliasActions.Type.ADD）、刪除別名（AliasActions.Type.REMOVE）和刪除索引（AliasActions.Type.REMOVE_INDEX）。

需要指出的是，索引別名不能重複，也不能和索引名稱重複；使用者可以增加、刪除別名，但不能修改別名。

AliasActions 還支援設定可選篩選器（filter）和可選路由（routing），對應的程式增加在 buildIndicatesAliasesRequest 方法內，如下所示：

```
// 建立別名操作，將索引的別名設為 indexAlias ADD
    AliasActions aliasAction = new AliasActions(AliasActions.Type.ADD).index
        (index).alias(indexAlias).filter("{\"term\":{\"year\":2019}}").routing
        ("niudong");
```

在建置 IndicatesAliasesRequest 時，使用者還可以設定可選參數，主要是所有節點確認索引操作的逾時、從節點連接到主節點的逾時。對應的程式增加在 IndexServiceImpl 類別中的 buildIndicatesAliasesRequest 方法內，如下所示：

```
// 可選參數設定
    // 所有節點確認索引操作的逾時
    request.timeout(TimeValue.timeValueMinutes(2));
    request.timeout("2m");
    // 從節點連接到主節點的逾時
    request.masterNodeTimeout(TimeValue.timeValueMinutes(1));
    request.masterNodeTimeout("1m");
```

2 執行索引別名請求

在 IndicatesAliasesRequest 建置後，即可執行索引別名請求。與建立索引請求類似，索引別名請求也有同步和非同步兩種執行方式。

同步方式

當以同步方式執行索引別名請求時，用戶端會等待 Elasticsearch 伺服器傳回的查詢結果 IndicatesAliasesResponse。在收到 IndicatesAliasesResponse 後，用戶端會繼續執行相關的邏輯程式。以同步方式執行的程式增加在

IndexServiceImpl 類別中，如下所示：

```
// 以同步方式執行 IndicatesAliasesRequest
  public void executeIndicatesAliasesRequest(String index, String indexAlias)
     {
   IndicesAliasesRequest request = buildIndicatesAliasesRequest(index,
      indexAlias);
   try {
     AcknowledgedResponse indicesAliasesResponse =
        restClient.indices().updateAliases(request, RequestOptions.DEFAULT);
     // 解析 AcknowledgedResponse
     processAcknowledgedResponse(indicesAliasesResponse);
   } catch (Exception e) {
     e.printStackTrace();
   } finally {
     // 關閉 Elasticsearch 連接
     closeEs();
   }
 }
```

☑ 非同步方式

當以非同步方式執行索引別名請求時，進階用戶端不必同步等待請求結果的傳回，可以直接向介面呼叫方傳回非同步介面執行成功的結果。

為了處理非同步傳回的回應資訊或處理在請求執行過程中引發的例外資訊，使用者需要指定監聽器。以非同步方式執行的核心程式如下所示：

```
client.indices().updateAliasesAsync(request, RequestOptions.DEFAULT, listener);
```

其中，listener 為監聽器。

在非同步請求處理後，如果請求執行成功，則呼叫 ActionListener 類別中的 onResponse 方法進行相關邏輯的處理；如果請求執行失敗，則呼叫 ActionListener 類別中的 onFailure 方法進行相關邏輯的處理。

以非同步方式執行的全部程式如下所示：

```
// 以非同步方式執行 IndicatesAliasesRequest
```

```
public void executeIndicatesAliasesRequestAsync(String index, String
    indexAlias) {
  IndicesAliasesRequest request = buildIndicatesAliasesRequest(index,
      indexAlias);
  // 建置監聽器
  ActionListener<AcknowledgedResponse> listener = new ActionListener
      <AcknowledgedResponse>() {
    @Override
    public void onResponse(AcknowledgedResponse indicesAliasesResponse) {
    }
    @Override
    public void onFailure(Exception e) {
    }
  };

  try {
    restClient.indices().updateAliasesAsync(request, RequestOptions.DEFAULT,
        listener);
  } catch (Exception e) {
    e.printStackTrace();
  } finally {
    // 關閉 Elasticsearch 連接
    closeEs();
  }
}
```

當然，在非同步請求執行過程中可能會出現例外，例外的處理與同步方式執行情況相同。

3 解析索引別名請求的回應結果

不論同步方式，還是非同步方式，當索引別名請求執行後，用戶端均需要對請求的回應結果 IndicatesAliaseResponse 進行處理和解析。

程式共分為三層，分別是 Controller 層、Service 層和 ServiceImpl 實現層。

在 Controller 層的 IndexController 類別中新增以下程式：

```
// 以同步方式執行 IndicatesAliasesRequest
@RequestMapping("/createAlias/sr")
public String executeIndicatesAliasesRequest(String indexName, String
    indexAliasName) {
  // 參數驗證
  if (Strings.isNullOrEmpty(indexName) || Strings.isNullOrEmpty
      (indexAliasName)) {
    return "Parameters are wrong!";
  }
  indexService.executeIndicatesAliasesRequest(indexName, indexAliasName);
  return "Execute IndicatesAliasesRequest success!";
}
```

在 Service 層的 IndexService 類別中新增以下程式：

```
// 以同步方式執行 IndicatesAliasesRequest
public void executeIndicatesAliasesRequest(String index, String indexAlias);
```

在 ServiceImpl 實現層的 IndexServiceImpl 類別中不用新增程式，可以重複使用前文提及的 processAcknowledgedResponse 方法。

隨後編輯專案，在專案根目錄下輸入以下指令：

```
mvn clean package
```

透過以下指令啟動專案服務：

```
java -jar ./target/esdemo-0.0.1-SNAPSHOT.jar
```

在專案服務啟動後，在瀏覽器中呼叫以下介面檢視索引 ultraman8 的別名命名情況：

```
http://localhost:8080/springboot/es/indexsearch/createAlias/sr?indexName=
ultraman8&indexAliasName=ultraman8_alias
```

請求執行後，如果在伺服器中輸出如下所示內容，則表示索引 ultraman8 的別名 ultraman8_alias 命名成功。

```
2019-09-07 15:01:47.053  INFO 46824 --- [nio-8080-exec-1] c.n.e.service.impl.IndexServiceImpl      : acknowledged is true
```

8.17 索引別名存在驗證

在 Elasticsearch 中，不僅提供了索引別名命名介面，還提供了檢視索引是否有別名的介面，即索引別名存在驗證介面。

1 建置索引別名存在驗證請求

在執行索引別名存在驗證請求前，需要建置索引別名存在驗證請求，即 GetAliasesRequest。GetAliasesRequest 的必要參數為待驗證的索引別名。對應的程式增加在 IndexServiceImpl 類別中，如下所示：

```
// 建置 GetAliasesRequest
public GetAliasesRequest buildGetAliasesRequest(String indexAlias) {
  GetAliasesRequest request = new GetAliasesRequest();
  GetAliasesRequest requestWithAlias = new GetAliasesRequest(indexAlias);
  GetAliasesRequest requestWithAliases =
      new GetAliasesRequest(new String[] {indexAlias, indexAlias});
  return request;
}
```

在建置 GetAliasesRequest 時，GetAliasesRequest 的可選參數主要有帶驗證存在性的別名、與別名連結的或多個索引名稱、是否本機尋找和 IndicesOptions。IndicesOptions 用於控制解析不可用索引及展開萬用字元運算式等。對應的程式增加在 IndexServiceImpl 類別的 buildGetAliasesRequest 方法內，如下所示：

```
// 設定可選參數
    // 帶驗證存在性的別名
    request.aliases(indexAlias);
    // 與別名連結的或多個索引
    request.indices(index);
    // 設定 IndicesOptions
    request.indicesOptions(IndicesOptions.lenientExpandOpen());
    // 本機標示 (預設為 false)，控制是否需要在本機叢集狀態或所選主節點持有的叢集狀態
    // 中查找別名
    request.local(true);
```

② 執行索引別名存在驗證請求

在 GetAliasesRequest 建置後，即可執行索引別名存在驗證請求。與索引別名請求類似，索引別名存在驗證請求也有同步和非同步兩種執行方式。

☑ 同步方式

當以同步方式執行索引別名存在驗證請求時，用戶端會等待 Elasticsearch 伺服器傳回的布林型查詢結果。在收到查詢結果後，用戶端會繼續執行相關邏輯程式。對應的程式增加在 IndexServiceImpl 類別中，如下所示：

```
// 以同步方式執行 GetAliasesRequest
 public void executeGetAliasesRequest(String indexAlias) {
   GetAliasesRequest request = buildGetAliasesRequest(indexAlias);
   try {
     boolean exists = restClient.indices().existsAlias(request,
         RequestOptions.DEFAULT);
     log.info("indexAlias exists is " + exists);
   } catch (Exception e) {
     e.printStackTrace();
   } finally {
     // 關閉 Elasticsecrch 連接
     closeEs();
   }
 }
```

☑ 非同步方式

當以非同步方式執行索引別存在驗證請求時，進階用戶端不必同步等待請求結果的傳回，可以直接向介面呼叫方傳回非同步介面執行成功的結果。

為了處理非同步傳回的回應資訊或處理在請求執行過程中引發的例外資訊，使用者需要指定監聽器。以非同步方式執行的核心程式如下所示：

```
client.indices().existsAliasAsync(request, RequestOptions.DEFAULT, listener);
```

其中，listener 為監聽器。

當非同步請求處理後，如果請求執行成功，則呼叫 ActionListener 類別中的 onResponse 方法進行相關邏輯的處理；如果請求執行失敗，則呼叫 ActionListener 類別中的 onFailure 方法進行相關邏輯的處理。

以非同步方式執行的全部程式增加在 IndexServiceImpl 類別中，如下所示：

```
// 以非同步方式執行 GetAliasesRequest
  public void executeGetAliasesRequestAsync(String indexAlias) {
    GetAliasesRequest request = buildGetAliasesRequest(indexAlias);
    // 建置監聽器
    ActionListener<Boolean> listener = new ActionListener<Boolean>() {
      @Override
      public void onResponse(Boolean exists) {
      }
      @Override
      public void onFailure(Exception e) {
      }
    };
    try {
      restClient.indices().existsAliasAsync(request, RequestOptions.DEFAULT,
          listener);
    } catch (Exception e) {
      e.printStackTrace();
    } finally {
      // 關閉 Elasticsearch 連接
      closeEs();
    }
  }
```

當然，在非同步請求執行過程中可能會出現例外，例外的處理與同步方式執行情況相同。

3 解析索引別名存在驗證請求的回應結果

不論同步方式，還是非同步方式，在索引別名存在驗證請求執行後，用戶端均需要對請求的回應結果進行處理和解析。

程式分為三層，分別是 Controller 層、Service 層和 ServiceImpl 實現層。

在 Controller 層的 IndexController 類別中新增以下程式：

```
// 以同步方式執行 GetAliasesRequest
@RequestMapping("/existsAlias/sr")
public String executeGetAliasesRequest(String indexAliasName) {
    // 參數驗證
    if (Strings.isNullOrEmpty(indexAliasName)) {
        return "Parameters are wrong!";
    }
    indexService.executeGetAliasesRequest(indexAliasName);
    return "Execute GetAliasesRequest success!";
}
```

在 Service 層的 IndexService 類別中新增以下程式：

```
// 以同步方式執行 GetAliasesRequest
public void executeGetAliasesRequest(String indexAlias);
```

無須在 ServiceImpl 實現層的 IndexServiceImpl 類別中新增程式。

隨後編輯專案，在專案根目錄下輸入以下指令：

```
mvn clean package
```

透過以下指令啟動專案服務：

```
java -jar ./target/esdemo-0.0.1-SNAPSHOT.jar
```

在專案服務啟動後，在瀏覽器中呼叫以下介面檢視索引別名 ultraman70_alias 和 ultraman7_alias 的存在驗證情況：

```
http://localhost:8080/springboot/es/indexsearch/existsAlias/sr?indexAliasName=
ultraman70_alias
http://localhost:8080/springboot/es/indexsearch/existsAlias/sr?indexAliasName=
ultraman7_alias
```

請求執行後，在伺服器中分別輸出如下所示內容，ultraman70_alias 索引不存在，ultraman7_alias 索引存在：

```
2019-09-07 15:23:18.452  INFO 3572 --- [nio-8080-exec-1] c.n.e.service.impl.IndexServiceImpl      : indexAlias exists is
false

2019-09-07 15:33:17.791  INFO 25752 --- [nio-8080-exec-1] c.n.e.service.impl.IndexServiceImpl      : indexAlias exists i
s true
```

8.18 取得索引別名

在 Elasticsearch 中，不僅提供了索引別名命名介面，還提供了取得索引別名介面。

1 建置取得索引別名請求

在執行取得索引別名請求前，需要建置取得索引別名請求，即 GetAliasesRequest。GetAliasesRequest 的必選參數為待驗證的索引別名，前面已經展示了 GetAliasesRequest 的建置方法，不再贅述。

2 執行取得索引別名請求

在 GetAliasesRequest 建置後，即可執行取得索引別名請求。與索引別名請求類似，取得索引別名請求也有同步和非同步兩種執行方式。

▨ 同步方式

當以同步方式執行取得索引別名請求時，用戶端會等待 Elasticsearch 伺服器傳回的查詢結果 GetAliasesResponse。在收到 GetAliasesResponse 後，用戶端會繼續執行相關的邏輯程式。對應的程式增加在 IndexServiceImpl 類別中，如下所示：

```
// 以同步方式執行 GetAliasesRequest
public void executeGetAliasesRequestForAliases(String indexAlias) {
  GetAliasesRequest request = buildGetAliasesRequest(indexAlias);
  try {
    GetAliasesResponse response = restClient.indices().getAlias(request,
        RequestOptions.DEFAULT);
    // 解析 GetAliasesResponse
    processGetAliasesResponse(response);
  } catch (Exception e) {
    e.printStackTrace();
  } finally {
    // 關閉 Elasticsearch 的連接
    closeEs();
```

```
      }
   }
```

☑ 非同步方式

當以非同步方式執行取得索引名請求時,進階用戶端不必同步等待請求結果的傳回,可以直接向介面呼叫方傳回非同步介面執行成功的結果。

為了處理非同步傳回的回應資訊或處理在請求執行過程中引發的例外資訊,使用者需要指定監聽器。以非同步方式執行的核心程式如下所示:

```
client.indices().getAliasAsync(request, RequestOptions.DEFAULT, listener);
```

其中,listener 為監聽器。

在非同步請求處理後,如果請求執行成功,則呼叫 ActionListener 類別中的 onResponse 方法進行相關邏輯的處理;如果請求執行失敗,則呼叫 ActionListener 類別中的 onFailure 方法進行相關邏輯的處理。

以非同步方式執行的全部程式如下所示:

```
// 以非同步方式執行 GetAliasesRequest
public void executeGetAliasesRequestForAliasesAsync(String indexAlias) {
    GetAliasesRequest request = buildGetAliasesRequest(indexAlias);
    // 建置監聽器
    ActionListener<GetAliasesResponse> listener = new ActionListener
        <GetAliasesResponse>() {
      @Override
      public void onResponse(GetAliasesResponse exists) {
      }
      @Override
      public void onFailure(Exception e) {
      }
    };

    try {
      restClient.indices().getAliasAsync(request, RequestOptions.DEFAULT,
          listener);
```

```
    } catch (Exception e) {
      e.printStackTrace();
    } finally {
      // 關閉 Elasticsearch 連接
      closeEs();
    }
  }
```

當然，在非同步請求執行過程中可能會出現例外，例外的處理與同步方式執行情況相同。

3 解析取得索引別名請求的回應結果

不論同步方式，還是非同步方式，在取得索引別名請求執行後，用戶端均需要對請求的回應結果 GetAliasesResponse 進行處理和解析。

程式共分為三層，分別是 Controller 層、Service 層和 ServiceImpl 實現層。

在 Controller 層的 IndexController 類別中新增以下程式：

```
// 以同步方式執行 GetAliasesRequest
  @RequestMapping("/getAlias/sr")
  public String executeGetAliasesRequestForAliases(String indexAliasName) {
    // 參數驗證
    if (Strings.isNullOrEmpty(indexAliasName)) {
      return "Parameters are wrong!";
    }
    indexService.executeGetAliasesRequestForAliases(indexAliasName);
    return "Execute GetAliasesRequestForAliases success!";
  }
```

在 Service 層的 IndexService 類別中新增以下程式：

```
// 以同步方式執行 GetAliasesRequest
  public void executeGetAliasesRequestForAliases(String indexAlias);
```

在 ServiceImpl 實現層的 IndexServiceImpl 類別中新增以下程式：

```
// 解析 GetAliasesResponse
  private void processGetAliasesResponse(GetAliasesResponse response) {
```

```
  // 檢索索引及其別名的對映
  Map<String, Set<AliasMetaData>> aliases = response.getAliases();
  // 如果為空，則傳回
  if (aliases == null || aliases.size() <= 0) {
    return;
  }
  // 檢查 Map
  Set<Entry<String, Set<AliasMetaData>>> set = aliases.entrySet();
  for (Entry<String, Set<AliasMetaData>> entry : set) {
    String key = entry.getKey();
    Set<AliasMetaData> metaSet = entry.getValue();
    if (metaSet == null || metaSet.size() <= 0) {
      return;
    }
    for (AliasMetaData meta : metaSet) {
      String aliaas = meta.alias();
      log.info("key is " + key + ";aliaas is " + aliaas);
    }
  }
}
```

隨後編輯專案，在專案根目錄下輸入以下指令：

```
mvn clean package
```

透過以下指令啟動專案服務：

```
java -jar ./target/esdemo-0.0.1-SNAPSHOT.jar
```

在專案服務啟動後，在瀏覽器中呼叫以下介面取得索引 ultraman7_alias 的別名情況：

```
http://localhost:8080/springboot/es/indexsearch/getAlias/sr?indexAliasName
=ultraman7_alias
```

在伺服器中輸出如下所示內容：

```
2019-09-07 15:44:09.168  INFO 11432 --- [nio-8080-exec-1] c.n.e.service.impl.IndexServiceImpl      : key is ultraman7;al
iaas is ultraman7_alias
```

8.19 索引原瞭解析

為了加強搜索效能，Elasticsearch 做了很多設計。下面介紹在 Elasticsearch 內部是如何索引文件的。由於第 5 章和第 6 章中曾提到部分文件的索引過程和分片過程，所以本節將從其他維度展開。

8.19.1 近即時搜索的實現

文件被索引動作和搜索該文件動作之間是有延遲的，因此，新的文件需要在幾分鐘後方可被搜索到，但這依然不夠快，其根本原因在於磁碟。

在 Elasticsearch 中，當提交一個新的段到磁碟時需要執行 fsync 操作，以確保段被物理地寫入磁碟，即使斷電資料也不會遺失。不過，fsync 很消耗資源，因此它不能在每個文件被索引時都觸發。

位於 Elasticsearch 和磁碟間的是檔案系統快取。記憶體索引快取中的文件被寫入新段的過程的資源消耗很低；之後文件會被同步到磁碟，這個操作的資源消耗很高。而一個檔案一旦被快取，它就可以被開啟和讀取。因此，Elasticsearch 利用了這一特性。

Lucene 允許新段在寫入後被開啟，以便讓段中包含的文件可被搜索，而不用執行一次全量提交。這是一種比提交更輕量的過程，可以經常操作，且不會影響效能。

在 Elasticsearch 中，這種寫入開啟一個新段的輕量級過程，就叫作 refresh。在預設情況下，每個分片每秒自動更新一次。這就是認為 Elasticsearch 是近即時搜索，但不是即時搜索的原因，即文件的改動不會立即被搜索，但是會在一秒內可見。

8.19.2 倒排索引的壓縮

Elasticsearch 為每個文件中的欄位分別建立了一個倒排索引。倒排索引示意圖如圖 2-6 所示。

隨著文件的不斷增加，倒排索引中的詞條和詞條對應的文件 ID 列表會不斷增大，進一步影響 Elasticsearch 的效能。

Elasticsearch 對詞條採用了 Term Dictionary 和 Term Index 的方式來簡化詞條的儲存和尋找；同時，Elasticsearch 對詞條對應的文件 ID 列表進行了必要的處理。

Elasticsearch 是如何處理這些文件 ID 列表的呢？答案很簡單，即透過增量編碼壓縮，將大數變小數，逐位元組儲存。

為了有效進行壓縮，詞條對應的文件 ID 列表是有序排列的。所謂增量編碼，就是將原來的大數變成小數，僅儲存增量值，範例如下。

假如原有的文件 ID 列表為：10233466100178。

增量編碼後文件 ID 列表為：1013113234 78。

所謂逐位元組儲存，即檢視增量編碼後的數字可以用幾位位元組儲存，而非全部用 int（4 位元組）儲存，進一步達到壓縮和節省記憶體的目的。當然，為了節省更多的記憶體，還可以對增量編碼後的文件 ID 列表進行分組，再分別計算每一個儲存需要的位元組數。

8.20 基礎知識連結

在 Elasticsearch 中，有兩種資料更新操作方式，即 Refresh 和 Flush。這兩種更新操作方式有什麼區別呢？

■ Refresh 方式

當我們索引文件時，文件是儲存在記憶體中，預設 1s 後會進入檔案系統快取。Refresh 操作本質上是對 Lucene Index Reader 呼叫了 ReOpen 操作，即對此時索引中的資料進行更新，使文件可以被使用者搜索到。

不過，此時文件還未儲存到磁碟上，如果 Elasticsearch 的伺服器當機了，那麼這部分資料就會遺失。如果要對這部分資料進行持久化，則需要呼叫消耗較大的 Lucene Commit 操作。因此，Elasticsearch 採用可以頻繁呼叫輕量級的 ReOpen 操作來達到近即時搜索的效果。

2 Flush 方式

雖然 Elasticsearch 採用了可以頻繁呼叫輕量級的 ReOpen 操作來達到近即時搜索的效果，但資料終究要持久化。

Elasticsearch 在寫入文件時，會寫一份 translog 記錄檔，用 translog 記錄檔可以恢復那些遺失的文件，在出現程式故障或磁碟例外時，保障資料的安全。

Flush 可高效率地觸發 Lucene Commit，同時清空 translog 記錄檔，使資料在 Lucene 層面持久化。

8.21 小結

本章主要介紹了索引 API 的使用，有關 18 個索引操作相關的介面，包含欄位索引分析、索引的增刪改查（建立索引、取得索引、刪除索引、索引存在驗證）、索引的開關（開啟索引、關閉索引）、索引的容量控制（縮小索引、拆分索引、強制合併索引）、索引資料更新（Refresh 更新、Flush 更新、同步 Flush 更新、清除索引快取、強制合併索引）和索引別名（建立索引別名、索引別名存在驗證、取得索引別名）。

Elasticsearch 外掛程式

東府買舟船
西府買器械

在 第二部分中，主要介紹了 Elasticsearch 實戰，本部分主要介紹
Elasticsearch 的生態圈。本章介紹 Elasticsearch 中的外掛程式生態。

外掛程式是使用者以自訂方式增強 Elasticsearch 功能的一種方法。
Elasticsearch 外掛程式包含增加自訂對映類型、自訂分析器、自訂指令稿引擎
和自訂發現等。

9.1 外掛程式簡介

Elasticsearch 外掛程式類型包含 jar 檔案、指令稿和設定檔，外掛程式必須安
裝在叢集中的每個節點上才能使用。安裝外掛程式後，必須重新啟動每個節
點，才能看到外掛程式。

在 Elasticsearch 官網上，外掛程式被歸納為兩大類，分別是核心外掛程式和社
區貢獻外掛程式。

① 核心外掛程式

核心外掛程式屬於 Elasticsearch 專案，外掛程式與 Elasticsearch 安裝套件同時提供，外掛程式的版本編號始終與 Elasticsearch 安裝套件的版本編號相同。這些外掛程式是由 Elasticsearch 團隊所維護。

在使用過程中，如果遇到問題，使用者可以在 GitHub 專案（開啟 GiHub 官網，搜索 Elasticsearch 即可檢視）的頁面上進行提交。

在 Elasticsearch 專案中，核心外掛程式列表如圖 9-1 所示。

Branch: master ▾	elasticsearch / plugins /		Create new file	Find file	History
👤 tlrx Fix usage of randomIntBetween() in testWriteBlobWithRetries (#46380) ...			Latest commit 0321073 Sep 6, 2019		
..					
📁 analysis-icu	Upgrade to Lucene 8.2.0 release (#44859)				Jul 26, 2019
📁 analysis-kuromoji	Add support for inlined user dictionary in the Kuromoji plugin (#45489)				Aug 20, 2019
📁 analysis-nori	Add support for inlined user dictionary in the Kuromoji plugin (#45489)				Aug 20, 2019
📁 analysis-phonetic	Upgrade to Lucene 8.2.0 release (#44859)				Jul 26, 2019
📁 analysis-smartcn	Upgrade to Lucene 8.2.0 release (#44859)				Jul 26, 2019
📁 analysis-stempel	Upgrade to Lucene 8.2.0 release (#44859)				Jul 26, 2019
📁 analysis-ukrainian	Upgrade to Lucene 8.2.0 release (#44859)				Jul 26, 2019
📁 discovery-azure-classic	Allow parsing the value of java.version sysprop (#44017)				Jul 22, 2019
📁 discovery-ec2	Enable caching of rest tests which use integ-test distribution (#43782)				Jul 3, 2019
📁 discovery-gce	Enable caching of rest tests which use integ-test distribution (#43782)				Jul 3, 2019
📁 examples	Update the schema for the REST API specification (#42346)				Aug 15, 2019
📁 ingest-attachment	Ingest Attachment: Upgrade tika to v1.22 (#45575)				Aug 19, 2019

圖 9-1

② 社區貢獻外掛程式

社區貢獻外掛程式屬於 Elasticsearch 專案外部的外掛程式。這些外掛程式由單一開發人員或私人公司提供，並擁有各自的授權及各自的版本控制系統。在使用社區貢獻外掛程式過程中，如果遇到問題，則可以在社區外掛程式的網站上進行提交。

9.2 外掛程式管理

Elasticsearch 提供了用於安裝、檢視和刪除外掛程式相關的指令，這些指令預設位於 $es_home / bin 目錄中。

使用者可以執行以下指令取得外掛程式指令的使用説明：

```
sudo bin / elasticsearch - plugin - h
```

1 外掛程式位置指定

當在根目錄中執行 Elasticsearch 時，如果使用 DEB 或 RPM 套件安裝了 Elasticsearch，則以根目錄執行 / usr / share / Elasticsearch/ bin / Elasticsearch-plugin，以便 Elasticsearch 可以寫入磁碟的對應檔案，否則需要以擁有所有 Elasticsearch 檔案的使用者身份執行 bin/ Elasticsearch 外掛程式。

當使用者自訂 URL 或檔案系統時，使用者可以透過指定 URL 直接從自訂位置下載外掛程式：

```
sudo bin / elasticsearch - plugin install [url]
```

其中，外掛程式名稱由其描述符號確定。如在 UNIX 環境下，可以透過以下指令進行外掛程式位置指定：

```
sudo bin / elasticsearch - plugin install file: ///path/to/plugin.zip
```

在 Windows 環境下，可以透過以下指令進行外掛程式位置指定：

```
bin\ elasticsearch - plugin install file: ///C:/path/to/plugin.zip
```

如果外掛程式檔案不在本機，則需要使用 HTTP 透過以下指令進行外掛程式位置指定：

```
sudo bin / elasticsearch - plugin install http: //some.domain/path/to/ plugin.zip
```

2 安裝外掛程式

在安裝外掛程式時，通常每個外掛程式的文件都包含該外掛程式的特定安裝説明。

下面以 Elasticsearch 的核心外掛程式為例,展示外掛程式的安裝。可以透過以下指令安裝 Elasticsearch 核心外掛程式:

```
sudo bin / elasticsearch - plugin install [plugin_name]
```

舉例來說,安裝核心 ICU 外掛程式,只需執行以下指令:

```
sudo bin / elasticsearch - plugin install analysis - icu
```

此指令將安裝與使用者的 Elasticsearch 版本相符合的外掛程式版本,並在下載時顯示進度指示器。

3 列出目前外掛程式

可以使用 list 指令檢索目前載入外掛程式的列表,指令如下所示:

```
sudo bin / elasticsearch - plugin list
```

4 刪除外掛程式

可以透過刪除 plugins / 下的對應目錄或使用公共指令手動刪除外掛程式,指令如下所示:

```
sudo bin / elasticsearch - plugin remove [pluginname]
```

需要指出的是,在刪除 Java 外掛程式之後,需要重新啟動節點,完成移除過程。

在預設情況下,外掛程式設定檔(如果有)會保留在磁碟上,以防止使用者在升級外掛程式時遺失設定。如果使用者希望在刪除外掛程式時清除設定檔,則使用 -p 或 --purge 指令。在刪除外掛程式後,可以使用此選項刪除所有延遲的設定檔。

5 更新外掛程式

外掛程式是為特定版本的 Elasticsearch 建置的,因此每次更新 Elasticsearch 時都必須重新安裝外掛程式。更新指令如下所示:

```
sudo bin / elasticsearch - plugin remove [pluginname]
sudo bin / elasticsearch - plugin install [pluginname]
```

6 其他命令列參數

外掛程式指令還支援許多其他命令列參數,如下所示:

(1)--verbose 參數,輸出更多偵錯資訊。

(2)--silent 參數,關閉包含進度指示器在內的所有輸出。

在指令執行過程中可能會傳回以下退出程式,各程式對應的含義如下所示:

(1)0: OK。

(2)64: 未知指令或選項參數不正確。

(3)74: I/O 錯誤。

(4)70: 有其他錯錯誤。

另外,某些外掛程式的安裝和執行需要更多的許可權。這些外掛程式將列出所需的許可權,並在繼續安裝之前要求使用者確認。

當基於安裝自動化指令稿執行外掛程式安裝指令稿時,外掛程式指令稿應檢測到沒有從主控台呼叫它,並跳過確認回應,自動授予所有請求的許可權。如果主控台檢測失敗,則使用者可以透過 -b 或 --batch 指令強制使用批次處理模式,指令如下所示:

```
sudo bin / elasticsearch - plugin install—batch [pluginname]
```

如果使用者的 elasticsearch.yml 設定檔位於自訂位置,則在使用外掛程式指令稿時需要指定設定檔的路徑:

```
sudo ES_PATH_CONF = /path/to / conf / dir bin / elasticsearch - plugin install
< plugin name >
```

當使用者透過代理安裝外掛程式時,則需要設定程式。可以用 Java 設定 http.proxyHost 和 http.proxyPort(https.proxyHost,https.proxyPort)將代理細節添到 ES_JAVA_OPTS 環境變數中,設定指令如下所示:

```
sudo ES_JAVA_OPTS = "-Dhttp.proxyHost=host_name -Dhttp.proxyPort=port_number
    -Dhttps.proxyHost=host_name -Dhttps.proxyPort=https_port_number"
bin / elasticsearch - plugin install analysis - icu
```

在 Windows 環境下，指令如下所示：

```
set ES_JAVA_OPTS = "-Dhttp.proxyHost=host_name -Dhttp.proxyPort=port_number
    -Dhttps.proxyHost=host_name -Dhttps.proxyPort=https_port_number"
bin\ elasticsearch - plugin install analysis - icu
```

9.3 分析外掛程式

9.3.1 分析外掛程式簡介

對分析器（Analyzer）而言，一般會接受一個字串作為輸入參數，分析器會將這個字串拆分成獨立的詞或語匯單元（也稱之為 token）。當然，在處理過程中會捨棄一些標點符號等字元，處理後會輸出一個語匯單元流（也稱之為 token stream）。

因此，一般分析器會包含三個部分：

（1）character filter：斷詞之前的前置處理，過濾 HTML 標籤、特殊符號轉換等。

（2）tokenizer：用於斷詞。

（3）token filter：用於標準化輸出。

Elasticsearch 為很多語言提供了專用的分析器，特殊語言所需的分析器可以由使用者根據需要以外掛程式的形式提供。Elasticsearch 內建的主要分析器有：

（1）Standard 分析器：預設的斷詞器。Standard 分析器會將詞匯單元轉換成小寫形式，並且去除了停用詞和標點符號，支援中文（採用的方法為單字切分）。停用詞指語氣助詞等修飾性詞語，如 the、an、的、這等。

（2）Simple 分析器：首先透過非字母字元分割文字資訊，並去除數字類型的字元，然後將詞匯單元統一為小寫形式。

（3）Whitespace 分析器：僅去除空格，不會將字元轉換成小寫形式，不支援中文；不對產生的詞匯單元進行其他標準化處理。

（4）Stop 分析器：與 Simple 分析器相比，增加了去除停用詞的處理。

（5）Keyword 分析器：該分析器不進行斷詞，而是直接將輸入作為一個單字
輸出。

（6）Pattern 分析器：該分析器透過正規表示法自訂分隔符號，預設是 "\
W+"，即把非字詞的符號作為分隔符號。

（7）Language 分析器：這是特定語言的分析器，不支援中文，支援如
English、French 和 Spanish 等語言。

通常來説，任何全文檢索的字串域都會預設使用 Standard 分析器，如果想要
一個自訂分析器，則可以按照以下方式重新製作一個「標準」分析器：

```
{
    "type":      "custom",
    "tokenizer": "standard",
    "filter":  [ "lowercase", "stop" ]
}
```

在這個自訂分析器中，主要使用了 Lowercase（小寫字母）和 Stop（停用詞）
詞彙單元篩檢程式。

什麼是 Standard 分析器呢？

一般來説，分析器會接受一個字串作為輸入。在工作時，分析器會將這個字
串拆分成獨立的詞或語匯單元（稱之為 token），當然也會捨棄一些標點符號
等字元，最後分析器輸出一個語匯單元流。這就是典型的 Standard 分析器的
工作模式。

分析器在識別詞彙時有多種演算法可供選擇。最簡單的是 Whitespace 斷詞演
算法，該演算法按空白字元，如空格、Tab、分行符號等，對敘述進行簡單的
拆分，將連續的不可為空格字元組成一個語匯單元。舉例來説，對下面的敘
述使用 Whitespace 斷詞演算法斷詞時，會獲得以下結果：

原文：You're the 1st runner home!

結果：You're、the、1st、runner、home!

而 Standard 分析器則使用 Unicode 文字分割演算法尋找單字之間的界限，並輸出所有界限之間的內容。Unicode 內包含的知識使其可以成功地對包含混合語言的文字進行斷詞。

一般來説，Standard 分析器是大多數語言斷詞的合理的起點。事實上，它組成了大多數特定語言分析器的基礎，如 English 分析器、French 分析器和 Spanish 分析器。另外，它還支援亞洲語言，只是有些缺陷，因此讀者可以考慮透過 ICU 分析外掛程式的方式使用 icu_tokenizer 進行取代。

9.3.2 Elasticsearch 中的分析外掛程式

分析外掛程式是一種外掛程式，我們可透過在 Elasticsearch 中增加新的分析器、標記化器、標記篩檢程式或字元篩檢程式等擴充 Elasticsearch 的分析功能。

Elasticsearch 官方提供的核心分析外掛程式如下。

（1）ICU 函數庫。

讀者可以使用 ICU 函數庫擴充對 Unicode 的支援，包含更進一步地分析亞洲語言、Unicode 規範化、支援 Unicode 的大小寫折疊、支援排序和音譯。

（2）Kuromoji 外掛程式。

讀者可以使用 Kuromoji 外掛程式對日語進行進階分析。

（3）Lucene Nori 外掛程式。

讀者可以使用 Lucene Nori 外掛程式對韓語進行分析。

（4）Phonetic 外掛程式。

讀者可以使用 Soundex、Metaphone、Caverphone 和其他編碼器 / 解碼器將標記分析為其語音相等物。

（5）SmartCN 外掛程式。

SmartCN 外掛程式可用於對中文或中英文混合文字進行分析。該外掛程式利用機率知識對簡化中文文字進行最佳斷詞。首先文字被分割成句子，然後每個句子再被分割成單字。

（6）Stempel 外掛程式。

Stempel 外掛程式為波蘭語提供了高品質的分析工具。

（7）Ukrainian 外掛程式。

Ukrainian 外掛程式可用於為烏克蘭語提供詞幹分析。

除官方的分析外掛程式外，Elasticsearch 技術社區也貢獻了不少分析外掛程式，比較常用且著名的有：

（1）IK Analysis Plugin。

IK 分析外掛程式將 Lucene IK Analyzer 整合到 Elasticsearch 中，支援讀者自訂字典。

（2）Pinyin Analysis Plugin。

Pinyin Analysis Plugin 是一款拼音分析外掛程式，該外掛程式可對中文字和拼音進行相互轉換。

（3）Vietnamese Analysis Plugin。

Vietnamese Analysis Plugin 是一款用於對越南語進行分析的外掛程式。

（4）Network Addresses Analysis Plugin。

Network Addresses Analysis Plugin 可以用於分析網路位址。

（5）Dandelion Analysis Plugin。

Dandelion Analysis Plugin 可譯為蒲公英分析外掛程式，該外掛程式提供了一個分析器（稱為「蒲公英 -A」），該分析器會從輸入文字中分析的實體進行語義搜索。

（6）STConvert Analysis Plugin。

STConvert Analysis Plugin 可對中文簡體和繁體進行相互轉換。

9.3.3 ICU 分析外掛程式

ICU 分析外掛程式將 Lucene ICU 模組整合到 Elasticsearch 中。

1 ICU 分析外掛程式的安裝

外掛程式必須安裝在叢集中的每個節點上，並且必須在安裝後重新啟動節點。在 Linux 環境下，可以透過以下指令安裝 ICU 分析外掛程式：

```
sudo bin/elasticsearch-plugin install analysis-icu
```

當需要刪除 ICU 分析外掛程式時，在 Linux 環境下，可以透過以下指令刪除 ICU 分析外掛程式：

```
sudo bin / elasticsearch - plugin remove analysis - icu
```

2 ICU 分析外掛程式簡介

ICU 分 析 外 掛 程 式 使 用 icu_normalizer char filter、icu_tokenizer 和 icu normalizer token filter 三個元件分別執行基底字元的規範化、字元標記和字元處理標準化操作。

ICU 分析外掛程式在工作時，需要 method 和 mode 兩個參數，其中，method 參數主要指的是歸一化方法，接受 NFKC、NFC 或 NFKC_CF（預設）方法；mode 參數主要指的是規範化模式，接受模式的組合或分解，預設是組合模式。

3 ICU 外掛程式的規範化字元篩檢程式

ICU 外掛程式的規範化字元篩檢程式主要用於規範化字元，可用於所有索引，無須進一步設定。規範化的類型可以由 name 參數指定，該參數接受 NFC、NFKC 和 NFKC_CF 方法，預設為 NFKC_CF 方法。模式參數還可以設定為分解，分別將 NFC 轉為 NFD，或將 NFKC 轉為 NFKD。

在使用時，可以透過指定 unicode_set_filter 參數控制規範化的字母，該參數接受 unicodeset。下面透過一個範例，展示字元篩檢程式的預設用法和自訂字元篩檢程式：

```
PUT icu_sample
{
  "settings": {
    "index": {
      "analysis": {
        "analyzer": {
```

```
      "nfkc_cf_normalized": { // 使用預設的 NFKC_CF 方法
        "tokenizer": "icu_tokenizer",
        "char_filter": [
          "icu_normalizer"
        ]
      },
      "nfd_normalized": {    // 使用自訂的 nfd_normalized 權杖篩檢程式，該篩檢
                             // 程式設定為使用帶分解的 NFC 方法
        "tokenizer": "icu_tokenizer",
        "char_filter": [
          "nfd_normalizer"
        ]
      }
    },
    "char_filter": {
      "nfd_normalizer": {
        "type": "icu_normalizer",
        "name": "nfc",
        "mode": "decompose"
      }
    }
  }
}
}
```

☑ ICU 外掛程式的使用方法

下面以 Windows 環境為例，展示 ICU 外掛程式的使用方法。

不同於 Linux 環境下的外掛程式安裝指令，在 Windows 環境下的安裝指令如
下所示：

```
bin\elasticsearch-plugin.bat install  file:///C:\Users\牛冬\Desktop\analysis-
icu-7.2.0.zip
```

其中，analysis-icu-7.2.0.zip 是下載到本機的檔案，設定方式如下：

```
file:/// 本機路徑
```

在執行安裝指令後，DOS 介面會顯示安裝過程和安裝結果，如圖 9-2 所示。

```
C:\elasticsearch-7.2.0-windows-x86_642\elasticsearch-7.2.0>bin\elasticsearch-plugin.bat install  file:///C:\Users\牛冬\D
esktop\analysis-icu-7.2.0.zip
warning: ignoring JAVA_TOOL_OPTIONS=-Dfile.encoding=UTF-8
future versions of Elasticsearch will require Java 11; your Java version from [C:\Program Files\Java\jdk1.8.0_171\jre] d
oes not meet this requirement
-> Downloading file:///C:\Users\牛冬\Desktop\analysis-icu-7.2.0.zip
[=============================================] 100%??
-> Installed analysis-icu
```

圖 9-2

可以透過以下指令檢視外掛程式的安裝結果，即安裝成功與否：

```
bin\elasticsearch-plugin.bat  list
```

在指令執行後，DOS 介面會輸出目前 Elasticsearch 中安裝的外掛程式，如圖 9-3 所示。

```
C:\elasticsearch-7.2.0-windows-x86_642\elasticsearch-7.2.0>bin\elasticsearch-plugin.bat  list
warning: ignoring JAVA_TOOL_OPTIONS=-Dfile.encoding=UTF-8
future versions of Elasticsearch will require Java 11; your Java version from [C:\Program Files\Java\jdk1.8.0_171\jre] d
oes not meet this requirement
analysis-icu
```

圖 9-3

透過圖 9-3 可以看出，目前安裝的外掛程式中含有 analysis-icu 外掛程式。

在啟動 Elasticsearch 時，在 Elasticsearch 的主控台也可以看到外掛程式的載入情況，如圖 9-4 所示。

```
2019-09-10T17:54:57,380][INFO ][o.e.p.PluginsService    ] [LAPTOP-1S8BALK3] loaded module [x-pack-sql]
2019-09-10T17:54:57,386][INFO ][o.e.p.PluginsService    ] [LAPTOP-1S8BALK3] loaded module [x-pack-watcher]
2019-09-10T17:54:57,397][INFO ][o.e.p.PluginsService    ] [LAPTOP-1S8BALK3] loaded plugin [analysis-icu]
```

圖 9-4

下面展示 ICU 外掛程式的使用方法。在下文提及的 Head 外掛程式中，使用 _analyze 指令，輸入以下參數對字串內容進行分析：

```
{
    "text": "我是牛冬，目前在好未來集團家長幫事業部做技術負責人 / 技術總監 ",
}
```

在使用 ICU 外掛程式前，為了比較分析效果，我們先用 Standard 分析器對字串文字進行分析。輸入如下所示內容：

```
{
    "text": " 我是牛冬，目前在好未來集團家長幫事業部做技術負責人 / 技術總監 ",
    "analyzer": "standard"
}
```

實際如圖 9-5 所示。

概览 | 索引 | 数据浏览 | 基本查询 [+] | 复合查询 [+]

▶ 历史记录
▼ 查询
http://localhost:9200/
analyze   POST ▼
{
    "text": "我是牛冬，目前在好未来集团家长帮事业部
做技术负责人/技术总监",
    "analyzer": "standard"
}

{
  ▼ "tokens": [
        {
            "token": "我",
            "start_offset": 0,
            "end_offset": 1,
            "type": "<IDEOGRAPHIC>",
            "position": 0
        }
        ,
        {
            "token": "是",
            "start_offset": 1,
            "end_offset": 2,
            "type": "<IDEOGRAPHIC>",
            "position": 1
        }
        ,
        {
            "token": "牛",
            "start_offset": 2,
            "end_offset": 3,
            "type": "<IDEOGRAPHIC>",
            "position": 2
        }
        ,
```

圖 9-5（編按：本圖為簡體中文介面）

在請求執行後，右側會輸出分析結果，分析結果如下所示：

```
{
  "tokens": [{
        "token": " 我 ",
        "start_offset": 0,
        "end_offset": 1,
        "type": "<IDEOGRAPHIC>",
        "position": 0
    },
    {
        "token": " 是 ",
        "start_offset": 1,
```
```

```
      "end_offset": 2,
      "type": "<IDEOGRAPHIC>",
      "position": 1
  },
  {

      "token": "牛",
      "start_offset": 2,
      "end_offset": 3,
      "type": "<IDEOGRAPHIC>",
      "position": 2
  },
  {

      "token": "冬",
      "start_offset": 3,
      "end_offset": 4,
      "type": "<IDEOGRAPHIC>",
      "position": 3
  },
  {

      "token": "目",
      "start_offset": 5,
      "end_offset": 6,
      "type": "<IDEOGRAPHIC>",
      "position": 4
  },
  {

      "token": "前",
      "start_offset": 6,
      "end_offset": 7,
      "type": "<IDEOGRAPHIC>",
      "position": 5
  },
  {

      "token": "在",
      "start_offset": 7,
      "end_offset": 8,
      "type": "<IDEOGRAPHIC>",
```

```
      "position": 6
},
{

      "token": " 好 ",
      "start_offset": 8,
      "end_offset": 9,
      "type": "<IDEOGRAPHIC>",
      "position": 7
},
{

      "token": " 未 ",
      "start_offset": 9,
      "end_offset": 10,
      "type": "<IDEOGRAPHIC>",
      "position": 8
},
{

      "token": " 來 ",
      "start_offset": 10,
      "end_offset": 11,
      "type": "<IDEOGRAPHIC>",
      "position": 9
},
{

      "token": " 集 ",
      "start_offset": 11,
      "end_offset": 12,
      "type": "<IDEOGRAPHIC>",
      "position": 10
},
{

      "token": " 團 ",
      "start_offset": 12,
      "end_offset": 13,
      "type": "<IDEOGRAPHIC>",
      "position": 11
},
```

```
    {
        "token": "家",
        "start_offset": 13,
        "end_offset": 14,
        "type": "<IDEOGRAPHIC>",
        "position": 12
    },
    {
        "token": "長",
        "start_offset": 14,
        "end_offset": 15,
        "type": "<IDEOGRAPHIC>",
        "position": 13
    },
    {
        "token": "幫",
        "start_offset": 15,
        "end_offset": 16,
        "type": "<IDEOGRAPHIC>",
        "position": 14
    },
    {
        "token": "事",
        "start_offset": 16,
        "end_offset": 17,
        "type": "<IDEOGRAPHIC>",
        "position": 15
    },
    {
        "token": "業",
        "start_offset": 17,
        "end_offset": 18,
        "type": "<IDEOGRAPHIC>",
        "position": 16
    },
    {
        "token": "部",
```

```
      "start_offset": 18,
      "end_offset": 19,
      "type": "<IDEOGRAPHIC>",
      "position": 17
   },
   {
      "token": " 做 ",
      "start_offset": 19,
      "end_offset": 20,
      "type": "<IDEOGRAPHIC>",
      "position": 18
   },
   {
      "token": " 技 ",
      "start_offset": 20,
      "end_offset": 21,
      "type": "<IDEOGRAPHIC>",
      "position": 19
   },
   {
      "token": " 術 ",
      "start_offset": 21,
      "end_offset": 22,
      "type": "<IDEOGRAPHIC>",
      "position": 20
   },
   {
      "token": " 負 ",
      "start_offset": 22,
      "end_offset": 23,
      "type": "<IDEOGRAPHIC>",
      "position": 21
   },
   {
      "token": " 責 ",
      "start_offset": 23,
      "end_offset": 24,
```

```
        "type": "<IDEOGRAPHIC>",
        "position": 22
    },
    {
        "token": " 人 ",
        "start_offset": 24,
        "end_offset": 25,
        "type": "<IDEOGRAPHIC>",
        "position": 23
    },
    {
        "token": " 技 ",
        "start_offset": 26,
        "end_offset": 27,
        "type": "<IDEOGRAPHIC>",
        "position": 24
    },
    {
        "token": " 術 ",
        "start_offset": 27,
        "end_offset": 28,
        "type": "<IDEOGRAPHIC>",
        "position": 25
    },
    {
        "token": " 總 ",
        "start_offset": 28,
        "end_offset": 29,
        "type": "<IDEOGRAPHIC>",
        "position": 26
    },
    {
        "token": " 監 ",
        "start_offset": 29,
        "end_offset": 30,
        "type": "<IDEOGRAPHIC>",
        "position": 27
```

```
        }
    ]
}
```

下面我們將 Standard 分析器換成 ICU 外掛程式分析器，此時 analyzer 的設定
參數變為 icu_analyzer，如下所示：

```
{
    "text": " 我是牛冬，目前在好未來集團家長幫事業部做技術負責人 / 技術總監 ",
    "analyzer": "icu-analyzer"
}
```

Head 外掛程式的設定情況如圖 9-6 所示。

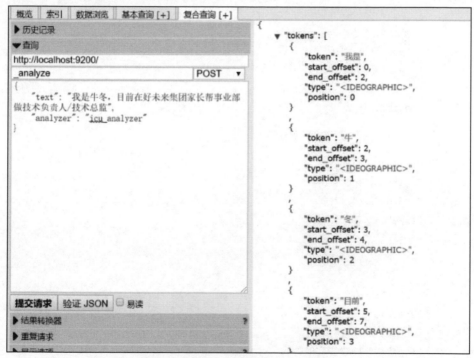

圖 9-6（編按：本圖為簡體中文介面）

在請求執行後，右側輸出的分析結果如下所示：

```
{
    "tokens": [{
```

```
      "token": " 我是 ",
      "start_offset": 0,
      "end_offset": 2,
      "type": "<IDEOGRAPHIC>",
      "position": 0
    },
    {
      "token": " 牛 ",
      "start_offset": 2,
      "end_offset": 3,
      "type": "<IDEOGRAPHIC>",
      "position": 1
    },
    {
      "token": " 冬 ",
      "start_offset": 3,
      "end_offset": 4,
      "type": "<IDEOGRAPHIC>",
      "position": 2
    },
    {
      "token": " 目前 ",
      "start_offset": 5,
      "end_offset": 7,
      "type": "<IDEOGRAPHIC>",
      "position": 3
    },
    {
      "token": " 在 ",
      "start_offset": 7,
      "end_offset": 8,
      "type": "<IDEOGRAPHIC>",
      "position": 4
    },
    {
      "token": " 好 ",
      "start_offset": 8,
```

```
      "end_offset": 9,
      "type": "<IDEOGRAPHIC>",
      "position": 5
},
{

      "token": " 未來 ",
      "start_offset": 9,
      "end_offset": 11,
      "type": "<IDEOGRAPHIC>",
      "position": 6
},
{

      "token": " 集團 ",
      "start_offset": 11,
      "end_offset": 13,
      "type": "<IDEOGRAPHIC>",
      "position": 7
},
{

      "token": " 家長 ",
      "start_offset": 13,
      "end_offset": 15,
      "type": "<IDEOGRAPHIC>",
      "position": 8
},
{

      "token": " 幫 ",
      "start_offset": 15,
      "end_offset": 16,
      "type": "<IDEOGRAPHIC>",
      "position": 9
},
{

      "token": " 事業 ",
      "start_offset": 16,
      "end_offset": 18,
      "type": "<IDEOGRAPHIC>",
```

```json
        "position": 10
    },
    {
        "token": " 部 ",
        "start_offset": 18,
        "end_offset": 19,
        "type": "<IDEOGRAPHIC>",
        "position": 11
    },
    {
        "token": " 做 ",
        "start_offset": 19,
        "end_offset": 20,
        "type": "<IDEOGRAPHIC>",
        "position": 12
    },
    {
        "token": " 技術 ",
        "start_offset": 20,
        "end_offset": 22,
        "type": "<IDEOGRAPHIC>",
        "position": 13
    },
    {
        "token": " 負責 ",
        "start_offset": 22,
        "end_offset": 24,
        "type": "<IDEOGRAPHIC>",
        "position": 14
    },
    {
        "token": " 人 ",
        "start_offset": 24,
        "end_offset": 25,
        "type": "<IDEOGRAPHIC>",
        "position": 15
    },
```

```
{
    "token": "技術",
    "start_offset": 26,
    "end_offset": 28,
    "type": "<IDEOGRAPHIC>",
    "position": 16
},
{
    "token": "總監",
    "start_offset": 28,
    "end_offset": 30,
    "type": "<IDEOGRAPHIC>",
    "position": 17
}
]
}
```

透過上述兩個分析器的比較，可以明顯看出 ICU 外掛程式的分析效果更好，更符合中文的語境。

9.3.4 智慧中文分析外掛程式

智慧中文分析外掛程式（SmartCN），本質上是將 Lucene 的智慧中文分析模組整合到 Elasticsearch 中。

SmartCN 提供了一個中文文字或中英文混合文字的分析器。該分析器利用機率知識對簡化中文文字進行最佳斷詞。首先文字被分成句子，然後每個句子再被分割成單字。

■ SmartCN 外掛程式的安裝

我們可以使用外掛程式管理員安裝 SmartCN 外掛程式，安裝指令如下所示：

```
sudo bin/elasticsearch-plugin install analysis-smartcn
```

需要指出的是，外掛程式必須安裝在叢集的每個節點上，並且必須在安裝後重新啟動節點。

我們還可以透過其他方式安裝（如 Windows 方式，在後面介紹）。從本機安裝時，需要先下載外掛程式到本機。

當需要刪除該外掛程式時，可以使用以下指令刪除：

```
sudo bin/elasticsearch-plugin remove analysis-smartcn
```

需要指出的是，在刪除外掛程式之前必須停止節點。

❷ SmartCN 外掛程式的簡介

SmartCN 外掛程式提供了不可設定的 SmartCN 分析器、SmartCN Tokenizer 和 SmartCN Stop Token 篩檢程式。

SmartCN 分析器是 Lucene 4.6 版本之後附帶的，中文斷詞效果不錯，但英文斷詞有問題。與 IKAnalyzer 分析器相比，SmartCN 分析器在斷詞時會帶來較多碎片，且目前不支援自訂詞函數庫。

SmartCN 分析器支援重新實現和擴充分析器。在使用時，可以將 SmartCN 分析器重新實現為自訂分析器，擴充和設定程式如下所示：

```
{
  "settings": {
    "analysis": {
      "analyzer": {
        "rebuilt_smartcn": {
          "tokenizer":  "smartcn_tokenizer",
          "filter": [
            "porter_stem",
            "smartcn_stop"
          ]
        }
      }
    }
  }
}
```

此外，SmartCN 分析器還支援使用者指定的自訂通用詞，設定方式如下所示：

```
{
  "settings": {
    "index": {
      "analysis": {
        "analyzer": {
          "smartcn_with_stop": {
            "tokenizer": "smartcn_tokenizer",
            "filter": [
              "porter_stem",
              "my_smartcn_stop"
            ]
          }
        },
        "filter": {
          "my_smartcn_stop": {
            "type": "smartcn_stop",
            "stopwords": [
              "_smartcn_",
              "stack",
              "的"
            ]
          }
        }
      }
    }
  }
}
```

❸ SmartCN 分析器實戰

前面介紹了 SmartCN 分析器在 Linux 環境下的安裝和移除，下面以 Windows 環境為例，展示該外掛程式的使用。

Windows 環境下的安裝指令如下所示：

```
bin\elasticsearch-plugin.bat install  file:///C:\Users\牛冬\Desktop\analysis-
smartcn-7.2.0.zip
```

其中，analysis-smartcn-7.2.0.zip 是下載到本機的檔案，設定方式如下：

```
file:/// 本機路徑
```

執行安裝指令後，DOS 介面會顯示安裝過程和安裝結果，如圖 9-7 所示。

```
C:\elasticsearch-7.2.0-windows-x86_642\elasticsearch-7.2.0>bin\elasticsearch-plugin.bat install  file:///C:\Users\牛冬\D
esktop\analysis-smartcn-7.2.0.zip
warning: ignoring JAVA_TOOL_OPTIONS=-Dfile.encoding=UTF-8
future versions of Elasticsearch will require Java 11; your Java version from [C:\Program Files\Java\jdk1.8.0_171\jre] d
oes not meet this requirement
-> Downloading file:///C:\Users\牛冬\Desktop\analysis-smartcn-7.2.0.zip
[=============================================] 100%??
-> Installed analysis-smartcn
```

圖 9-7

可以透過以下指令檢視外掛程式的安裝結果，即安裝成功與否：

```
bin\elasticsearch-plugin.bat list
```

指令執行後，DOS 介面會輸出目前 Elasticsearch 中安裝的外掛程式，內容如圖 9-8 所示。

```
C:\elasticsearch-7.2.0-windows-x86_642\elasticsearch-7.2.0>bin\elasticsearch-plugin.bat list
warning: ignoring JAVA_TOOL_OPTIONS=-Dfile.encoding=UTF-8
future versions of Elasticsearch will require Java 11; your Java version from [C:\Program Files\Java\jdk1.8.0_171\jre] d
oes not meet this requirement
analysis-icu
analysis-smartcn
```

圖 9-8

透過圖 9-8 可以看到，在目前 Elasticsearch 安裝的外掛程式中含有 analysis-smartcn 外掛程式。

在啟動 Elasticsearch 時，在 Elasticsearch 的主控台可以看到外掛程式的載入情況，如圖 9-9 所示。

```
[2019-09-10T18:04:41,869][INFO ][o.e.p.PluginsService     ] [LAPTOP-1S8BALK3] loaded plugin [analysis-icu]
[2019-09-10T18:04:41,869][INFO ][o.e.p.PluginsService     ] [LAPTOP-1S8BALK3] loaded plugin [analysis-smartcn]
```

圖 9-9

下面介紹 SmartCN 外掛程式的使用。在 Head 外掛程式中，使用 _analyze 指令輸入以下參數，對字串內容進行分析：

```
{
    "text": " 我是牛冬，目前在好未來集團家長幫事業部做技術負責人 / 技術總監 ",
}
```

此時 "analyzer" 參數設定為 "smartcn"，即使用 SmartCN 外掛程式對字串進行分析。如圖 9-10 所示。

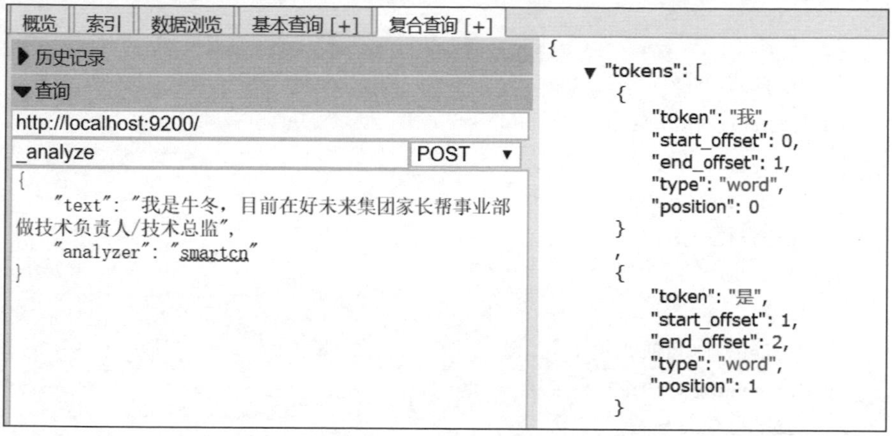

圖 9-10 （編按：本圖為簡體中文介面）

在請求執行後，右側會顯示使用 SmartCN 外掛程式對字串進行分析的結果，如下所示：

```
{
  "tokens": [{
      "token": " 我 ",
      "start_offset": 0,
      "end_offset": 1,
      "type": "word",
      "position": 0
  },
  {
      "token": " 是 ",
      "start_offset": 1,
      "end_offset": 2,
      "type": "word",
      "position": 1
  },
  {
      "token": " 牛 ",
      "start_offset": 2,
```

```
            "end_offset": 3,
            "type": "word",
            "position": 2
        },
        {

            "token": " 冬 ",
            "start_offset": 3,
            "end_offset": 4,
            "type": "word",
            "position": 3
        },
        {

            "token": " 目前 ",
            "start_offset": 5,
            "end_offset": 7,
            "type": "word",
            "position": 5
        },
        {

            "token": " 在 ",
            "start_offset": 7,
            "end_offset": 8,
            "type": "word",
            "position": 6
        },
        {

            "token": " 好 ",
            "start_offset": 8,
            "end_offset": 9,
            "type": "word",
            "position": 7
        },
        {

            "token": " 未來 ",
            "start_offset": 9,
            "end_offset": 11,
            "type": "word",
```

```
      "position": 8
   },
   {
      "token": " 集團 ",
      "start_offset": 11,
      "end_offset": 13,
      "type": "word",
      "position": 9
   },
   {
      "token": " 家長 ",
      "start_offset": 13,
      "end_offset": 15,
      "type": "word",
      "position": 10
   },
   {
      "token": " 幫 ",
      "start_offset": 15,
      "end_offset": 16,
      "type": "word",
      "position": 11
   },
   {
      "token": " 事業 ",
      "start_offset": 16,
      "end_offset": 18,
      "type": "word",
      "position": 12
   },
   {
      "token": " 部 ",
      "start_offset": 18,
      "end_offset": 19,
      "type": "word",
      "position": 13
   },
```

```
    {
      "token": " 做 ",
      "start_offset": 19,
      "end_offset": 20,
      "type": "word",
      "position": 14
    },
    {
      "token": " 技術 ",
      "start_offset": 20,
      "end_offset": 22,
      "type": "word",
      "position": 15
    },
    {
      "token": " 負責人 ",
      "start_offset": 22,
      "end_offset": 25,
      "type": "word",
      "position": 16
    },
    {
      "token": " 技術 ",
      "start_offset": 26,
      "end_offset": 28,
      "type": "word",
      "position": 18
    },
    {
      "token": " 總監 ",
      "start_offset": 28,
      "end_offset": 30,
      "type": "word",
      "position": 19
    }
  ]
}
```

與前面兩個分析器（Standard 分析器和 ICU 分析器）相比，可以看出 SmartCN 的分析效果顯然更好，更符合中文的語境。

9.4 API 擴充外掛程式

如果 Elasticsearch 內建的介面不夠用，則可以使用 API 擴充外掛程式。

API 擴充外掛程式透過增加新的、與搜索有關的 API 或功能，實現對 Elasticsearch 新功能的增加。Elasticsearch 社區人員陸陸續續貢獻了不少 API 擴充外掛程式編輯器，整理如下。

（1）Carrot2 Plugin。
該外掛程式用於結果分群，讀者可存取 GitHub 官網，搜索 elasticsearch-carrot2，檢視搭配程式。

（2）Elasticsearch Trigram Accelerated Regular Expression Filter。
該外掛程式包含查詢、篩檢程式、原生指令稿、評分函數，以及使用者最後建立的任意其他內容。透過該外掛程式可以讓搜索變得更好。讀者可存取 GitHub 官網，搜索 search-extra 取得外掛程式。

（3）Elasticsearch Experimental Highlighter。
該外掛程式是用 Java 撰寫的，用於文字反白顯示。讀者可存取 GitHub 官網，搜索 search-highlighter 取得外掛程式。

（4）Entity Resolution Plugin。
該外掛程式使用 Duke（Duke 是一個用 Java 撰寫的、快速靈活的、刪除重復資料的引擎）進行重複檢測。讀者可存取 GitHub 官網，搜索 elasticsearch-entity-resolution 取得外掛程式。

（5）Entity Resolution Plugin(zentity)。
該外掛程式用於即時解析 Elasticsearch 中儲存的物理資訊。讀者可存取 GitHub 官網，搜索 zentity 取得外掛程式。

（6）PQL language Plugin。

該外掛程式允許使用者使用簡單的管線查詢語法對 Elasticsearch 進行查詢。讀者可存取 GitHub 官網，搜索 elasticsearch-pql 取得外掛程式。

（7）Elasticsearch Taste Plugin。

該外掛程式以 Mahout Taste 為基礎的協作過濾演算法實現。讀者可存取 GitHub 官網，搜索 elasticsearch-taste 取得外掛程式。

（8）WebSocket Change Feed Plugin。

該外掛程式允許用戶端建立到 Elasticsearch 節點的 WebSocket 連接，並從資料庫接收更改的提要。讀者可存取 GitHub 官網，搜索 es-change-feed-plugin 取得外掛程式。

9.5 監控外掛程式

在使用 Elasticsearch 的過程中，CPU 的使用率可能會意外增加，這會導致應用服務的回應時間變長；也可能會出現 HTTP 傳回碼（如 503 的錯誤）的數量迅速上升，此時，Elasticsearch 的索引速度會直線下降。

幸好 Elasticsearch 有監控和警告功能，使得使用者可以即時了解 Elasticsearch 的狀態。

監控外掛程式允許使用者監視 Elasticsearch 索引，並在違反設定值時觸發警示。那麼警示是如何觸達使用者的呢？

Elasticsearch 內建了電子郵件、PagerDuty、Slack 和 HipChat 的相關功能，使用者可以在多項警告選項中自由選擇。此外，Elasticsearch 還有強大的 WebHook 輸出功能，可以與使用者現有的監控基礎設施或任意協力廠商系統整合。

另外，使用者可以對警告功能進行設定，將搜索中的相關資料封包含在通知內，警告功能還支援簡易範本。

同時，使用者還可以透過 Kibana 聯通 Elasticsearch 的監控和警告功能。

Elasticsearch 中內建了監控和警告外掛程式，即 X-Pack。X-Pack 允許使用者根據資料中的更改需求採取操作，其設計原則是，如果使用者在 Elasticsearch 中查詢某些內容，就對其發出警示。因此，使用者只需定義一個查詢、條件、計畫和將要採取的操作，X-Pack 就可以完成剩下的工作。

此外，Elasticsearch 社區還提供了一些知名的外掛程式，如 Head 外掛程式和 Cerebro 外掛程式，它們都能對 Elasticsearch 進行監控。這兩個外掛程式將在後面介紹。

9.6 資料分析外掛程式

當我們不想或不能透過 API 介針對 Elasticsearch 中儲存資料時，可以使用資料分析外掛程式向 Elasticsearch 中寫入資料。

Elasticsearch 中內建的核心資料分析外掛程式主要如下。

（1）附件分析外掛程式。

附件分析外掛程式允許使用者使用 Apache 文字分析函數庫 Tika 以通用格式（如 PPT、XLS 和 PDF 等）分析檔案中的資料。

（2）Geoip 資料分析外掛程式。

Geoip 資料分析外掛程式預設在 Geoip 欄位中。根據來自 MaxMind 資料庫的資料增加有關 IP 位址、地理位置等資訊。目前，該外掛程式已經內建在 Elasticsearch 中。

（3）分析使用者代理 user_agent 外掛程式。

該外掛程式用於從使用者代理頭值中分析詳細資訊。目前，該外掛程式已經內建在 Elasticsearch 中。

此外，Elasticsearch 技術社區也貢獻了一些資料分析外掛程式，如分析 CSV 格式檔案資料的外掛程式等。

由於附件分析外掛程式尚未成為 Elasticsearch 內建的外掛程式，因此在使用時，使用者需要自行安裝。在 Linux 環境下的安裝指令如下所示：

```
sudo bin/elasticsearch-plugin install ingest-attachment
```

需要指出的是，外掛程式必須安裝在叢集的每個節點上，並且必須在安裝後重新啟動節點。

如果不是即時下載安裝，而是離線下載後安裝，則可以從 Elastcsearch 官網下載外掛程式。

當外掛程式不再使用時，可以使用以下指令刪除外掛程式：

```
sudo bin/elasticsearch-plugin remove ingest-attachment
```

需要指出的是，在刪除外掛程式之前必須停止節點。

9.7 常用外掛程式實戰

本節介紹 Head 外掛程式和 Cerebro 外掛程式的使用。這兩個外掛程式都能監控 Elasticsearch 叢集中每個節點的情況，也都提供了 API 搜索相關的功能。下面一一介紹。

9.7.1 Head 外掛程式

Head 外掛程式，全稱為 elasticsearch-head，是一個介面化的叢集操作和管理工具，可以對叢集進行「傻瓜式」操作。既可以把 Head 外掛程式整合到 Elasticsearch 中，也可以把 Head 外掛程式當成一個獨立服務。

Head 外掛程式主要有三方面的功能：

（1）顯示 Elasticsearch 叢集的拓撲結構，能夠執行索引和節點等級的操作。
（2）在搜索介面能夠查詢 Elasticsearch 叢集中原始 JSON 或表格格式的資料。
（3）能夠快速存取並顯示 Elasticsearch 叢集的狀態。

1 Head 外掛程式的安裝

Head 外掛程式的安裝方式有兩種，方式一是透過 Elasticsearch 附帶的 plugin 指令進行安裝，如下所示：

```
elasticsearch/bin/elasticsearch-plugin -install mobz/elasticsearch-head
```

透過指令安裝時，會即時下載安裝套件進行安裝。

方式二是離線安裝，即使用者將 Head 外掛程式下載到本機後再進行安裝。讀者可存取 GitHub 官網，搜索 elasticsearch-head，取得 Head 外掛程式。

在 Head 外掛程式頁面，可以看到如圖 9-11 所示內容。

圖 9-11

可以使用 git 指令下載外掛程式：

```
git clone https://github.com/mobz/elasticsearch-head.git
```

也可以點擊圖 9-11 中所示的 "Download ZIP" 按鈕，進行下載。

將 Head 外掛程式下載到本機後，需將其解壓縮到某一目錄下。解壓縮後的目錄如圖 9-12 所示。

📄 _site		2019/8/14 19:27
📄 crx		2019/8/14 19:27
📄 proxy		2019/8/14 19:27
📄 src		2019/8/14 19:27
📄 test		2019/8/14 19:27
📄 .dockerignore		2019/8/14 19:27
📄 .gitignore		2019/8/14 19:27
📄 .jshintrc		2019/8/14 19:27
📄 Dockerfile		2019/8/14 19:27
📄 Dockerfile-alpine		2019/8/14 19:27
📄 elasticsearch-head.sublime-project		2019/8/14 19:27
📄 grunt_fileSets.js		2019/8/14 19:27
📄 Gruntfile.js		2019/8/14 19:27
📄 index.html		2019/8/14 19:27
📄 LICENCE		2019/8/14 19:27
📄 package.json		2019/8/14 19:27
📄 plugin-descriptor.properties		2019/8/14 19:27
📄 README.textile		2019/8/14 19:27

圖 9-12

隨後，開啟 Elasticsearch 服務，在瀏覽器中輸入以下 URL：

```
http://localhost:9200/_plugin/head/
```

上述安裝方式在 Linux 系統環境下和在 Windows 系統環境下均相同，不同的
是，在 Windows 環境下，可以直接點擊圖 9-12 中的 index.html 開啟對應的頁
面，如圖 9-13 所示。

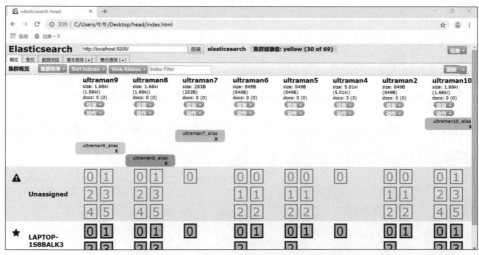

圖 9-13（編按：本圖為簡體中文介面）

在預設情況下，Head 外掛程式將立即嘗試連接位於 http://localhost:9200/ 的叢集節點。如果使用者需要更改連接節點的資訊，則可以在「連接」框中輸入正確的 Elasticsearch 其他節點位址，並點擊「連接」按鈕進行連接。

在 Windows 環境下，如果直接開啟 Head 外掛程式首頁，則在瀏覽器的開發者模式下可以看到跨域的顯示出錯，因此需要對 Elasticsearch 跨域請求存取進行設定。設定是透過修改 elasticsearch/config/elasticsearch.yml 檔案實現的，內容如下所示：

```
http.cors.enabled: true
http.cors.allow-origin: "*"
```

設定後，即可正常存取 Head 外掛程式首頁。在首頁中，可以看到如圖 9-14 所示的幾部分內容。

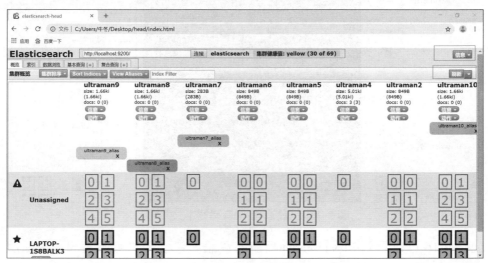

圖 9-14（編按：本圖為簡體中文介面）

從圖 9-14 可以看出，Head 外掛程式首頁由 4 部分組成：節點位址輸入區域、資訊更新區域、導覽列和概覽中的叢集資訊整理。

第一部分是節點位址輸入區域。

第二部分是資訊更新區域，可以檢視 Elasticsearch 相關的資訊和更新外掛程式的資訊。其中，外掛程式提供的資料更新方式如圖 9-15 所示。

圖 9-15（編按：本圖為簡體中文介面）

使用者可以選擇手動更新、快速更新、每 5 秒更新或每 1 分鐘更新。

此外，在「資訊」按鈕部分，可以檢視 Elasticsearch 相關的資訊，如圖 9-16 所示。

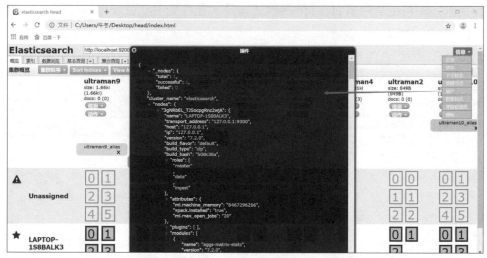

圖 9-16（編按：本圖為簡體中文介面）

從圖 9-16 可以看到 Elasticsearch 相關的資訊，包含叢集節點資訊、節點狀態、叢集狀態、叢集資訊、叢集健康值等內容。點擊對應的按鈕，即可檢視對應的資訊。

第三部分是導覽列，從圖 9-16 中可以看到概覽、索引、資料瀏覽、基本查詢和複合查詢五個 Tab 導覽，預設為概覽。

第四部分是概覽中的叢集資訊整理。我們可以看到 Elasticsearch 已經建立的索引，這些索引資料封包含了索引的名稱、索引的大小和索引的資料量，並且透過「資訊」和「動作」兩個按鈕可以檢視索引資訊，或給索引建立別名。

以 ultraman9 為例，點擊索引 ultraman9 下的「資訊」→「索引狀態」選項，可以看到如圖 9-17 所示內容。

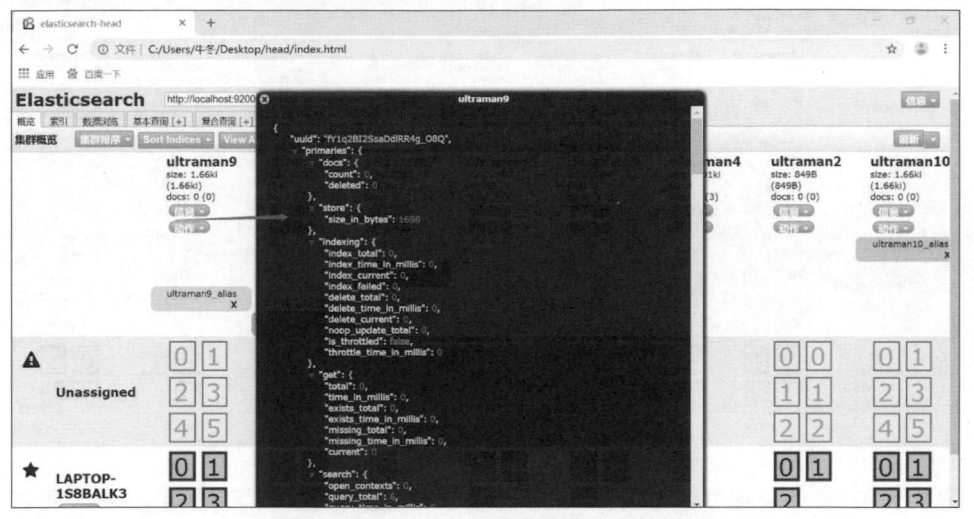

圖 9-17（編按：本圖為簡體中文介面）

點擊索引 ultraman9 下的「資訊」→「索引資訊」選項，可以看到如圖 9-18 所示內容。

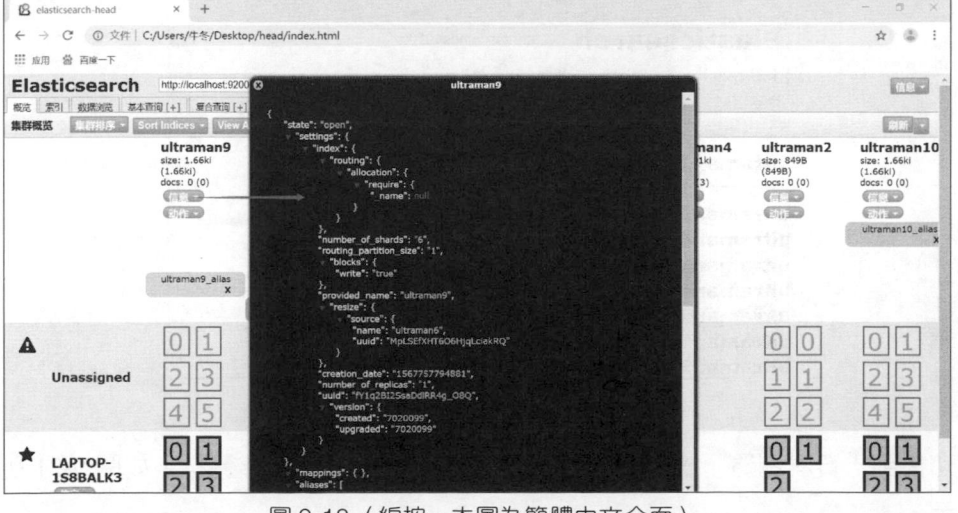

圖 9-18（編按：本圖為簡體中文介面）

而「動作」選項下的資訊有建立別名、快照、關閉等功能，如圖 9-19 所示。

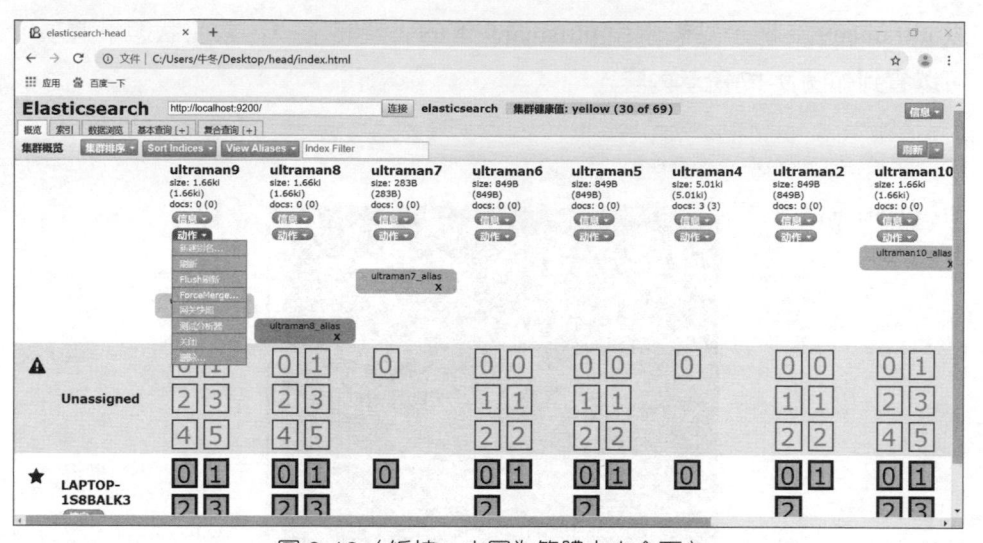

圖 9-19（編按：本圖為簡體中文介面）

圖 9-17 中帶有驚嘆號的 Unassigned 表示未分配的節點，帶有星星的表示主節點，其節點名稱叫作 LAPTOP-1S8BALK3。

切換到「索引」標籤頁，可以檢視目前 Elasticsearch 叢集中的索引情況，如圖 9-20 所示。

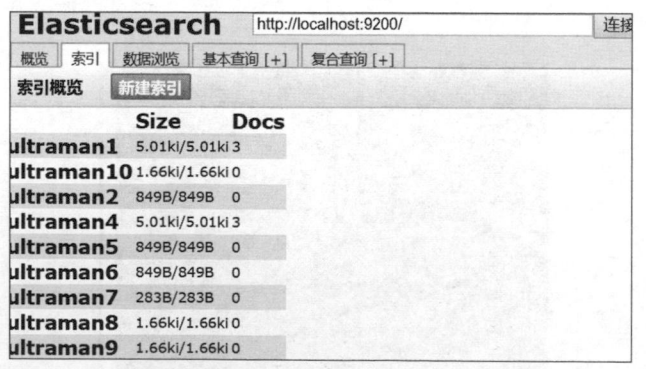

圖 9-20（編按：本圖為簡體中文介面）

切換到「資料瀏覽」標籤頁，可以檢視特定索引下的儲存資料，如圖 9-21 所示。

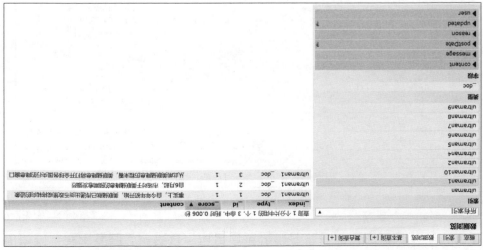

圖 9-21（編按：本圖為展儒中文小圖）

在圖 9-21 中，我們除了可以檢視索引 ultraman1 中儲存的資料，也可以直接從欄位進行資料編輯，如圖 9-22 所示。

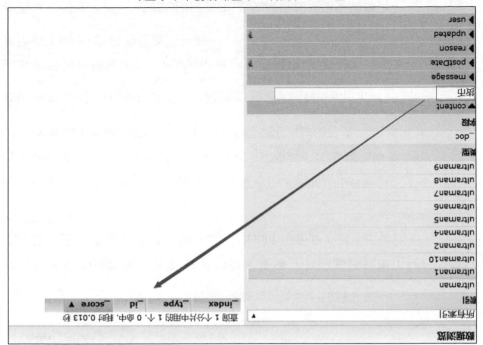

圖 9-22（編按：本圖為展儒中文小圖）

切換到「基本查詢」標籤頁,用自由連接條件進行簡單的資料查詢,如圖 9-23 所示。

圖 9-23(編按:本圖為簡體中文介面)

當用多個查詢準則進行搜索或查詢時,需要注意多個查詢準則間的比對方式。比對方式主要有 3 種,即 must、should 和 mus_tnot。三種比對方式的説明如下所示。

(1)must 子句:文件必須比對 must 查詢準則,相當於 "="。

(2)should 子句:文件應該比對 should 子句查詢的或多個條件。

(3)must_not 子句:文件不能比對該查詢準則,相當於 " ! ="。

在圖 9-23 中選擇 match 方式,常用的比對方式還有 term、text 和 range 等。

term 表示的是精確比對,wildcard 表示的是萬用字元比對,prefix 表示的是字首比對,range 表示的是區間查詢。

在圖 9-23 中的 "+"、"−" 按鈕用於增加查詢準則或減少查詢準則。

在查詢結果展示區域中，使用者可以設定資料的呈現形式，如 table、JSON、CVS 表格等，還可以選取「顯示查詢敘述」選項，呈現透過表單內容連接的搜索敘述。

當切換到「複合查詢」標籤頁時，可以自由連接條件，進行複雜的資料查詢。

「複合查詢」標籤頁提供給使用者了撰寫 RESTful 介面風格的請求，使用者可以使用 JSON 進行複雜的查詢，例如發送 PUT 請求新增及更新索引，使用 delete 請求刪除索引等。總之，在「複合查詢」頁面，使用者可對 Elasticsearch 中的資料或索引進行各種增刪改查等操作請求。如圖 9-24 所示。

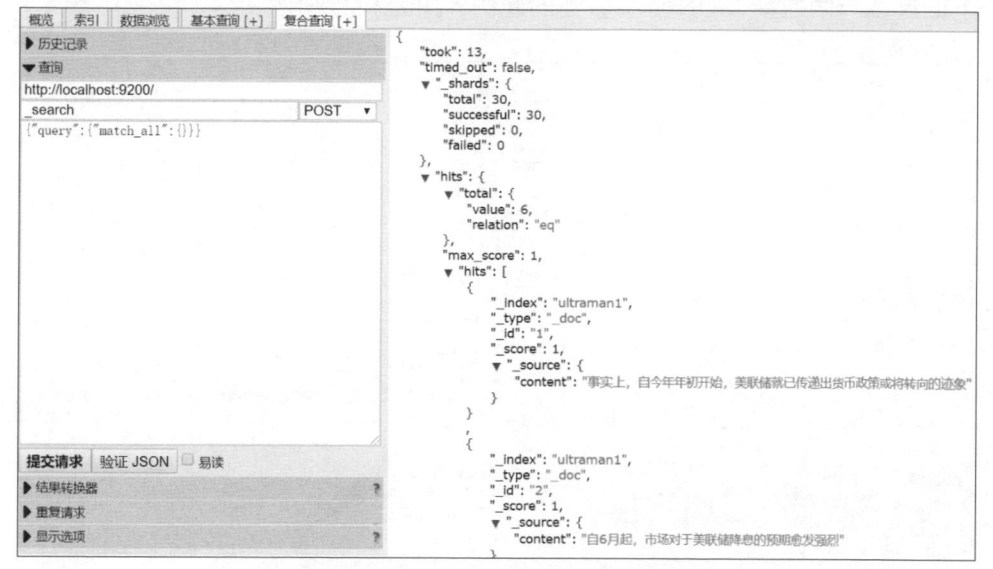

圖 9-24（編按：本圖為簡體中文介面）

從圖 9-24 可以看到，頁面中有一個輸入視窗，允許使用者任意呼叫 Elasticsearch 的 RESTful API。

在 Elasticsearch 中，RESTful API 的基本格式如下所示：

```
http://ip:port/ 索引 / 類型 / 文件 ID
```

在 Elasticsearch 中，以 POST 方法自動產生 ID，而 PUT 方法需要指明 ID。

設定介面包含以下四個選項：

（1）請求方法，與 HTTP 的請求方法相同，如 GET、PUT、POST、DELETE 等。還可以設定查詢 JSON 請求資料、請求對應的 Elasticsearch 節點和請求路徑。

（2）支援設定 JSON 驗證器對使用者輸入的 JSON 請求資料進行 JSON 格式驗證。

（3）支援重複請求計時器設定重複請求的頻率和時間。

（4）在結果轉換器中支援使用 JavaScript 運算式轉換結果。

下面展示一個連接 JSON 查詢敘述的範例，如圖 9-25 所示。

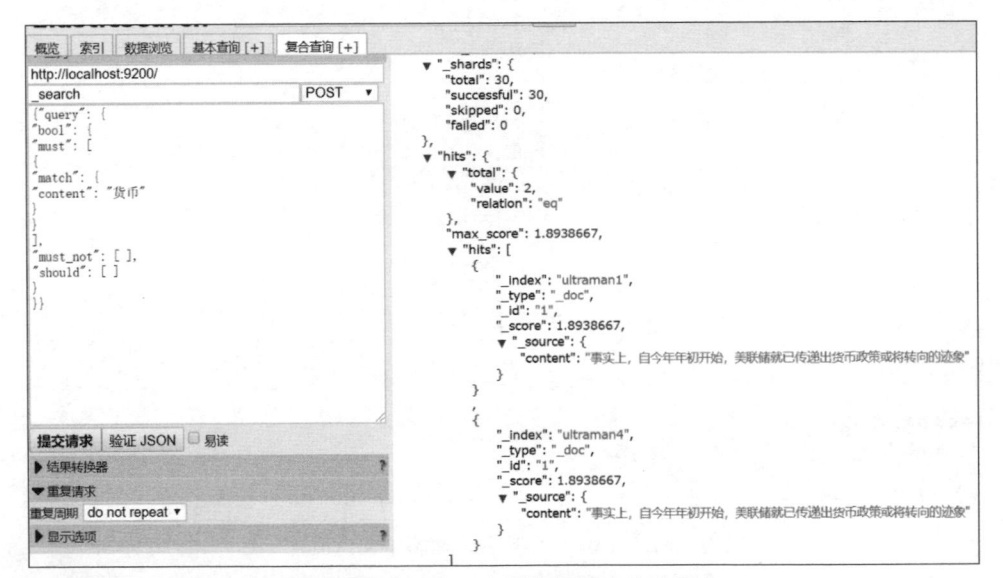

圖 9-25（編按：本圖為簡體中文介面）

下面以索引 ultraman1 為例，展示對相關索引的操作。如檢視索引 ultraman1 中編號為 1 的文件，頁面設定如圖 9-26 所示。

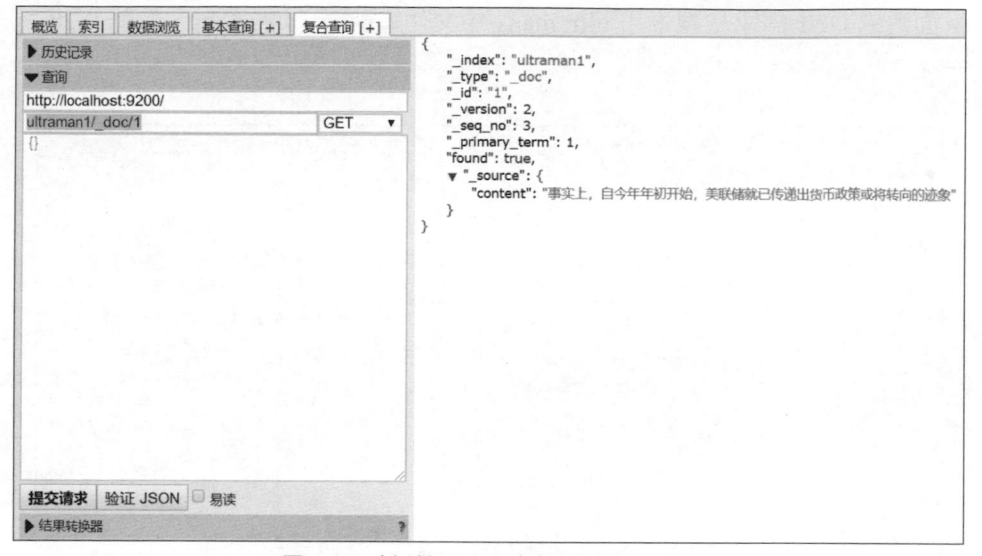

圖 9-26（編按：本圖為簡體中文介面）

從圖 9-26 可見，文件 1 已經被搜索並展示在右側視窗。

在索引 ultraman1 中提交資料的範例如圖 9-27 所示，增加文件編號為 10 的內容。

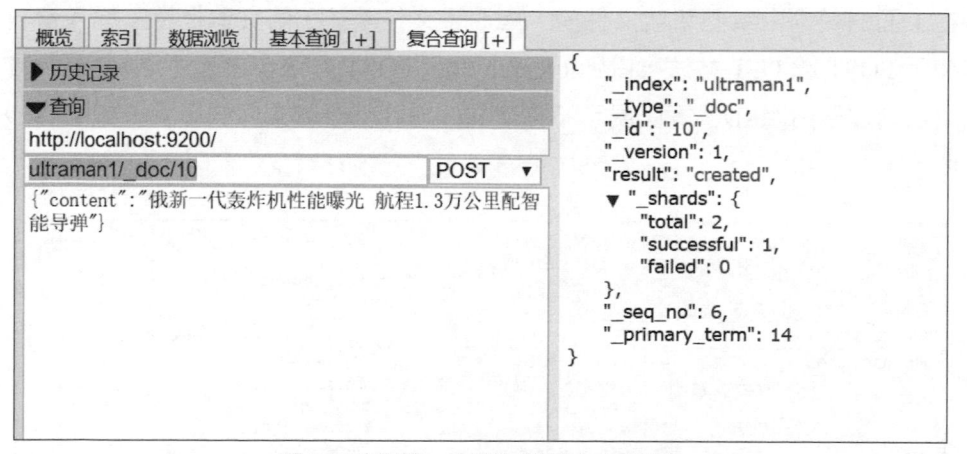

圖 9-27（編按：本圖為簡體中文介面）

從圖 9-27 可見，新增文件編號為 10 的請求已經執行成功，文件編號為 10，目前文件版本編號為 1。

下面透過 GET 請求檢視索引 ultraman1 中文件編號為 10 的資料，如圖 9-28 所示。

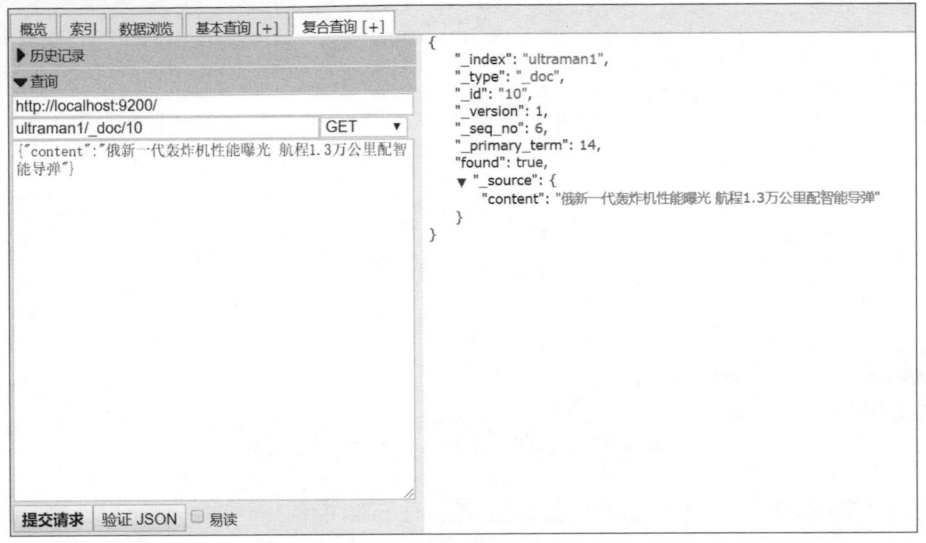

圖 9-28(編按：本圖為簡體中文介面)

從圖 9-28 可見，文件編號為 10 的資料確實已經被索引到 ultraman1 中了。

在上面的範例中我們新增文件時，指明了文件編號。前文曾提及，插入資料使用 POST 或 PUT 方法，Elasticsearch 會用 POST 方法自動產生 ID，而 PUT 方法則需要指明 ID。下面展示不指明文件編號就新增文件的場景，如圖 9-29 所示。

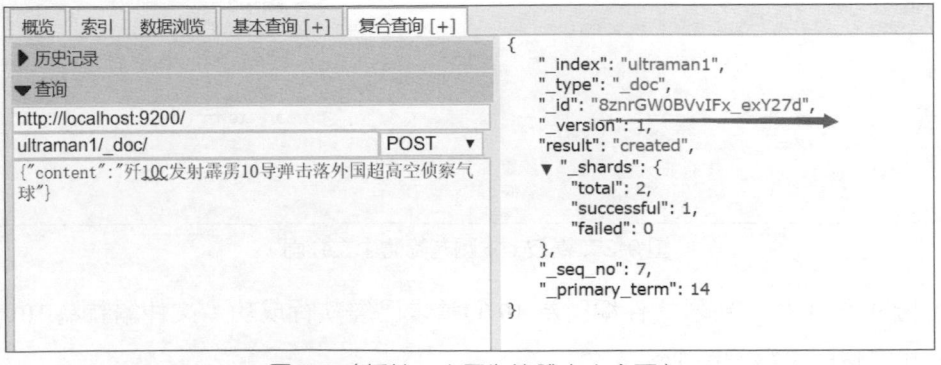

圖 9-29(編按：本圖為簡體中文介面)

從圖 9-29 可見，文件的編號是個隨機值。那麼使用 PUT 方法時的場景呢？我們依然不指明文件編號，執行請求後，結果如圖 9-30 所示。

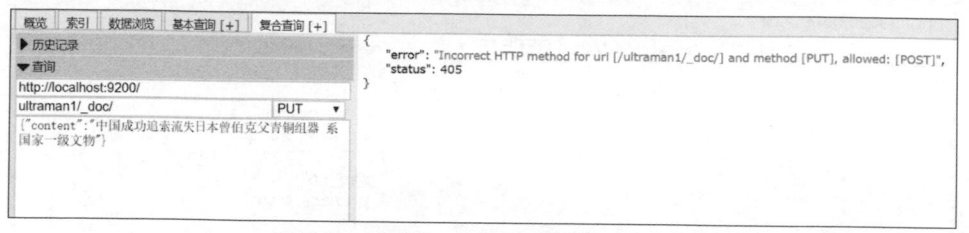

圖 9-30（編按：本圖為簡體中文介面）

從圖 9-30 可見，在索引 ultraman1 中使用 PUT 方法且不指明文件編號為新增文件時會失敗。

下面展示在索引 ultraman1 中使用 PUT 方法且指明文件編號的新增文件的場景，如圖 9-31 所示。

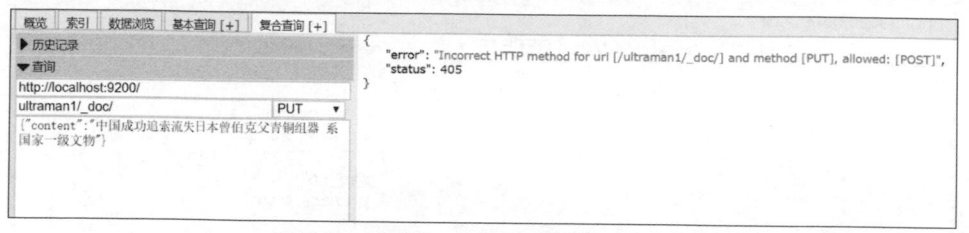

圖 9-31（編按：本圖為簡體中文介面）

從圖 9-31 可見，在索引 ultraman1 中使用 PUT 方法，且指明文件編號的新增文件執行成功。

下面展示刪除文件的方法，如圖 9-32 所示。

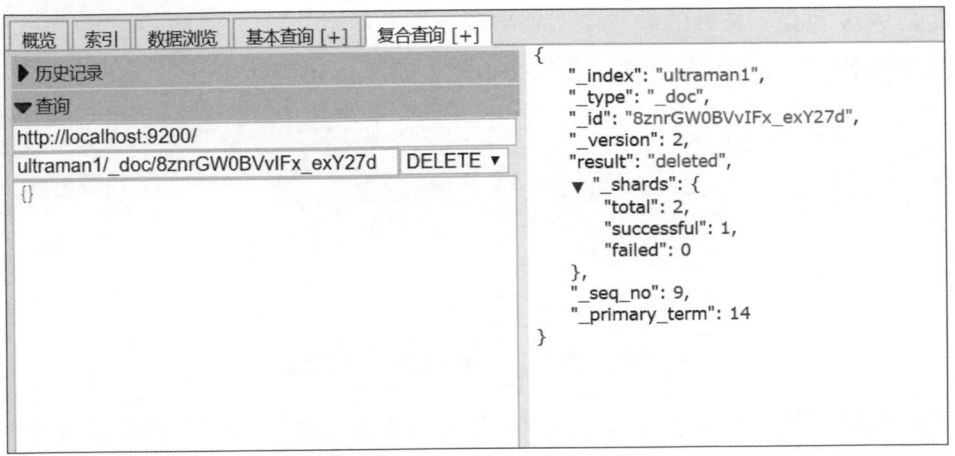

圖 9-32（編按：本圖為簡體中文介面）

從圖 9-32 可見，文件刪除請求執行成功。

需要指出的是，由於 Head 外掛程式可以對資料進行增刪改查，因此在實際開發時儘量不要使用，如果一定要用，則至少要限制 IP 位址。

刪除文件之後還能查到嗎？我們嘗試一下，查詢結果如圖 9-33 所示。

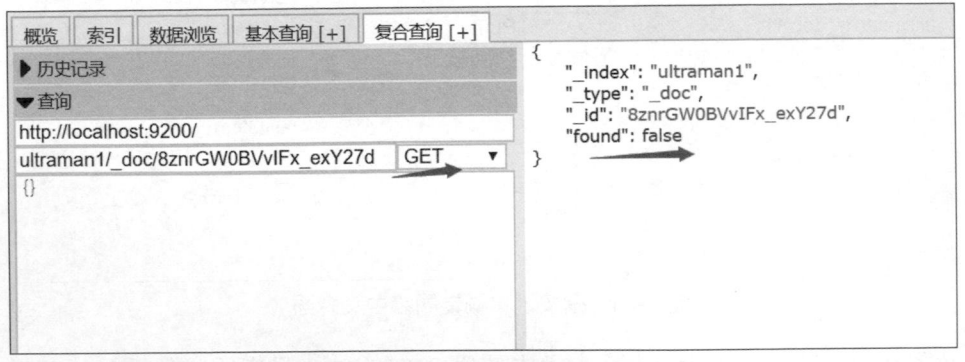

圖 9-33（編按：本圖為簡體中文介面）

從圖 9-33 可見，在索引 ultraman1 中查詢已刪除文件的請求執行失敗。也就是說，已刪除的文件不能再被查詢。

除了可以刪除對單一索引單一文件的請求之外，還可以對單一索引單一類型的所有資料進行查詢，如圖 9-34 所示。

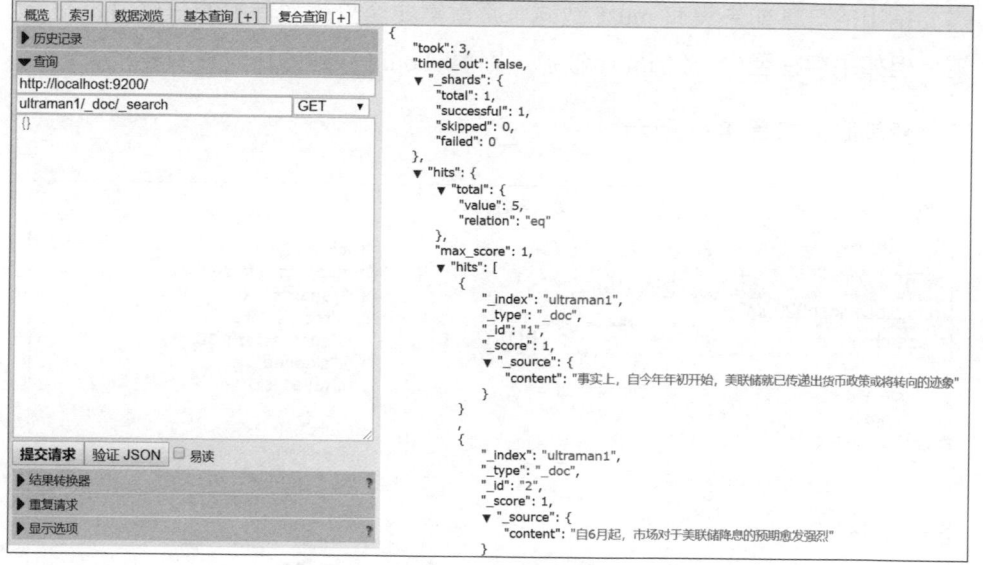

圖 9-34（編按：本圖為簡體中文介面）

在簡單查詢頁面可以進行多個查詢準則的比對；同理，在複雜查詢頁面，也能進行相關查詢，如使用布林查詢。布林查詢的基本格式如下所示：

```
{
    "bool": {
        "must": [],
        "should": [],
        "must_not": [],
        "filter": {}
    }
}
```

（1）must。文件必須比對這些條件才能被搜索出來。

（2）must_not。文件必須不符合這些條件才能被搜索出來。

（3）should。如果滿足這些敘述中的任意敘述，則將增加搜索排名結果_score；否則對查詢結果沒有任何影響。其主要作用是修正每個文件的相關性得分。

（4）filter。表示必須比對，但它是以不評分的過濾模式進行。這些敘述對評分沒有貢獻，只是根據過濾標準排除或包含文件。

需要指出的是，如果沒有 must 敘述，那麼需要至少比對其中的一條 should 敘述。但如果存在至少一條 must 敘述，則對 should 敘述的比對沒有要求。

布林查詢範例如圖 9-35 所示。

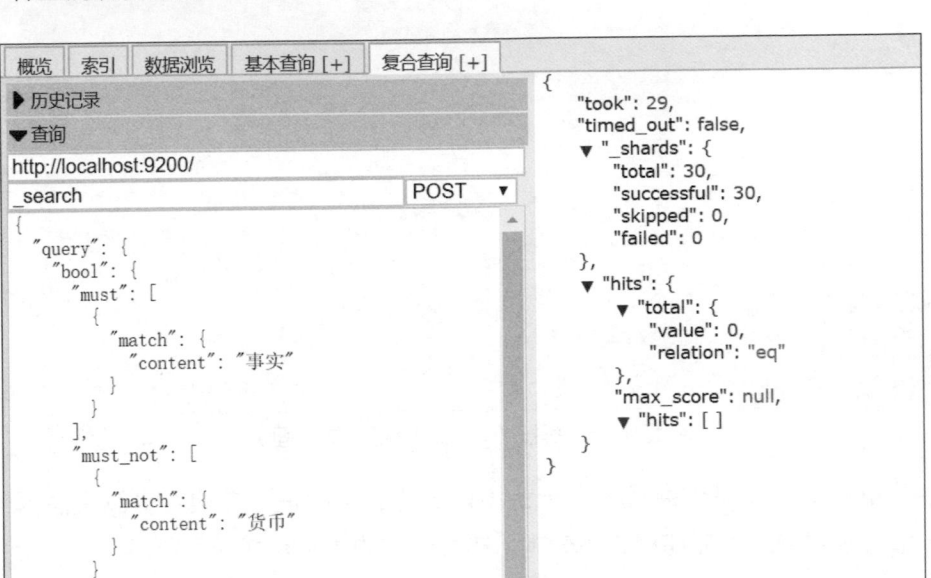

圖 9-35（編按：本圖為簡體中文介面）

由圖 9-35 可知，查詢比對「事實」、且不符合「貨幣」的文件為空，這與儲存在所有索引中的資料相比對。

9.7.2 Cerebro 外掛程式

Cerebro 外掛程式是外掛程式工具 kopf 的升級版本。Cerebro 外掛程式中包含了 kopf 的功能，如監控工具，並包含了 Head 外掛程式的部分功能，可以圖形化地進行新增索引等操作。

被取代的 kopf 早已不再更新；而 Cerebro 外掛程式則在不斷維護和升級中。

讀者可以選擇對應作業系統環境的版本進行下載。如在 Windows 環境下，可以下載 cerebro-0.8.4.tgz。下載到本機後，將檔案解壓縮至根目錄，根目錄內容如圖 9-36 所示。

📁	bin	2019/7/23 19:54
📁	conf	2019/7/23 19:54
📁	lib	2019/7/23 19:54
📁	logs	2019/7/23 19:54
📄	cerebro.db	2019/9/11 18:08
📄	README.md	2019/7/23 19:54

圖 9-36

在使用 Cerebro 外掛程式之前，可以進行必要的設定。當然不進行設定，採用預設設定也可以。設定檔為 cerebro-0.8.4\cerebro-0.8.4\conf\application.conf。application.conf 的設定內容如下所示：

```
# Secret will be used to sign session cookies, CSRF tokens and for other
encryption utilities.
# It is highly recommended to change this value before running cerebro in
production.
secret = "ki:s:[[@=Ag?QI`W2jMwkY:eqvrJ]JqoJyi2axj3ZvOv^/KavOT4ViJSv?6YY4[N"
# Application base path
basePath = "/"
# Defaults to RUNNING_PID at the root directory of the app.
# To avoid creating a PID file set this value to /dev/null
#pidfile.path = "/var/run/cerebro.pid"
pidfile.path=/dev/null
# Rest request history max size per user
rest.history.size = 50 // defaults to 50 if not specified
# Path of local database file
#data.path: "/var/lib/cerebro/cerebro.db"
data.path = "./cerebro.db"
es = {
  gzip = true
}
# Authentication
auth = {
  # either basic or ldap
```

```
type: ${?AUTH_TYPE}
settings {
  # LDAP
  url = ${?LDAP_URL}
  # OpenLDAP might be something like "ou=People,dc=domain,dc=com"
  base-dn = ${?LDAP_BASE_DN}
  # Usually method should  be "simple" otherwise, set it to the SASL
  # mechanisms to try
  method = ${?LDAP_METHOD}
  # user-template executes a string.format() operation where
  # username is passed in first, followed by base-dn. Some examples
  #  - %s => leave user untouched
  #  - %s@domain.com => append "@domain.com" to username
  #  - uid=%s,%s => usual case of OpenLDAP
  user-template = ${?LDAP_USER_TEMPLATE}
  // User identifier that can perform searches
  bind-dn = ${?LDAP_BIND_DN}
  bind-pw = ${?LDAP_BIND_PWD}
  group-search {
    // If left unset parent's base-dn will be used
    base-dn = ${?LDAP_GROUP_BASE_DN}
    // Attribute that represent the user, for example uid or mail
    user-attr = ${?LDAP_USER_ATTR}
    // Define a separate template for user-attr
    // If left unset parent's user-template will be used
    user-attr-template = ${?LDAP_USER_ATTR_TEMPLATE}
    // Filter that tests membership of the group. If this property is empty
    // then there is no group membership check
    // AD example => memberOf=CN=mygroup,ou=ouofthegroup,DC=domain,DC=com
    // OpenLDAP example => CN=mygroup
    group = ${?LDAP_GROUP}
  }
  # Basic auth
  username = ${?BASIC_AUTH_USER}
  password = ${?BASIC_AUTH_PWD}
 }
}
```

```
# A list of known hosts
hosts = [
  #{
  #  host = "http://localhost:9200"
  #  name = "Localhost cluster"
  #  headers-whitelist = [ "x-proxy-user", "x-proxy-roles", "X-Forwarded-
  #  For" ]
  #}
  # Example of host with authentication
  #{
  #  host = "http://some-authenticated-host:9200"
  #  name = "Secured Cluster"
  #  auth = {
  #    username = "username"
  #    password = "secret-password"
  #  }
  #}
]
```

主要參數的設定説明如下。

（1）pidfile.path：服務執行的 pid 的儲存位置。如果要避免產生 pid，則可以使用 /dev/null。

（2）data.path：Cerebro 儲存資料的位置，預設為 Cerebro 安裝目錄。

（3）auth -> settings -> username：Cerebro Web 服務的帳號。

（4）auth -> settings -> password：Cerebro Web 服務的密碼。

（5）hosts -> host：Elasticsearch 叢集的 host 地址。

（6）hosts -> name：Elasticsearch 叢集的名稱。

設定 hosts 的程式如下所示：

```
hosts = [
  {
    host = "http:// localhost:9200"
    name = "Test Elasticsearch Cluster for niudong"
  }
```

設定後，切換到 cerebro-0.8.4 目錄下並啟動 Cerebro 服務，切換指令和啟動指令如圖 9-37 所示。

```
PS C:\Users\牛冬> cd C:\Users\牛冬\Desktop\cerebro-0.8.4\cerebro-0.8.4
PS C:\Users\牛冬\Desktop\cerebro-0.8.4\cerebro-0.8.4> bin/cerebro
Picked up JAVA_TOOL_OPTIONS: -Dfile.encoding=UTF-8
[info] play.api.Play - Application started (Prod) (no global state)
[info] p.c.s.AkkaHttpServer - Listening for HTTP on /0:0:0:0:0:0:0:0:9000
```

圖 9-37

當 Cerebro 服務啟動後，即可在瀏覽器中存取以下 URL，開啟 Cerebro 圖形介面：

```
http://localhost:9000/
```

從圖 9-37 可以看出，Cerebro 會預設連接 http://localhost:9200/。使用者可以根據自己環境的情況連接對應的 Elasticsearch 節點。設定後，點擊 "Connect" 按鈕進行連接，如圖 9-38 所示。

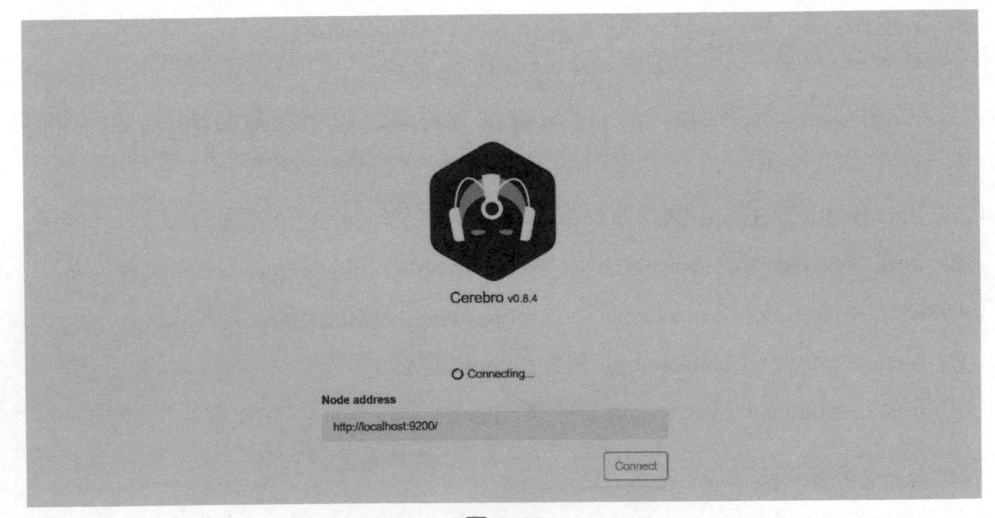

圖 9-38

連接 Elasticsearch 節點成功後，會進入 Cerebro 的首頁，如圖 9-39 所示。

從圖 9-39 可見，Cerebro 首頁中有 "Overview"、"nodes"、"rest" 和 "more" 四個標籤頁。預設目前是 "Overview" 頁面。

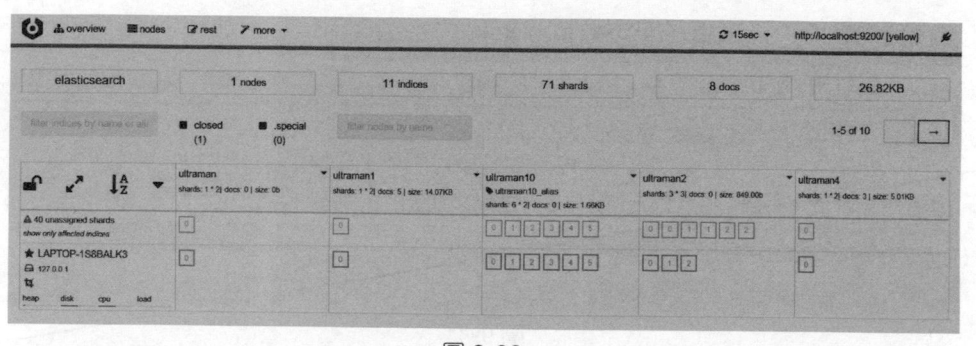

圖 9-39

在 "Overview" 頁面中，可以看到 Elasticsearch 叢集的各個 Node 節點的詳細
資訊。該標籤頁分為三部分，頂端的線條、Elasticsearch 叢集內各種資訊的統
計、各個節點的資訊。其中：

（1）頂端的線條：顏色釋義與 Head 外掛程式中的顏色釋義相同，共有綠色、
　　　紅色、黃色三種，綠色代表叢集工作正常。

（2）Elasticsearch 叢集內各種資訊的統計：包含叢集名稱、節點數量、索引數
　　　量、分片數量、文件數量和索引所佔儲存空間的大小等資訊。

（3）各個節點的資訊：在最下方表格中，每行代表一個節點，每列代表一個
　　　索引。

點擊某個索引，即可檢視該索引下的資訊，如圖 9-40 所示。

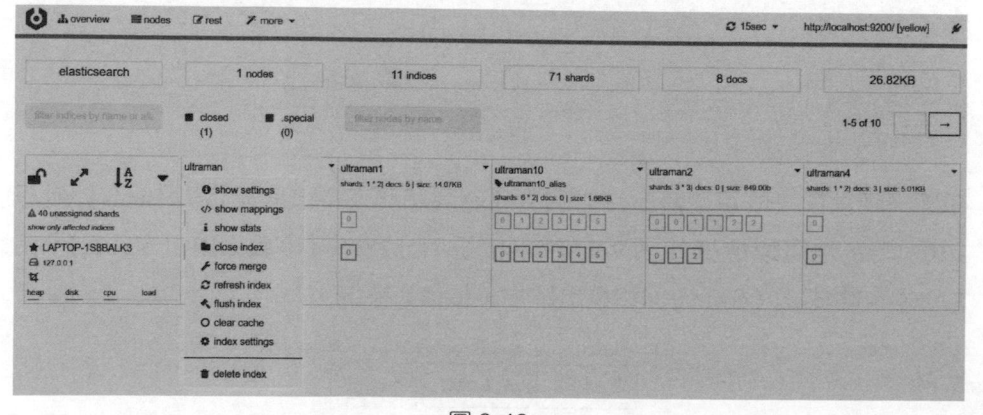

圖 9-40

切換到 "Nodes" 標籤頁，可以看到各節點的資源使用情況，如圖 9-41 所示。

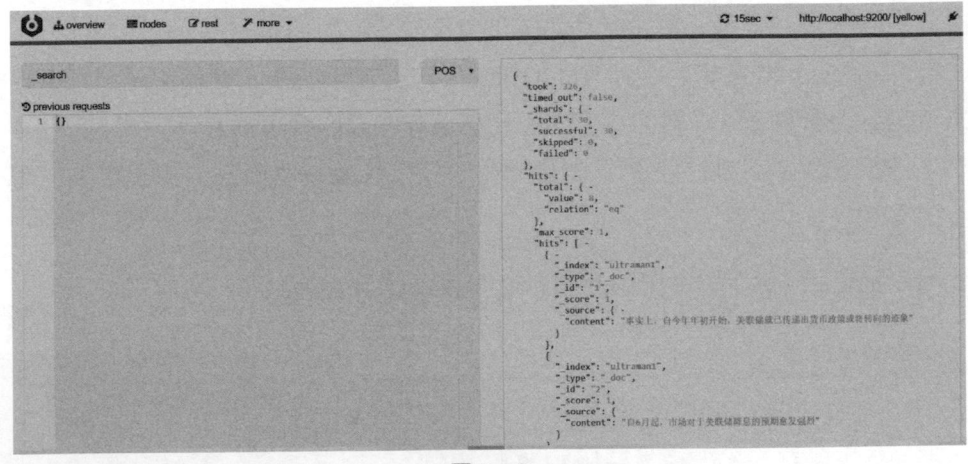

圖 9-41

從圖 9-41 可見，各節點的資源使用情況主要有 CPU、堆、磁碟使用、資料更新時間等。

切換到 "Rest" 標籤頁，使用者可以向 Elasticsearch 叢集發出 RESTful 格式的 API 請求，如圖 9-42 所示。

圖 9-42

在提交請求前，既可以設定請求的更新時間和頻率（這些設定與 Head 外掛程式類似），也可以透過如圖 9-43 所示按鈕，分別進行複製 URL（cURL）、請求 JSON 格式化（format）和發出請求（send）操作。

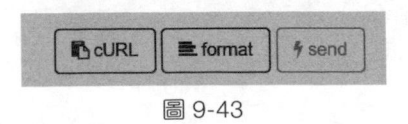

圖 9-43

可以按下 "More" 接鈕，可以做更多的操作，如圖 9-44 所示。

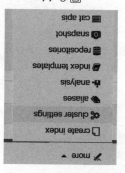

圖 9-44

點擊 "create index" 接鈕，即可建立索引，欄名如圖 9-45 所示。

name　　index name
settings　　│ X │
number of shards　　number of replicas
　　# of shards　　# of replicas
load settings from existing index

圖 9-45

填寫相關欄位後可即可建立新增的索引。

點擊 "cluster settings" 接鈕，便可以針對 Elasticsearch 叢集進行設定，設定參數如圖 9-46 所示。

☰ show static settings 🔍

ACTION

action.auto_create_index	action.destructive_requires_name	action.search.shard_count.limit
true	false	9223372036854775807

CLUSTER

cluster.blocks.read_only	cluster.blocks.read_only_allow_delete	cluster.indices.close.enable
false	false	true
cluster.info.update.interval	cluster.info.update.timeout	cluster.routing.allocation.allow_rebalance
30s	15s	indices_all_active
cluster.routing.allocation.awareness.attributes	cluster.routing.allocation.balance.index	cluster.routing.allocation.balance.shard
	0.55	0.45

圖 9-46

點擊 "aliases" 選項，可以為索引維護其別名資訊，如圖 9-47 所示。

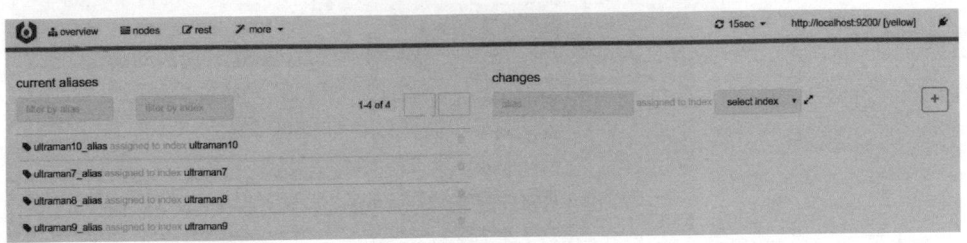

圖 9-47

點擊 "analysis" 選項，可以進行字串分析操作，如圖 9-48 所示。

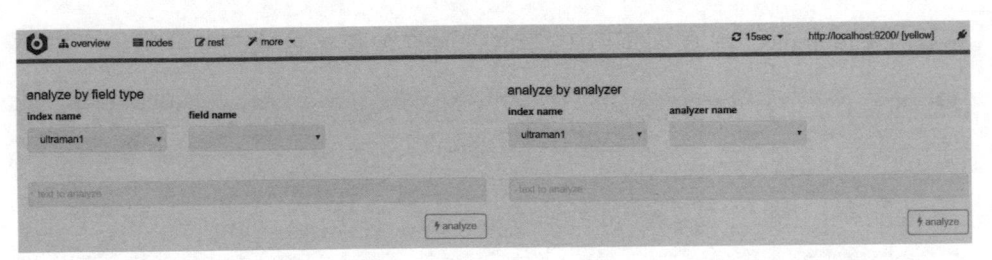

圖 9-48

9.8　基礎知識連結

外掛程式，即 Plug-in，又稱為 addin、add-in、addon 或 add-on，在遊戲或手機等場景中被稱為「外掛」，是一個多稱謂的詞彙。

一般來說，外掛程式是一種程式碼的統稱。外掛程式的撰寫會遵循一定標準，特別是載體的應用程式介面。因此，外掛程式只能執行在程式規定的系統平台下，像 Java 一樣「一次編譯、到處執行」的外掛程式幾乎是不存在的。

此外，外掛程式一般不能脫離指定的平台單獨執行，畢竟外掛程式需要由載體系統提供必要的本機函數程式庫和資料。與之相反的是，應用程式不用依賴外掛程式即可執行，因而當外掛程式載入到應用程式上時，動態更新不會對應用程式造成影響。

最早的外掛程式出現在 20 世紀 70 年代中期的 EDT 文字編輯器中。EDT 文字編輯器在 Univac 90/60 系列大型主機上執行 Unisys VS/9 作業系統時，可以提供執行外掛程式的功能，該功能允許外掛程式進入編輯器的緩衝，允許外掛程式染指記憶體中正在編輯的工作。外掛程式使得編輯器可以在緩衝區上進行文字編輯，而這個緩衝區是編輯器和外掛程式共同享用的。

在個人電腦時代，第一個帶有外掛程式的應用軟體是 1987 發行、安裝在蘋果電腦上的 HyperCard 和 QuarkXPress。

現在，我們常用的瀏覽器中就有各種各樣的外掛程式，如廣告過濾、智慧填寫表單，甚至是火車票搶票外掛程式等。

外掛程式的工作原理是什麼呢？

一般來說，已公開的應用程式介面提供一個標準的介面或函數程式庫，允許其他人撰寫外掛程式與應用程式互動，並提供一個穩定的應用程式介面，允許其他外掛程式正常執行。

使用外掛程式的好處很多，特別是在產品功能擴充、生態建置過程中。

其一，一般外掛程式的載體程式會提供類似匯流排的架構，介面清晰，易於開發者了解；多個外掛程式安裝後也是獨立執行、互不干擾。

其二，外掛程式和載體程式是可抽換關係，組合很靈活，因而外掛程式和載體程式都容易修改，方便升級和維護。

其三，外掛程式的撰寫一般都是單一職責原則，因此外掛程式的可攜性較強、重複使用的機會多。也正是因為基於單一職責原則，所以多個外掛程式之間是鬆散耦合關係。

需要指出的是，與外掛程式概念類別相似的概念是元件。元件和外掛程式有明顯的區別，外掛程式屬於程式介面；元件是一種控制項、物件，重複使用程度更高。

9.9 小結

本章主要介紹了 Elasticsearch 的外掛程式生態，包含官方維護的外掛程式和技術社區維護的外掛程式；介紹了外掛程式的統一管理方法，如安裝和移除等。

本章先後介紹了分析外掛程式、API 擴充外掛程式、監控外掛程式和資料分析外掛程式，重點介紹了分析外掛程式中的 ICU 分析外掛程式和智慧中文分析外掛程式、監控外掛程式中的 Head 外掛程式和 Cerebro 外掛程式。

Elasticsearch 生態圈

天際浮雲入思深
物情生態看銷沉

第 9 章介紹了 Elasticsearch 的外掛程式生態，外掛程式生態是依附於 Elasticsearch 內部的，屬於一種相對狹義、微觀的生態；本章主要介紹 Elasticsearch 的巨觀生態。

10.1 ELK

提到 Elasticsearch 生態，很多人第一反應就是 ELK Stack。什麼是 ELK Stack 呢？

很簡單，ELK Stack 指的就是 Elastic Stack。

10.1.1 Elastic Stack

"ELK" 是三個開放原始碼專案的字首縮寫，這三個專案分別是：Elasticsearch、Logstash 和 Kibana，如圖 10-1 所示。當然，這並非是 Elastic Stack 的全部，讀者可以根據需要在生態中增加 Redis、Kafka、Filebeat 等軟體。

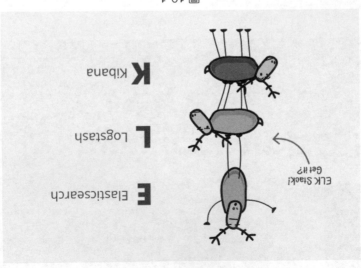

圖 10-1

目前的 Elastic Stack 其實就是 ELK Stack 的改朝換代產品。2015 年，ELK Stack 中加入了一系列輕量型的產品——功能資料蒐集器，並把它們叫作 Beats。

Beats 加入 ELK 家族後，舊叫 ELK 顯然就不合適了，那麼新的家族該叫什麼呢？BELK？ELKB？

其實，Elastic 官方也曾對此進行過讓瀰漫況用字其縮寫的方式，但 Elastic 又相繼收購了 APM 公司 Opbeat、機器學習公司 Prelert、SaaS 服務公司 Found，後來還收購了 Swiftype、終端安全公司 Endgame 等來擴大自己的業務版圖。

對於 Elastic 擴充業務如此之快的步伐來說，一直採用字母其縮寫的搭配方式不是久之計。於是，Elastic 為這個「一家族」取的名字就誕生了。

10.1.2 Elastic Stack 版本的由來

Elasticsearch 的版本編號從 2 直接升到 5 是否有回事呢？

在最初的 Elastic Stack 產品中，Elasticsearch、Logstash、Kibana 和 Beats 有各自的版本編號，如當 Elasticsearch 和 Logstash 的版本編號是 V2.3.4 件，Kibana 的版本編號是 V4.5.3，而 Beats 的版本編號是 V1.2.3。

因此，Elastic Stack 官方將產品版本編號也進行了統一，從 V5.0 開始。因為當時的 Kibana 版本編號已經是 4.x 了，其下個版本只能是 5.0，所以其他產品的版本編號也隨之「跳級」，於是 V5.0 版本的 Elastic Stack 在 2016 年就面世了。

10.1.3 ELK 實戰的背景

在實際使用過程中，什麼場景適合使用 ELK 呢？

在實戰中，我們既可以用 ELK 管理和分析記錄檔，也可以用 ELK 分析索引中的資料。

在目前的軟體開發過程中，業務發展節奏越來越快，伺服器整理越來越多，隨之而來的就是各種存取記錄檔、應用記錄檔和錯誤記錄檔。隨著時間的流逝，記錄檔的累積也越來越多。

此時，會出現這樣的問題：運行維護人員無法極佳地管理記錄檔；開發人員排除業務問題時需要到伺服器上查詢大量記錄檔；當營運人員需要一些業務資料時，需要到伺服器上分析記錄檔。

在上述場景中，通常意義上的 "awk" 和 "grep" 指令已經力不從心，而且效率很低。這時 ELK 就可以「隆重登場」啦！

ELK 的三個元件是如何分工協作的呢？

首先，我們使用 Logstash 進行記錄檔的搜集、分析和過濾。一般工作方式為 C/S 架構，Client 端會被安裝在需要收集記錄檔的主機上，Server 端則負責收集的各節點的記錄檔資料，並進行過濾、修改和分析等操作，前置處理過的資料會一平行處理到 Elasticsearch 上。

隨後將 Kibana 連線 Elasticsearch，並為 Logstash 和 Elasticsearch 提供記錄檔分析人性化的 Web 介面，幫助使用者整理、分析和搜索重要資料的記錄檔。

10.1.4 ELK 的部署架構變遷

ELK 架構為資料分散式儲存、視覺化查詢和記錄檔解析建立了一個功能強大的管理鏈。ELK 架構為使用者建立了集中式記錄檔收集系統，將所有節點上的記錄檔統一收集、管理和存取。三者相互配合，截長補短，共同完成分散式大數據處理工作。

目前官方推薦的 ELK 部署架構並非一步合格，而是經過反覆運算演進發展而來的。下面簡單介紹 ELK 架構的發展歷程。

最簡單的一種 ELK 部署架構方式如圖 10-2 所示。

首先由分佈於各個服務節點上的 Logstash 搜集相關記錄檔和資料，經過 Logstash 的分析和過濾後發送給遠端伺服器上的 Elasticsearch 進行儲存。Elasticsearch 將資料以分片的形式壓縮儲存，並提供多種 API 供使用者進行查詢操作。使用者還可以透過設定 Kibana Web Portal 對記錄檔進行查詢，並根據資料產生報表。

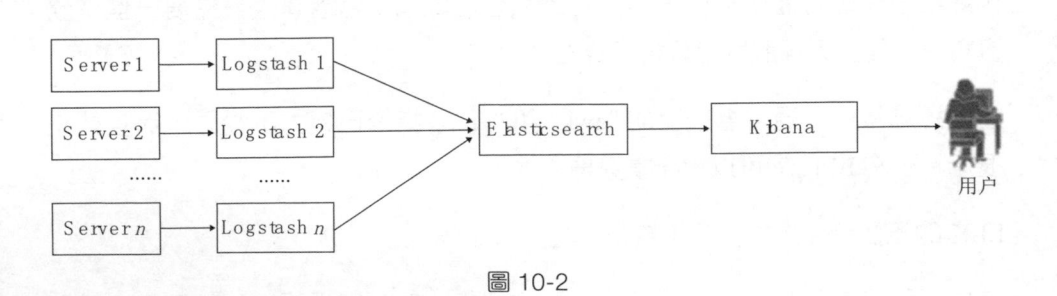

圖 10-2

該架構最顯著的優點是架設簡單，易於上手。但缺點同樣很突出，因為 Logstash 消耗資源較大，所以在執行時期會佔用很多的 CPU 和記憶體。並且系統中沒有訊息佇列快取等持久化方法，因而存在資料遺失隱憂。因此，一般這種部署架構通常用於學習和小規模叢集。

以第一種 ELK 部署架構為基礎的優缺點，第二種架構引用了訊息佇列機制，如圖 10-3 所示。

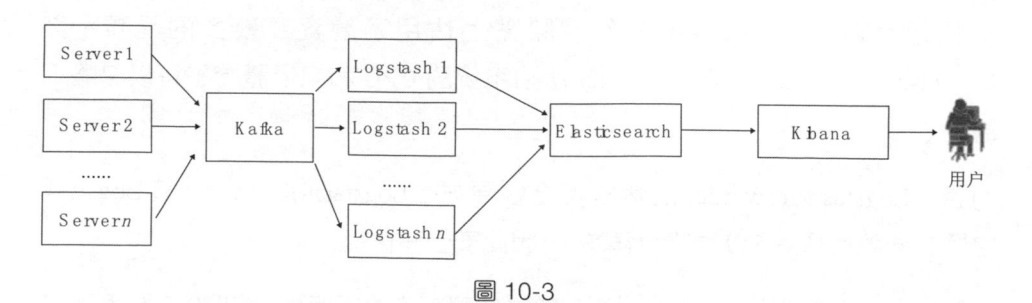

圖 10-3

位於各個節點上的 Logstash 用戶端先將資料和記錄檔等內容傳遞給 Kafka，當然，也可以用其他訊息機制，如各種 MQ（Message Queue）和 Redis 等。

Kafka 會將佇列中的訊息和資料傳遞給 Logstash，經過 Logstash 的過濾和分析等處理後，傳遞給 Elasticsearch 進行儲存。最後由 Kibana 將記錄檔和資料呈現給使用者。

在該部署架構中，Kafka 的引用使得即使遠端 Logstash 因故障而停止執行，資料也會被儲存下來，進一步避免資料遺失。

第二種部署架構解決了資料的可用性問題，但 Logstash 的資源消耗依然較多，因而引出第三種架構。第三種架構引用了 Logstash-forwarder，如圖 10-4 所示。

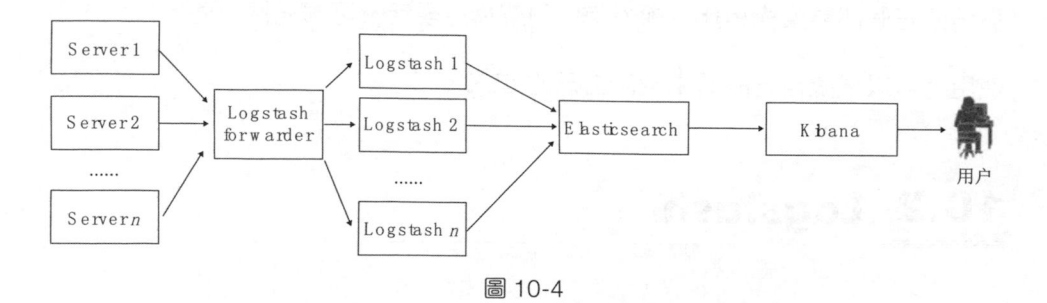

圖 10-4

Logstash-forwarder 將記錄檔資料搜集並統一後發送給主節點上的 Logstash，Logstash 在分析和過濾記錄檔資料後，把記錄檔資料發送至 Elasticsearch 進行儲存，最後由 Kibana 將資料呈現給使用者。

這種架構解決了 Logstash 在各電腦點上佔用系統資源較多的問題。與 Logstash 相比，Logstash-forwarder 所佔系統的 CPU 和記憶體幾乎可以忽略不計。

而且，Logstash-forwarder 的資料安全性更好。Logstash-forwarder 和 Logstash 之間的通訊是透過 SSL 加密傳輸的，因此安全有保障。

隨著 Beats 元件引用 ELK Stack，第四種部署架構應運而生，如圖 10-5 所示。

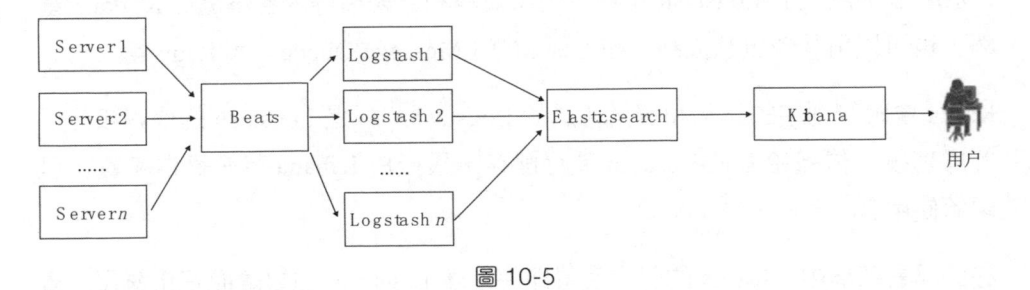

圖 10-5

在實際使用中，Beats 平台在滿負荷狀態時所耗系統資源和 Logstash-forwarder 相當，但其擴充性和靈活性更好。Beats 平台目前包含 Packagebeat、Topbeat 和 Filebeat 三個產品，均為 Apache 2.0 License。同時使用者可以根據需要進行延伸開發。

與前面三個部署架構相比，顯然第四種架構更靈活，可擴充性更強。

使用者可以根據自己的需求架設自己的 ELK。

10.2 Logstash

10.2.1 Logstash 簡介

Logstash 由三部分組成，即輸入模組（INPUTS）、篩檢程式模組（FILTERS）和輸出模組（OUTPUTS），如圖 10-6 所示。

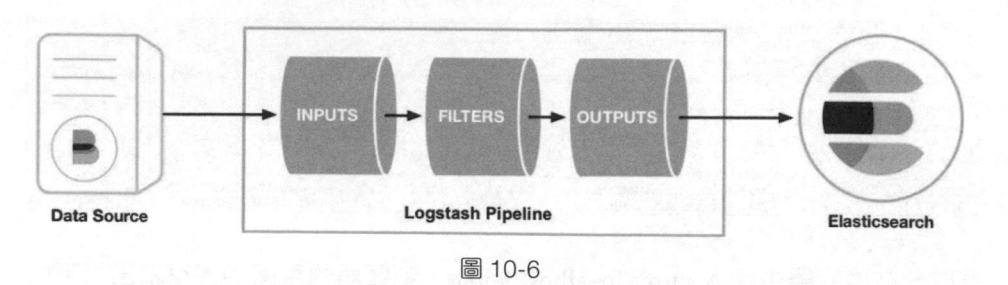

圖 10-6

Logstash 能夠動態地擷取、轉換和傳輸資料，不受格式或複雜度的影響。利用 Grok 從非結構化資料中衍生出結構，從 IP 位址解碼出地理座標，匿名化或排除敏感欄位，並簡化整體處理過程。

從官網下載 Logstash 安裝套件。下載完成後，在本機解壓縮。解壓縮後的根目錄內容如 10-7 所示。

bin	2019/9/16 15:44
config	2019/9/16 15:44
data	2019/9/16 15:57
lib	2019/9/16 15:44
logs	2019/9/16 15:57
logstash-core	2019/9/16 15:44
logstash-core-plugin-api	2019/9/16 15:44
modules	2019/9/16 15:44
tools	2019/9/16 15:44
vendor	2019/9/16 15:45
x-pack	2019/9/16 15:45
CONTRIBUTORS	2019/9/6 16:46
Gemfile	2019/9/6 16:46
Gemfile.lock	2019/9/6 16:46
LICENSE.txt	2019/9/6 16:46
NOTICE.TXT	2019/9/6 16:46

圖 10-7

根目錄下有 bin、config、data、lib、logstash core 和 tools 等內容。

在 Logstash 啟動後，會自動建立 logs 目錄。隨後設定 config 目錄下的 logstash.conf 檔案。第一次設定時可參考同目錄的 logstash-simple.conf 範例進行設定。設定後，執行 bin/logstash -f logstash.conf 即可啟動 Logstash 服務，如下所示：

```
PS C:\elasticsearch-7.2.0-windows-x86_642\logstash-7.3.2> bin/logstash -f logstash.conf
"warning: ignoring JAVA_TOOL_OPTIONS=$JAVA_TOOL_OPTIONS"
Thread.exclusive is deprecated, use Thread::Mutex
Sending Logstash logs to C:/elasticsearch-7.2.0-windows-x86_642/logstash-7.3.2/logs which is now configured via log4j2.p
roperties
[2019-09-16T16:00:27,478][WARN ][logstash.config.source.multilocal] Ignoring the 'pipelines.yml' file because modules or
 command line options are specified
[2019-09-16T16:00:27,532][INFO ][logstash.runner          ] Starting Logstash {"logstash.version"=>"7.3.2"}
[2019-09-16T16:00:29,268][INFO ][logstash.config.source.local.configpathloader] No config files found in path {:path=>"C
:/elasticsearch-7.2.0-windows-x86_642/logstash-7.3.2/logstash.conf"}
[2019-09-16T16:00:29,287][ERROR][logstash.config.sourceloader] No configuration found in the configured sources.
[2019-09-16T16:00:29,911][INFO ][logstash.agent           ] Successfully started Logstash API endpoint {:port=>9600}
```

此時，在瀏覽器中輸入 http://localhost:9600/，瀏覽器的頁面中即可輸出以下內容：

```
{
  "host": "LAPTOP-1S8BALK3",
  "version": "7.3.2",
  "http_address": "127.0.0.1:9600",
  "id": "e0903a8b-e533-4836-98bb-79d905720920",
  "name": "LAPTOP-1S8BALK3",
  "ephemeral_id": "a96193e8-2ce9-4eea-8a79-90972c92085c",
  "status": "green",
  "snapshot": false,
  "pipeline": {
    "workers": 4,
    "batch_size": 125,
    "batch_delay": 50
  },
  "build_date": "2019-09-06T16:42:57+00:00",
  "build_sha": "1071165b526bcd43b475caed0b4289e3d32dc52f",
  "build_snapshot": false
}
```

需要指出的是，Logstash 資料夾儲存的路徑中不能有中文命名的資料夾，否則會列出以下錯誤訊息：

```
Logstash - java.lang.IllegalStateException: Logstash stopped processing
because of an error: (ArgumentError) invalid byte sequence in US-ASCII
```

10.2.2 Logstash 的輸入模組

Logstash 的輸入模組用於擷取各種樣式、大小和來源的資料。一般來說，資料常常以各種各樣的形式，或分散或集中地儲存於很多系統中。Logstash 支援各種輸入選擇，可以在同一時間從許多常用來源捕捉事件，能夠以資料流方式，輕鬆地從使用者的記錄檔、指標、Web 應用、資料儲存及各種 AWS 服務中擷取資料。

為了支援各種資料登錄，Logstash 提供了很多輸入外掛程式，整理如下。

（1） azure_event_hubs：該外掛程式從微軟 Azure 事件中心接收資料。讀者可存取 GitHub 官網，搜索 logstash-input-azure_event_hubs 取得外掛程式。

（2） beats：該外掛程式從 Elastic Beats 架構接收資料。讀者可存取 GitHub 官網，搜索 logstash-input-beats 取得外掛程式。

（3） cloudwatch：該外掛程式從 Amazon Web Services CloudWatch API 中分析資料。讀者可存取 GitHub 官網，搜索 logstash-input-cloudwatch 取得外掛程式。

（4） couchdb_changes：該外掛程式從 CouchDB 更改 URI 的流式處理事件中取得資料。讀者可存取 GitHub 官網，搜索 logstash-input-couchdb_changes 取得外掛程式。

（5） dead_letter_queue：該外掛程式從 logstash 的 dead letter 佇列中讀取資料。讀者可存取 GitHub 官網，搜索 logstash-input-dead_letter_queue 取得外掛程式。

（6） elasticsearch：該外掛程式從 ElasticSearch 叢集中讀取查詢結果。讀者可存取 GitHub 官網，搜索 logstash-input-elasticsearch 取得外掛程式。

（7） exec：該外掛程式將 shell 指令的輸出捕捉為事件，並取得資料。讀者可存取 GitHub 官網，搜索 logstash-input-exec 取得外掛程式。

（8） file：該外掛程式從檔案流式處理中取得資料。讀者可存取 GitHub 官網，搜索 logstash-input-file 取得外掛程式。

（9） ganglia：該外掛程式透過 UDP 資料封包讀取 ganglia 中的資料封包來取得資料。讀者可存取 GitHub 官網，搜索 logstash-input-ganglia 取得外掛程式。

（10）gelf：該外掛程式從 graylog2 中讀取 gelf 格式的訊息，並取得資料。讀者可以存取 GitHub 頁面，搜索 logstash-input-gelf 取得此外掛程式。

（11）http：該外掛程式經過 HTTP 或 HTTPS 接收事件並取得資料。讀者可以存取 GitHub 頁面，搜索 logstash-input-http 取得此外掛程式。

（12）jdbc：該外掛程式經過 JDBC 介面從資料庫中取得資料。讀者可以存取 GitHub 頁面，搜索 logstash-input-jdbc 取得此外掛程式。

（13）kafka：該外掛程式從 Kafka 主題中讀取事件，進一步取得資料。讀者可以存取 GitHub 頁面，搜索 logstash-input-kafka 取得此外掛程式。

（14）log4j：該外掛程式經過 TCP 通訊端從 Log4j SocketAppender 物件中讀取資料。讀者可以存取 GitHub 頁面，搜索 logstash-input-log4j 取得此外掛程式。

（15）rabbitmq：該外掛程式從 RabbitMQ 資料交換中分析取得資料。讀者可以存取 GitHub 頁面，搜索 logstash-input-rabbitmq 取得此外掛程式。

https://github.com/logstash-plugins/logstash-input-rabbitmq

（16）redis：該外掛程式從 redis 實例中讀取資料。讀者可以存取 GitHub 頁面，搜索 logstash-input-redis 取得此外掛程式。

10.2.3 Logstash 篩檢程式

Logstash 篩檢程式用於即時解析和轉換資料。

在資料從來源傳輸到儲存庫的過程中，Logstash 篩檢程式能夠解析各個事件，識別已命名的欄位，並將它們轉換為通用格式，以便更輕鬆、更快速地進行分析，實現商業價值。

Logstash 篩檢程式有以下幾種：

（1）利用 Grok 從非結構化資料中分析出結構。

（2）從 IP 位址解讀出地理座標。

（3）將 PII 資料匿名化，完全排除敏感欄位。

（4）簡化整體處理，不受資料來源、格式或架構的影響。

為了處理各種各樣的資料來源，Logstash 提供了豐富多樣的篩檢程式外掛程式庫，常用的篩檢程式外掛程式整理如下。

（1）aggregate：該外掛程式用於從一個工作的多個事件中蒐集資訊。讀者可以存取 GitHub 官網，搜索 logstash-filter-aggregate 取得外掛程式。

（2）alter：該外掛程式對 mutate 篩檢程式不處理的欄位執行正常整理。讀者可存取 GitHub 官網，搜索 logstash-filter-alter 取得外掛程式。

（3）bytes：該外掛程式以電腦儲存單位表示的各式字串形式，如 "123MB" 或 "5.6GB"，解析為以位元組為單位的數值。讀者可以存取 GitHub 官網，搜索 logstash-filter-bytes 取得外掛程式。

（4）cidr：該外掛程式從網路位址中檢查簡單看看 IP 位址。讀者可以存取 GitHub 官網，搜索 logstash-filter-cidr 取得外掛程式。

（5）cipher：該外掛程式用於對事件中欄位的值進行加密或解密。讀者可以存取 GitHub 官網，搜索 logstash-filter-cipher 取得外掛程式。

（6）clone：該外掛程式用於複製事件。讀者可以存取 GitHub 官網，搜索 logstash-filter-clone 取得外掛程式。

（7）csv：該外掛程式用於將逗點分隔的值資料解析為單一欄位。讀者可以存取 GitHub 官網，搜索 logstash-filter-csv 取得外掛程式。

（8）date：該外掛程式用於分析欄位中的日期，為以後事件記錄標準中儲存存的時間戳記。讀者可以存取 GitHub 官網，搜索 logstash-filter-date 取得外掛程式。

（9）dns：該外掛程式用於正向或反向解析 DNS 條款。讀者可以存取 GitHub 官網，搜索 logstash-filter-dns 取得外掛程式。

（10）elasticsearch：該外掛程式用於將 Elasticsearch 記錄檔中的欄位複製到目前事件中。讀者可以存取 GitHub 官網，搜索 logstash-filter-elasticsearch 取得外掛程式。

（11）geoip：該外掛程式用於增加有關 IP 位址的地理資訊。讀者可以存取 GitHub 官網，搜索 logstash-filter-geoip 取得外掛程式。

（12）json：該外掛程式用於解析 JSON 事件。讀者可存取 GitHub 官網，搜索 logstash-filter- json 取得外掛程式。

（13）kv：該外掛程式用於分析鍵值對。讀者可存取 GitHub 官網，搜索 logstash-filter-kv 取得外掛程式。

（14）memcached：該外掛程式用於提供與 memcached 中資料的整合。讀者可存取 GitHub 官網，搜索 logstash-filter- memcached 取得外掛程式。

（15）split：該外掛程式用於將多行訊息拆分為不同的事件。讀者可存取 GitHub 官網，搜索 logstash-filter-split 取得外掛程式。

10.2.4 Logstash 的輸出模組

Logstash 的輸出模組用於將目標資料匯出到使用者選擇的儲存函數庫。

在 Logstash 中，儘管 Elasticsearch 是 Logstash 官方首選的，但它並非唯一選擇。

Logstash 提供許多輸出選擇，使用者可以將資料發送到指定的地方，並且能夠靈活地解鎖許多下游使用案例。

（1）csv：該外掛程式以 CVS 格式將結果資料寫入磁碟。讀者可存取 GitHub 官網，搜索 logstash-output-csv 取得外掛程式。

（2）mongodb：該外掛程式將結果資料寫入 MongoDB。讀者可存取 GitHub 官網，搜索 logstash-output-mongodb 取得外掛程式。

（3）elasticsearch：該外掛程式將結果資料寫入 Elasticsearch。讀者可存取 GitHub 官網，搜索 logstash-output-elasticsearch 取得外掛程式。

（4）email：該外掛程式將結果資料發送到指定的電子郵件。讀者可存取 GitHub 官網，搜索 logstash-output- email 取得外掛程式。

（5）kafka：該外掛程式將結果資料寫入 Kafka 的 Topic 主題。讀者可存取 GitHub 官網，搜索 logstash-output- kafka 取得外掛程式。

（6）file：該外掛程式將結果資料寫入磁碟上的檔案。讀者可存取 GitHub 官網，搜索 logstash-output- file 取得外掛程式。

（7）redis：該外掛程式使用 redis 中的 rpush 指令將分散式資料發送到 redis 佇列作為，讓其可作為 logstash-output- redis 接著，透過存取 GitHub 官網，取得外掛程式。

10.3 Kibana

Kibana 是一個以 Web 為基礎的圖形介面，可以讓使用者在 Elasticsearch 中使用圖形和圖表將資料進行視覺化。

在實際使用過程中，Kibana 一般用於搜尋、分析和視覺化儲存在 Elasticsearch 拔棒中的即時資料。Kibana 利用 Elasticsearch 的 REST 介面檢索資料，不僅使使用者建立立自己的資料的訂製儀表板檢視，還允許他們以特殊的方式查詢和過濾資料。可以說從頭到尾、有條到列了解運用資料和如何流經整個應用，Kibana 都能輕鬆完成。

10.3.1 Kibana 簡介

Kibana 提供了基本分析服務、位置分析服務、時間序列服務、機器學習服務，以及圖表和網路服務。

（1）基本分析服務：指的是 Kibana 核心產品中搭載的一批經典功能，如直方圖、線圖、圓形圖、資料分佈圖、區域圖和日曆圖等。

（2）位置分析服務：主要借助 Elastic Maps 搜素地理資料。另外，還可以連結第三方資料來源、繪製自訂圖層和向量形狀進行視覺化。

（3）時間序列服務：借助 Kibana 圖形化精選的時序資料 UI，對使用者所用的 Elasticsearch 中的資料執行進階時間序列分析。圖中，使用者可以利用功能強大、簡單易學的運算式來描述查詢、轉換和視覺化。

（4）機器學習服務：主要是借助非監督機器學習功能檢測隱藏在所使用的 Elasticsearch 資料中的例外情況，並探索那些影響它們的有關屬性的關係。

圖表。

（5）圖表和網路服務：憑藉搜尋引擎的相關性功能，結合 Graph 連結分析，揭示使用者所用 Elasticsearch 資料中極其常見的關係。

此外，Kibana 還支援使用者把 Kibana 視覺化內容分享給他人，如團隊成員、老闆、客戶、符合規範經理或承包商等，進而讓每個人都感受到 Kibana 的便利。

除分享連結外，Kibana 還有其他內容輸出形式，如嵌入儀表板，匯出為 PDF、PNG 或 CSV 等格式檔案，以便把這些檔案作為附件發送給他人。

我們可從官網下載 Kibana，下載完成後，即可在本機進行解壓縮。解壓縮後的 Kibana 的根目錄如圖 10-8 所示。

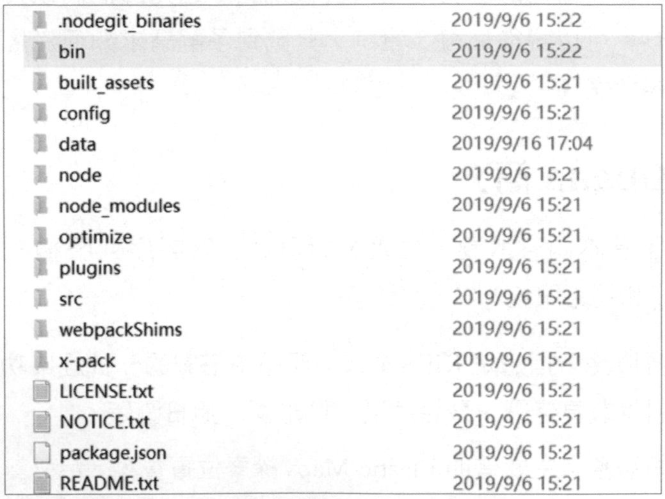

📁 .nodegit_binaries	2019/9/6 15:22	
📁 bin	2019/9/6 15:22	
📁 built_assets	2019/9/6 15:21	
📁 config	2019/9/6 15:21	
📁 data	2019/9/16 17:04	
📁 node	2019/9/6 15:21	
📁 node_modules	2019/9/6 15:21	
📁 optimize	2019/9/6 15:21	
📁 plugins	2019/9/6 15:21	
📁 src	2019/9/6 15:21	
📁 webpackShims	2019/9/6 15:21	
📁 x-pack	2019/9/6 15:21	
📄 LICENSE.txt	2019/9/6 15:21	
📄 NOTICE.txt	2019/9/6 15:21	
📄 package.json	2019/9/6 15:21	
📄 README.txt	2019/9/6 15:21	

圖 10-8

10.3.2 連接 Elasticsearch

由於 Kibana 服務需要連接到 Elasticsearch，因此在啟動 Kibana 前，需要先啟動 Elasticsearch，不然 Kibana 在啟動過程中會顯示如下所示錯誤：

```
[warning][admin][elasticsearch] Unable to revive connection: http://
localhost:9200/
```

在啟動 Elasticsearch 後，需要對 config/kibana.yml 檔案進行設定。主要是設定 elasticsearch.hosts 屬性，該屬性在無設定的情況下預設連接到 http://localhost:9200。

在設定好 elasticsearch.hosts 屬性後，即可透過以下指令啟動 Kibana：

```
bin/kibana
```

如果是在 Windows 環境下啟動，則使用以下指令啟動 Kibana：

```
bin\kibana.bat
```

當 Kibana 啟動成功後，輸出內容如下所示：

```
Ready
  log   [09:38:23.046] [info][status][plugin:snapshot_restore@7.2.0] Status changed from uninitialized to yellow - Waiti
ng for Elasticsearch
  log   [09:38:23.061] [info][status][plugin:data@7.2.0] Status changed from uninitialized to green - Ready
  log   [09:38:23.442] [info][status][plugin:timelion@7.2.0] Status changed from uninitialized to green - Ready
  log   [09:38:23.452] [info][status][plugin:ui_metric@7.2.0] Status changed from uninitialized to green - Ready
  log   [09:38:24.983] [info][status][plugin:elasticsearch@7.2.0] Status changed from yellow to green - Ready
  log   [09:38:25.049] [info][license][xpack] Imported license information from Elasticsearch for the [data] cluster: mo
de: basic | status: active
  log   [09:38:25.073] [info][status][plugin:xpack_main@7.2.0] Status changed from yellow to green - Ready
  log   [09:38:25.074] [info][status][plugin:graph@7.2.0] Status changed from yellow to green - Ready
  log   [09:38:25.075] [info][status][plugin:searchprofiler@7.2.0] Status changed from yellow to green - Ready
  log   [09:38:25.079] [info][status][plugin:ml@7.2.0] Status changed from yellow to green - Ready
  log   [09:38:25.079] [info][status][plugin:tilemap@7.2.0] Status changed from yellow to green - Ready
  log   [09:38:25.083] [info][status][plugin:watcher@7.2.0] Status changed from yellow to green - Ready
  log   [09:38:25.083] [info][status][plugin:grokdebugger@7.2.0] Status changed from yellow to green - Ready
  log   [09:38:25.084] [info][status][plugin:logstash@7.2.0] Status changed from yellow to green - Ready
  log   [09:38:25.085] [info][status][plugin:beats_management@7.2.0] Status changed from yellow to green - Ready
  log   [09:38:25.086] [info][status][plugin:index_management@7.2.0] Status changed from yellow to green - Ready
  log   [09:38:25.087] [info][status][plugin:index_lifecycle_management@7.2.0] Status changed from yellow to green - Rea
dy
  log   [09:38:25.117] [info][status][plugin:rollup@7.2.0] Status changed from yellow to green - Ready
  log   [09:38:25.118] [info][status][plugin:remote_clusters@7.2.0] Status changed from yellow to green - Ready
  log   [09:38:25.119] [info][status][plugin:cross_cluster_replication@7.2.0] Status changed from yellow to green - Read
y
  log   [09:38:25.120] [info][status][plugin:snapshot_restore@7.2.0] Status changed from yellow to green - Ready
  log   [09:38:25.122] [info][kibana-monitoring][monitoring] Starting monitoring stats collection
  log   [09:38:25.140] [info][status][plugin:maps@7.2.0] Status changed from yellow to green - Ready
  log   [09:38:30.029] [        ][reporting] Generating a random key for xpack.reporting.encryptionKey. To prevent pendin
g reports from failing on restart, please set xpack.reporting.encryptionKey in kibana.yml
  log   [09:38:30.043] [info][status][plugin:reporting@7.2.0] Status changed from uninitialized to green - Ready
  log   [09:38:30.143] [        ][task_manager] This Kibana instance defines an older template version (7020099) than is
currently in Elasticsearch (7030299). Because of the potential for non-backwards compatible changes, this Kibana instanc
e will only be able to claim scheduled tasks with "kibana.apiVersion" <= 1 in the task metadata.
  log   [09:38:30.196] [        ][task_manager] This Kibana instance defines an older template version (7020099) than is
currently in Elasticsearch (7030299). Because of the potential for non-backwards compatible changes, this Kibana instanc
e will only be able to claim scheduled tasks with "kibana.apiVersion" <= 1 in the task metadata.
  log   [09:38:30.379] [        ][task_manager] This Kibana instance defines an older template version (7020099) than is
currently in Elasticsearch (7030299). Because of the potential for non-backwards compatible changes, this Kibana instanc
e will only be able to claim scheduled tasks with "kibana.apiVersion" <= 1 in the task metadata.
  log   [09:38:30.638] [info][migrations] Creating index .kibana_1.
  log   [09:38:31.049] [info][migrations] Pointing alias .kibana to .kibana_1.
  log   [09:38:31.136] [info][migrations] Finished in 508ms.
  log   [09:38:31.139] [info][listening] Server running at http://localhost:5601
  log   [09:38:31.863] [info][status][plugin:spaces@7.2.0] Status changed from yellow to green - Ready
```

我們可以在瀏覽器的網址列中輸入 http://localhost:5601，開啟 Kibana 頁面。此時，開啟的頁面是一個歡迎頁面。

我們既可以點擊 "Try our sample data" 按鈕體驗 Kibana 的功能，也可以點擊 "Explore on my own" 按鈕使用自己的資料體驗 Kibana 的功能。

下面以點擊 "Try our sample data" 按鈕為例，體驗 Kibana 的功能。

點擊 "Try our sample data" 按鈕，進入如圖 10-9 所示頁面。

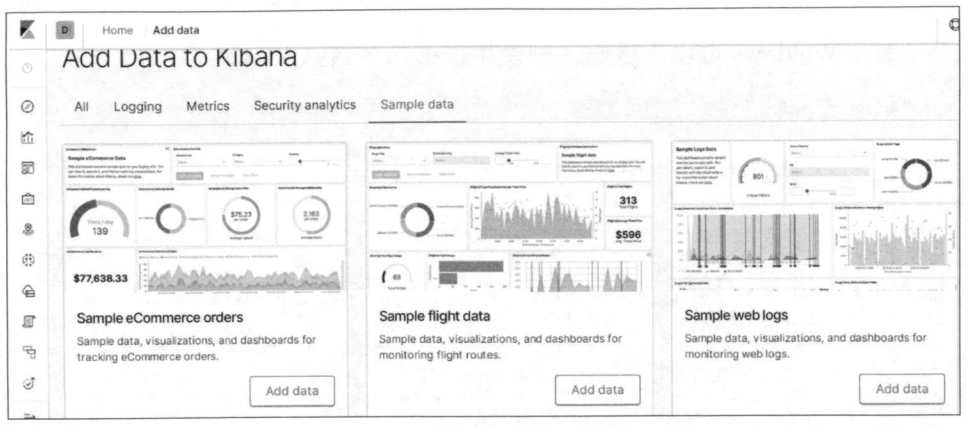

圖 10-9

我們選擇第一種展現類型作為範例，點擊 "Add Data" 按鈕後，Kibana 開始對資料進行載入，這個過程會持續數十秒。當資料載入後，"Add Data" 按鈕會變成如圖 10-10 所示的 "View Data" 按鈕。

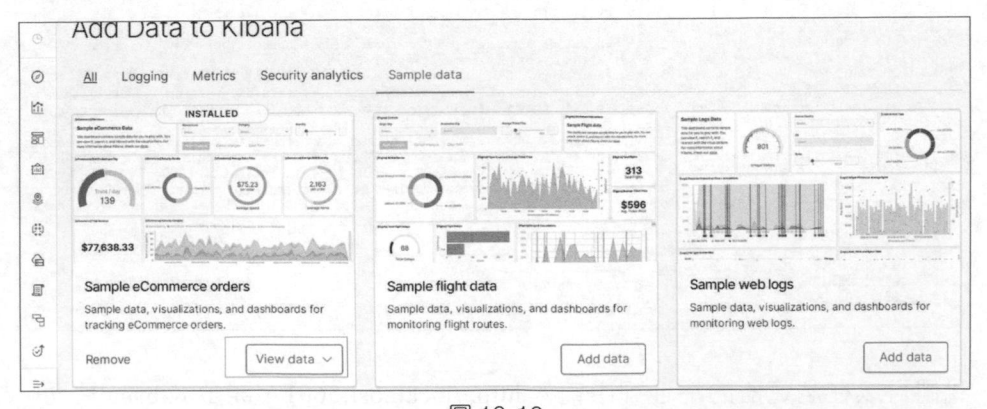

圖 10-10

點擊 "View Data" 按鈕，進入如圖 10-11 所示頁面。

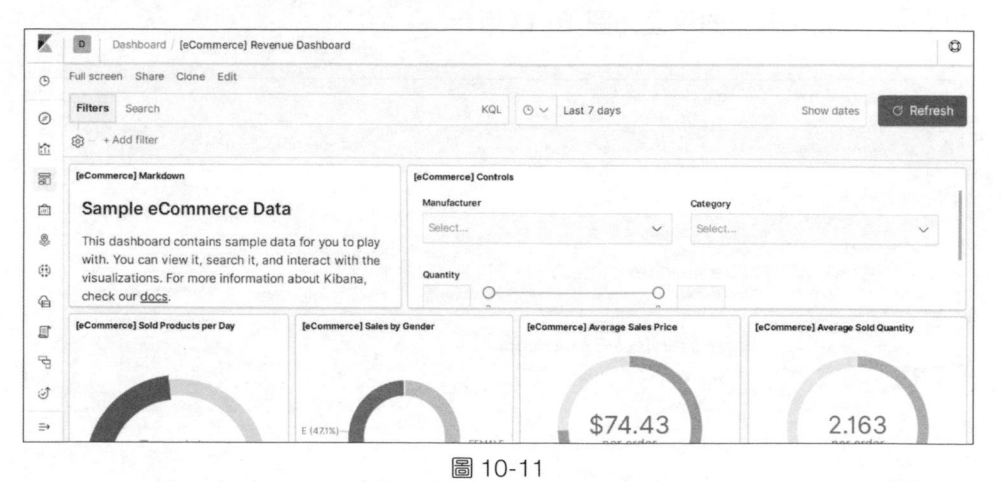

圖 10-11

在圖 10-11 所示頁面中，可以透過篩選資料，檢視細粒度的呈現。主要有關鍵字和時間兩個篩選維度。點擊 "Discover" 按鈕，切換到如圖 10-12 所示頁面，檢視資料的時間軸資訊和資料詳情資訊。

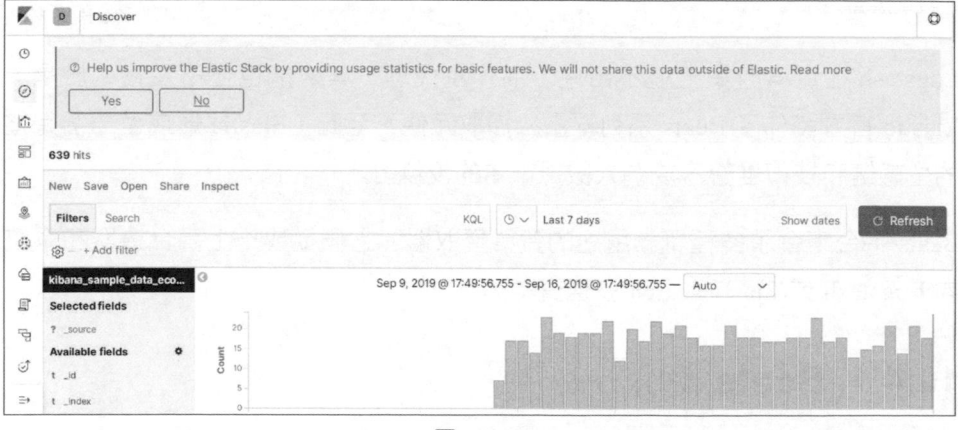

圖 10-12

在實際使用中，需要設定必要的 Elasticsearch 索引資訊，以便與 Kibana 進行資料聯通。設定索引的頁面如圖 10-13 所示。

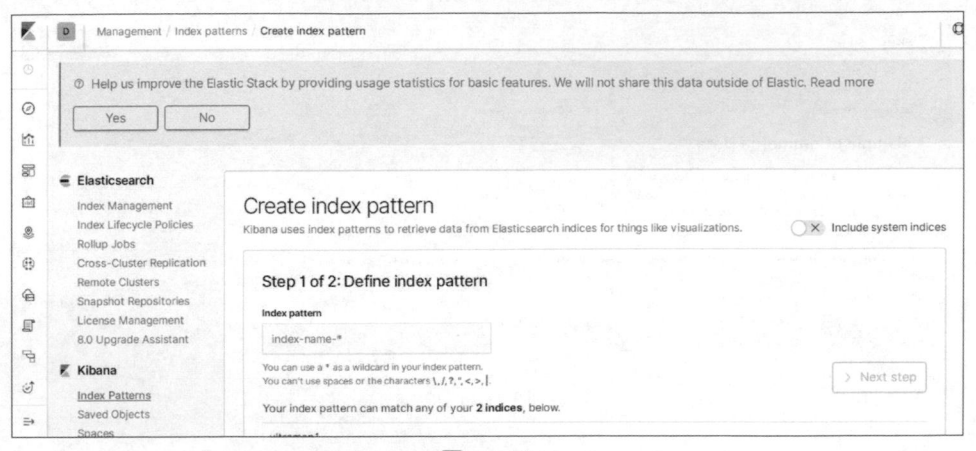

圖 10-13

10.4 Beats

Logstash 在資料收集上並不出色，而作為代理，其效能也並不及格。於是，Elastic 官方發佈了 Beats 系列輕量級擷取元件。至此，Elastic 形成了一個完整的生態鏈和技術堆疊，成為大數據市場的佼佼者。

Beats 平台集合了多種單一用途的資料獲取器。它們從成千上萬台機器和系統向 Logstash 或 Elasticsearch 發送資料。

10.4.1 Beats 簡介

Beats 是一組輕量級擷取程式的統稱，如圖 10-14 所示。

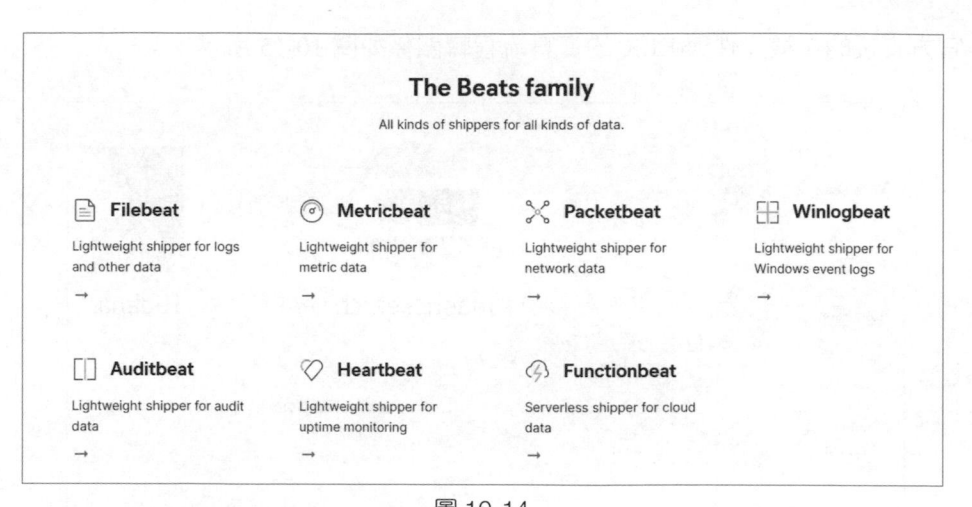

圖 10-14

Beats 中包含但不限於以下元件：

（1）Filebeat：該元件會進行檔案和目錄的擷取，主要用於收集記錄檔資料。

（2）Metricbeat：該元件會進行指標擷取。這裡說的指標可以是系統的，也可以是許多中介軟體產品的。主要用於監控系統和軟體的效能。

（3）Packetbeat：該元件透過網路封包截取和協定分析，對一些請求回應式的系統通訊進行監控和資料收集，可以收集到很多正常方式無法收集到的資訊。

（4）Winlogbeat：該元件專門針對 Windows 的 event log 進行資料獲取。

（5）Audibeat：該元件用於稽核資料場景，收集稽核記錄檔。

（6）Heartbeat：該元件用於系統間連通性檢測，如 ICMP、TCP、HTTP 等的連通性監控。

（7）Functionbeat：該元件用於不需要伺服器的擷取器。

以上是 Elastic 官方支援的 7 種元件，事實上，借助開放原始碼的力量，網際網路上早已創造出大大小小幾十甚至上百種元件，只有我們沒想到，沒有 Beats 做不到。官方不負責維護的 Beats，社區統一稱之為 Community Beats。

下方來講解的 7 種元件前 ELK 的資料流程轉圖係如圖 10-15 所示。

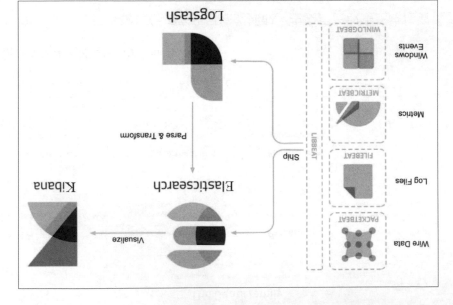

圖 10-15

更多元件可以從 Elasticsearch 官網下載。

10.4.2 Beats 輕量級設計的考量

如果擴充，Beats 是一組輕量級資料擷取方式的系統，那麼 Beats 是如何做到輕量級設計的呢？

（1）資料佔用資源少。在蒐集以及處理，Beats 並不進行複雜的運算的資料處理，只是將資料蒐集需要的組織儲存並上轉給上游系統。

（2）平行處理性好、便於擴展。Beats 採用 Go 語言開發而成，當所採用 Go 語言是一種並行程式語言，能夠在不依賴外部機器的情況下執行，包括撰其具有一種並行程式語言，Beats 與 Go 語言保持一致，支援多種作業系統，如 Linux、Windows、FreeBSD 和 macOS。

因此，Beats 的效能便明顯優於 Logstash。

10.4.3 Beats 的架構

Beats 之所以有上乘的效能及優秀的可攜充性，能讓得如此還是因此的底層統構支撐，其接下面我們就來介紹一起探討 Beats 的架構式架構。

Beats 的架構設計如圖 10-16 所示。

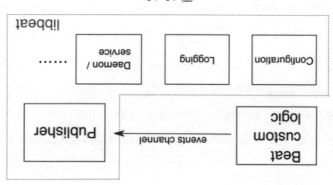

圖 10-16

libbeat 是 Beats 的核心元件。

在 Beats 架構中，有輸出模組（Publisher）、資料收集模組（Logging）、設定模組（Configuration）、記錄模組模組和守護應置程序模組（Daemon/service）。

其中，輸出模組有其將收集到的資料發送至 Logstash 或 Elasticsearch。

因為 Go 語言天然就有 channel，所以它在資料的邏輯模式下前端出模組都是經通過 channel 通訊的。也就是說，每個模組的構片在都是一個收集器時，完全不需要知道輸出模組的方在，當程式執行時時，自然就「神不知鬼覺地接收服務完了」。

除此之外，記錄模組模組、記錄模組置理模組、守護應模組程序模組等功能模組的發存讓充了 Beats 的功能提供了相當大的空間。

10.5 基礎知識連結

本章主要介紹了 Elasticsearch 的生態圈，不難看出其生態圈十分繁榮。不僅官方在維護 Elasticsearch 的生態，社區技術人員也在積極地貢獻自己的力量，正所謂「團結力量大」！

而這背後正是生態思維。

在技術圈中，Java 技術堆疊中的 Spring 生態發展得也是花團錦簇。從 Spring Boot 到 Eureka、Hystrix、Zuul、Archaius、Consul、Sleuth、Spring Cloud ZooKeeper、Feign、Ribbon……終於，以 Spring Boot 為核心的 Spring Cloud 微服務生態建立起來，並持續發展。

Android 系統、微信等也都在積極建立屬於自己的生態系統。Android 系統生態中數以萬計的開發者貢獻了許多的 App。微信在擴充和增強功能的同時，透過小程式的生態打造，連接了許多的線下場景，孕育了無限想像空間。

技術如此，商業如此，同理，技術人員的發展也是如此。

求職之路漫漫，因此在求職或跳槽時不能光盯著眼前的利益，更要看重未來的個人「生態」：找到適合自己的公司，找到有能力、有意願、有方法培養自己的主管，找到一群「技術派」的同事，找到一個技術氣氛好的團隊，等等。

10.6 小結

本章主要介紹了 Elasticsearch 的生態圈，即 ELK Stack。先後介紹了 ELK Stack 的背景、ELK 的實戰部署架構設計，以及 Logstash、Kibana 和 Beats。